ECONOMIC GROWTH AND DEMOGRAPHIC TRANSITION IN THIRD WORLD NATIONS

A Chaos and Complexity Theory Perspective

ECONOMIC GROWTH AND DEMOGRAPHIC TRANSITION IN THIRD WORLD NATIONS

A Chaos and Complexity Theory Perspective

Edited by

Şefika Şule Erçetin, PhD

Nilanjan Ray, PhD

Saurabh Sen, PhD

AAP | APPLE ACADEMIC PRESS

Apple Academic Press Inc.
3333 Mistwell Crescent
Oakville, ON L6L 0A2
Canada USA

Apple Academic Press Inc.
1265 Goldenrod Circle NE
Palm Bay, Florida 32905
USA

First issued in paperback 2021

Exclusive worldwide distribution by CRC Press, a member of Taylor & Francis Group

ISBN 13: 978-1-77463-430-1 (pbk)
ISBN 13: 978-1-77188-745-8 (hbk)

Library and Archives Canada Cataloguing in Publication

Title: Economic growth and demographic transition in the third world nations : a chaos and complexity theory perspective / edited by Şefika Şule Erçetin, PhD, Nilanjan Ray, PhD, Saurabh Sen, PhD.

Names: Erçetin, Şefika Şule, editor. | Ray, Nilanjan, 1984- editor. | Sen, Saurabh, 1978- editor.

Description: Includes bibliographical references and index.

Identifiers: Canadiana (print) 20190092572 | Canadiana (ebook) 20190092645 | ISBN 9781771887458 (hardcover) | ISBN 9780429433894 (eBook)

Subjects: LCSH: Demographic transition—Economic aspects—Developing countries. | LCSH: Economic development—Developing countries. | LCSH: Chaotic behavior in systems. | LCSH: Critical theory. | LCSH: Developing countries—Economic conditions.

Classification: LCC HB887 .E36 2019 | DDC 304.6/2091724—dc23

CIP data on file with US Library of Congress

Apple Academic Press also publishes its books in a variety of electronic formats. Some content that appears in print may not be available in electronic format. For information about Apple Academic Press products, visit our website at **www.appleacademicpress.com** and the CRC Press website at **www.crcpress.com**

About the Editors

Şefika Şule Erçetin, PhD, Professor, is currently chairwomen of the Educational Administration, Supervising, Planning and Economics Department at Hacettepe University in Ankara, Turkey, and was also head of the department during the years 2007–2010 years. As project coordinator, she has directed many international and national projects on organizational intelligence, municipalities, and vocational education. She is the president, board member, and founding member of the International Science Association (ISCASS). Along with ISCASS, Dr. Erçetin has organized six international symposiums called Chaos Complexity and Leadership. She is also editor-in-chief of several national and international journals and has many published articles. She has written seven national books, over 20 international book chapters, and edited seven international books. She is one of rare academicians in Turkey who has been interested in "chaos, complexity, quantum leadership" since 1990. Her other main research fields include comparative education policies; research methods on organizational intelligence, stupidity and wisdom; leadership; migration; fuzzy logic; children at risk; women and peace studies; social media; plasma leadership; quantum leadership; and women leadership models.

Nilanjan Ray, PhD, is from Kolkata, India and presently associated as an Associate Professor of Marketing Management in Adamas University in School of Management, West Bengal, India. Dr. Ray has obtained certified *Accredited Management Teacher Award from All India Management Association*, New Delhi, India. He has obtained his PhD(Mkt); M.Com (Mkt); MBA (Mkt), STC FMRM (IIT-Kgp). He has 10 years teaching experience in BBA, MBA, BCom and 6 years research experience and guided around 56 postgraduate students' project. Dr. Ray has contributed over 60 research papers in reputed national and International referred, peer- reviewed journals, proceedings and 8 edited research handbooks from springer, IGI-Global USA, and Apple Academic Publisher CRC Press (A Taylor & Francis Group), USA. He has also associated himself as a reviewer of *Tourism Management Elsevier, Journal of Service Marketing Emerald Group Publishing Limited, Journal of Business and Economics, Research Journal of Business and Management Accounting,* and as an Editorial Board

Member of several referred journals. He has also chaired in a technical session at IJAS Conference 2012, at Harvard University, Boston, USA. Dr. Ray is a life-member of the International Business Studies Academia, Fellow Member of Institute of Research Engineers and Doctors Universal Association of Arts and Management Professionals (UAAMP) New York, USA.

Saurabh Sen, PhD, has authored many research papers and articles published in international and national journals. He also participates and presents research papers at many national and international seminars and conferences. He is a life member of the Indian Accounting Association and the founding member of its Varanasi Chamber. He qualified UGC-NET for eligibility for assistant professorship in commerce. Dr. Sen acquired his master's degree in commerce with a specialization in accounts and finance and a PhD in commerce from Banaras Hindu University (BHU), India.

Contents

DEDICATION

To all young women scientists, including my beloved Dr. Nihan Potas and Şuay Nilhan Açıkalın, MSc, who devote their life to science and who enrich their fields with beauty.

— Prof. Dr. Şefika Şule Erçetin

First and foremost, I would like to thank my parents, Sri Nirmalendu Ray and Smt. Rina Ray, for their unending inspiration and for standing beside me throughout my career and while I was writing this book. I would like to thank my wife, Trina Ray, for providing me moral support and inspiration as well as the necessary time and resource toward the applications and maintaining databases. I would like to especially thank to my doctoral supervisor and my teacher, Dr. Dillip Kumar Das, Asst. Professor and Head of the Department, Department of Tourism, The University of Burdwan, India, for guiding me and developing the applications for this project, for giving me the freedom to manage my projects, and for providing the necessary time and resources toward the applications and databases.

— Dr. Nilanjan Ray

In loving memory of my beloved parents.

— Dr. Saurabh Sen

Biographies

Şuay Nilhan Açıkalın
Şuay Nilhan Açikalin graduated from Bilkent University in 2012. She had MA degree in International Relations Department of Middle East Technical University and currently she is a PhD candidate in same department and visiting researcher in Humbolt University of Berlin, Germany. She has some international articles, international book parts. Her main interests are: Chaos and Complexity, Turkish Foreign Policy, EU-Turkey Relations and Leadership. On the other hand, she has worked in many national and international projects as a member of project team and project coordinator. She is vice president and founder member of the International Science Association (ISCASS). Under association, she has been member of organizing committee of International Symposium on Chaos Complexity and Leadership.

Yener Akman
Dr. Yener Akman graduated from the Division of Computer Education and Instructional Technologies at Uludağ University. He completed master's degree at Gazi University and PhD degree at Hacettepe University at the Division of Educational Administration and Supervision. The author, who works as a teacher in the Ministry of National Education since 2007, has published several of articles in academic journals. His main research fields are organizational behavior, educational policies, multiculturalism, and research methods.

Salih Aydemir
Salih Aydemir did his BA degree in Islamic studies at the Faculty of Divinity in Ankara University. He also continued to do his postgraduate studies in the same university. Aydemir did an MA and PhD in sociology of religion. The title of his PhD thesis is "Sufi orders as a community based on the relationship between sheikh and disciples: The case of the community of Sheikh Sudeysi Order in Sanliurfa." He also deals with issues including terror organizations and violent ideologies with religious references, immigration, and small religious communities, on which he directed and finished two national projects. Aydemir has written numerous articles and chapters in various academic journals and books. He is currently working as an Assistant Professor in the

department of philosophy and religious sciences at the Faculty of Theology, Harran University in Sanliurfa, Turkey and teaches the sociology of religion, the history of religions, and the religious communities and theories in the sociology of religion at undergraduate and postgraduate levels.

Debasish Batabyal

Dr. Debasish Batabyal, PhD, is currently teaching at Amity University, Kolkata, India. Dr. Debasish regularly speaks at academic conferences around the world and shares his views on various research themes, such as sustainable tourism, tourism for economic development, relationship marketing, and services marketing. He has contributed to articles and chapters in various prestigious journals and edited books, respectively, at international and national levels. As a key member, he also worked with various research institutes and team members, including committee for the production of development reports by UNDP, District Gazetteer Development committee, etc. He was also a guest/invited speaker in many Indian Universities and institutes of high reputes, for example, Indira Gandhi National Tribal University, Sikkim Central University, and Burdwan University. He has more than 10 years of experience of teaching business courses in India at several tourism-based schools globally. He participates and designs various short-term training courses for the academicians and industry professionals. Recently, he has completed a UNESCO sponsored MOOC from the Universita della Svizerra Italiana, Switzerland. Besides, he has authored a book "*Development of Hill Tourism in Sikkim: Aspects & Challenges*."

Selda Çelik

She was born on March 15, 1984 in Tunceli. After her primary and secondary education in Elazığ, she graduated from İzmir Dokuz Eylül University in 2005 with a major in Foreign Language Education. In 2016, she completed her master's degree at Hacettepe University in the field of education management inspection planning and economics. In 2017, she started her PhD in educational administration at Hacettepe University. She currently works as an English teacher at Mehmet Akif Ersoy Primary School in Elazığ.

Mukund Deshpande

Prof. Dr. Mukund Deshpande is currently an independent researcher, who is graduated in engineering in 1972 from Laxminarayan Institute of Technology, Nagpur (India) and having worked thereafter in corporate sector, for diverse industrial projects, since then in India and gained practical management experience. He later turned his interests in management and obtained PhD in organization management from University of Pune. He has more than 25 years industrial experience and 14 years in teaching as Professor and training in industrial and academic institutions. He has been a consultant for SMEs in and around Pune, India. Currently, a visiting Professor at the Department of Management of accredited business schools in Pune having keen interest in academics and research. He has presented research papers and written case studies and chapters in his domain of policy designing at national and international conferences. He has contributed to peer-reviewed paper and chapter publications in leading journals and books published in India, the United Kingdom and the United States. He can be contacted at mvdeshpande49@yahoo.co.in.

Hasan Dinçer

Dr. Hasan Dincer is an Associate Professor of finance at Istanbul Medipol University, Faculty of Economics and Administrative Sciences, Istanbul-Turkey. Dr. Dincer has BAs in Financial Markets and Investment Management at Marmara University. He received PhD in Finance and Banking with his thesis entitled "The Effect of Changes on the Competitive Strategies of New Service Development in the Banking Sector." He has work experience in the finance sector as a portfolio specialist and has his major academic studies focusing on financial instruments, performance evaluation, and economics. He is the executive editor of the *International Journal of Finance and Banking Studies* (*IJFBS*) and the founder member of the Society for the Study of Business and Finance (SSBF).

Şefika Şule Erçetin

Şefika Şule Erçetin, Professor, is currently chairwomen of the Educational Administration, Supervising, Planning and Economics Department at Hacettepe University in Ankara, Turkey, and was also head of the department during the years 2007–2010 years. As project coordinator, she has directed many international and national projects on organizational intelligence, municipalities, and vocational education. She is the president, board member, and founding member of the International Science Association (ISCASS). Along with ISCASS, Dr. Erçetin has organized six international symposiums called Chaos Complexity and Leadership.

She is also editor-in-chief of several national and international journals and has many published articles. She has written seven national books, over

20 international book chapters, and edited seven international books. She is one of rare academicians in Turkey who has been interested in "chaos, complexity, quantum leadership" since 1990. Her other main research fields include comparative education policies; research methods on organizational intelligence, stupidity and wisdom; leadership; migration; fuzzy logic; children at risk; women and peace studies; social media; plasma leadership; quantum leadership; and women leadership models.

Niyazi Erdoğan

Niyazi Erdoğan graduated from Atatürk University's Faculty of Arts and Sciences at the Department of French Language and Literature in 1970. He graduated from Gazi University in 1977 with a master's degree in financial management and in 1993 with the PhD degree in business administration from Gazi University Social Sciences Institute. Dr. Niyazi Erdogan began his career in 1972 as an officer at Akbank Central Branch. In 1984, he was appointed as the Director of Ankara Branch of Interbank A.Ş. He worked as the Coordinator of Cukurova Holding AŞ between 1993 and 1999, as Executive Director of Inter Lease AŞ between 1998 and 1999, and as Vice Chairman of Inter Yatırım AS between 1999 and 2001. From 2001 to 2003, he served as T.C. Ziraat Bankası AŞ Chief Executive Officer (CEO) and General Manager and Board Member of Board of Directors, Kazakhstan Ziraat Intl Bank/ALMATY Board of Directors, Başak Hayat Sigorta AŞ/İstanbul Board of Directors, Başak Sigorta AŞ Board Member and Deutsche Turkish Bank/Germany. He also served as a member of the Board of Directors and as the Chairman of the Board of Savings Protection Fund between April 2001 and May 2003. He carried out the Public Banks Structuring Process and T. Emlak Bankası and T.C. Ziraat Bank merger process successfully. Dr. Niyazi Erdoğan has served as Vice Chairman of the Executive Committee of Tubin Group (Construction-Machinery and Prefabricated) between 2004 and 2006, Chairman of Guris Construction and Engineering Inc. between 2006 and 2010 and Chairman of the Board of Directors of Çelik Holding AŞ. He has been serving as Chairman of Güriş Holding AŞ since 2010. Dr. Niyazi Erdoğan also served as a part-time instructor at Gazi University Faculty of Economics and Administrative Sciences between 1997 and July 2012. He has been serving as an Associate Professor at Ufuk University, Faculty of Economics and Administrative Sciences International Trade Department. There are many articles and three books published in financial matters to his credit.

Behiye Ertaş
She graduated from Cumhuriyet University Primary School Teaching Department in 2007 and took her master's degree from Cumhuriyet University Educational Administration Department in 2012. She has been attending the PhD program in Hacettepe University Educational Administration Department.

Manuel Alberto M. Ferreira
Manuel Alberto M. Ferreira is electrotechnical engineer and Master in Applied Mathematics by Lisbon Technical University, PhD in Management-Quantitative Methods, and Habilitation in Quantitative Methods by ISCTE-Lisbon University Institute. He is Former Chairman of the Board of Directors and Vice President of ISCTE-Lisbon University Institute, Full Professor at ISCTE-Lisbon University Institute, and Former Director of Department of Mathematics in ISTA-School of Technology and Architecture. His research interests are: mathematics; statistics; stochastic processes-queues and applied probabilities; game theory; chaos theory; Bayesian statistics: application to forensic identification; applications to economics, management, business, marketing, finance, and social problems in general. He has published more than 420 papers in scientific journals and conference proceedings, and also 40 book chapters. He has presented 172 communications in international scientific conferences. https://ciencia.iscte-iul.pt/public/person/mamf.

José António Filipe
José António Filipe is graduated in Economics by ISEG/UTL—Instituto Superior de Economia e Gestão, Universidade Técnica de Lisboa; has a master's degree in management sciences by ISCTE-IUL (Instituto Universitário de Lisboa) and a PhD in quantitative methods (operations research) by ISCTE-IUL. Assistant Professor with Habilitation at ISCTE-IUL; Invited Professor in several European universities (Spain, Slovakia, Greece, and Estonia); Member of BRU-IUL and ISTAR-IUL research groups. Former Subdirector of Department of Mathematics in ISTA—School of Technology and Architecture, at ISCTE-IUL. Research interests (among others): finance, economics, management, marketing, mathematics; statistics; stochastic processes—queues and applied probabilities; game theory; application of quantitative methods to economics, management, finance and social problems; environmental and natural resource economics, tourism, etc. He has published more than 250 papers in scientific journals and conference proceedings, and more than 30 book chapters, 10 books, and 2 edited books. He has presented more than 100 communications in international scientific

conferences. He is Member of more than 20 international organizations, mentor and patron of some of them and Country Head for Portugal of an International Academic Organization. He is associated to many world organizations, particularly in Academia in European Countries, India, the United States, and Brazil. He is examiner of PhD theses in Portugal and abroad, participant in PhD theses progress evaluation panels, examiner of master theses, and President of Board of some of them. He is Supervisor of PhD theses (finance, management, economics, quantitative methods) and master theses (finance, management, economics, quantitative methods) of students from Portugal and abroad. https://ciencia.iscte-iul.pt/authors/jose-antonio-candeias-bonito-filipe/cv.

Saheli Guha Neogi Ghatak

Dr. Saheli Guga Neogi Ghatak has graduated from Lady Brabourne College, Kolkata, West Bengal and postgraduated from University of Calcutta and did her PhD from University of Calcutta. Her area of research includes sociology of aging, quality of life, rural and urban sociology. She has more than 6 national and international publications and more than 20 paper presentations. She was invited as resource person in National TV channel and UGC sponsored seminars in colleges and universities. She has more than 3 years of experience of teaching in West Bengal State University and Vidyasagar College for women.

Halime Güngör

Halime Güngör was born on April 17, 1979 in Bursa, Turkey. She was graduated from Boğaziçi University, Faculty of Education Guidance and Psychological Counseling Department in 2003. Then she received her master's degree in Department of Educational Administration and Supervision from Sakarya University, Turkey in May 2013. Her thesis was entitled, "Relationship between Stigma Tendecies of School Principals' and Teachers' and Organizational Culture." She has started to attend educational administration, supervision, planning and economics PhD program since September 2013, in Hacettepe University, Turkey. She is a total quality management leader and psychological counselor, and also has managed lots local and international projects on education.

Ümit Hacıoğlu

Dr. Ümit Hacıoğlu is an Associate Professor of finance at İstanbul Medipol University, Faculty of Economics and Administrative Sciences, İstanbul,

Turkey. Dr. Hacıoğlu has BAs in business and international business from Beykent University. He received his MBA from Beykent University and his PhD in finance and banking from Kadir Has University. He has published extensively in major academic journal focusing on financial markets, behavioral finance, performance evaluation, and economics of markets. He edits and serves on several journals related to his core research areas—behavioral finance and financial decision-making.

Gülşah İmamoğlu Akman

Gülşah Imamoğlu Akman graduated from the Division of Elementary Mathematics Education at Abant İzzet Baysal University in 2007. She had master's degree at the Division of Educational Administration and Supervision at Gazi University in 2011. She is currently continuing her PhD education at the Division of Educational Administration and Supervision at Hacettepe University. Since 2007, she has been working as a teacher in the Ministry of National Education and is interested in the fields of organizational behavior and management theories.

Gökhan Kılıçoğlu

Gökhan Kılıçoğlu is currently working as an Assistant Professor at Eskişehir Osmangazi University. He completed his master's degree in educational administration at Ege University. He received his PhD degree from Eskişehir Osmangazi University. He has several national and international articles published in journals such as *Management in Education*, *Urban Review*, *Leadership and Policy in Schools* in the field of educational administration. Dr. Kılıçoğlu has presented at numerous national and international conference meetings including AERA, BERA, and EERA. He has also given seminars in various governmental organizations such as the Ministry of National Education and Turkish Air Forces in the fields of leadership, communication and classroom management. His research interests have been focused on organizational behavior, change management, leadership in educational organizations, and educational sociology.

Sevda Kubilay

Sevda Kubilay is a graduate of Bogaziçi University Foreign Language Education Department and worked as an English teacher in primary and secondary schools of the Turkish Ministry of Education. After 5 years of teaching experience, she moved to Belgium and continued her career as a teacher of Turkish language and culture. Now she has been working as an English instructor in a State University of Turkey. She completed her master's degree on educational administration, inspection, economy, and planning in 2013, and she has been doing her PhD in Educational Administration department of Hacettepe University. She is interested in education policy of refugee students and social inclusion.

Duran Mavi

He was born on June 12, 1985 in Kahramanmaraş Elbistan. After completing his primary and secondary education in Elbistan, he graduated from the Department of Primary School Teaching at Selçuk University Faculty of Education in 2008. He worked as a primary school teacher in Kahramanmaras during 2008–2011. He has been working as assistant school principal in MONE in Kahramanmaraş, Elbistan since 2014. In 2015, he completed his graduate program in education management, inspection, planning, and economics in Gaziantep. He has been a PhD student since the autumn of 2016, in Hacettepe University Educational Sciences Institute Educational Administration Program.

İlknur Maya

Maya graduated from the Elementary School Teaching Department of Dokuz Eylül University in 1996. The researcher completed the Master of Arts in the Educational Management and Supervision Department of the Institute of Educational Sciences of the same University. In 2008, she graduated from the Educational Management, Supervision, Planning and Economy Department of the Institute of Educational Sciences of Hacettepe University, and was granted the title of Doctor. İlknur Maya, PhD (2008), is currently an Associate Professor in the Department of Educational Sciences at Canakkale Onsekiz Mart University. Maya started her career as a teacher. In 2010 she wrote a book about risk management in educational organizations which was published in Turkey. Her main fields of interests are risk, risk management and educational organizations, education systems of EU countries, women, leadership and change, values and educational organizations, and sex inequality in educational organizational. She has published several articles in many scientific journals about risk management, crisis management, educational organizations and entrepreneurship, education systems of EU countries, women, and leadership.

P. K. Mishra

P. K. Mishra, PhD is currently Associate Professor in the Department of Economic Studies, Central University of Punjab, Bathinda, India. He is having more than 21 years of teaching experience at undergraduate and postgraduate levels. He has to his credit more than 100 research papers published in refereed journals of national and international repute. His principal research areas include issues concerning socioeconomic development of India and economics of finance and tourism.

S. K. Mishra

S. K. Mishra, PhD is currently Assistant Professor in the Department of Humanities and Social Sciences, Siksha 'O' Anusandhan (deemed to be university), Bhubaneswar, Odisha, India. He has more than 14 years of teaching experience at undergraduate level. He has to his credit more than 20 research papers published in various journals of national and international repute. His principal research areas include issues concerning socioeconomic development of India and her States .

Daniels Aide Okun

Daniels Aide Okun is a cosmopolitan humanist. He has lived in 23 countries, and is an educator. He has a bachelors of arts degree in international studies and diplomacy. He has a master's degrees in international trade and finance and in international security. He is enthusiastically passionate about the intellectual force of education and knowledge in the development of humanity. His aspirations and goals are envisioned around the forces of intellectualism.

Mustafa Özmusul

Mustafa Özmusul is an Associate Professor in the Faculty of Education at Harran University in Şanlıurfa, Turkey, and is also head of the department of elementary education. He completed his PhD on educational administration, supervising, and planning and economics in 2013 at Hacettepe University, Turkey. While conducting two European Union school projects, he observed various school systems in the European countries, and received eTwinning Label award. During the years 2003–2013, he was a school teacher; he participated in various professional development activities mainly on improving the school education, project development, creativity, and sociopsychological issues in Turkey and EU countries. He is also member of scientific advisory committees of several international conferences, and an editorial board member of two international-refereed journals. Along with the author of a Turkish book written on theoretical issues on school management, and the many articles published in both international and national journals, he is mainly interested in the research areas of school leadership, school policies and implementations, international dimension, and indicators in education.

Nihan Potas

Nihan Potas graduated from Başkent University, Department of Statistics and Computer Sciences in 2005. She has been working as assistant professor at Ankara Hacı Bayram Veli University, Faculty of Economics and Administrative Sciences, Department of Healthcare Management. She had completed her master education at Gazi University Department of Statistics, Faculty of Science. During her PhD, she was enrolled TUBITAK's international doctoral research fellowship programme. Due to her fellowship and doctoral research, she is now research scholar at University of North Carolina at Chapel Hill in the department of Biostatistics. She got her PhD degree from Ankara University Department of Statistics, Faculty of Science. She has worked in many national and international projects as member of project team and expert of statistic. She is in the board and founder member board of the International Science Association. She has written one national and seventeen international book chapters. Her main interests are non-parametric statistics, categorical data analysis, survival analysis and statistical applications.

Nilanjan Ray

Dr. Nilanjan Ray is from Kolkata, India and presently associated as an Associate Professor of Marketing Management in Adamas University in School of Management, West Bengal, India. Dr. Ray has obtained certified Accredited Management Teacher Award from All India Management Association, New Delhi, India. He has obtained his PhD(Mkt); HYPERLINK "http://m.com/" \t "_blank"M.Com (Mkt); MBA (Mkt), STC FMRM (IIT-Kgp). He has 10 years teaching experience in BBA, MBA, BCom and 6 years Research experience and guided around 56 Post Graduate students' project. Dr. Ray has contributed over 60 research papers in reputed National and International Referred, Peer Reviewed Journals, Proceedings and 8 Edited Research Hand Books from Springer, IGI-Global USA and Apple Academic Publisher CRC Press (A Taylor & Francis Group), USA. He has also associated himself as a reviewer of Tourism Management Elsevier, Journal of Service Marketing Emerald Group Publishing Limited, Journal of Business and Economics, Research Journal of Business and Management Accounting and as an

Editorial Board Member of several referred Journals. He has also chaired in a technical session at IJAS Conference 2012, at Harvard University, Boston, USA. Dr. Ray is a life-member of the International Business Studies Academia, Fellow Member of Institute of Research Engineers and Doctors Universal Association of Arts and Management Professionals (UAAMP) New York, USA.

Debojyoti Roy

Dr. Debojyoti Roy has over 11 years of work experience as data scientist. He has managed multiple projects working, namely, for Hewlett Packard, US Army, Credit Suisse Bank, and Charter communication. He has completed Doctor of Management (Strategy) from Webster University St Louis, MO, USA. He completed his double major on leadership and information technology from Webster University. His areas of research are corporate social responsibility and business ethics. He has previously worked at Globsyn Business School as visiting faculty. He has worked as corporate trainer addressing client's (Big and Small) concern regarding data mining.

Saurabh Sen

Dr. Saurabh Sen was born on October 25, 1978. He completed his Higher Secondary from a government school. Dr. Sen acquired MCom with specialization in accounts and finance; his appetite for knowledge and excellence drove him to research. The virtue of his sincere effort and hard work awarded him PhD in Commerce by Banaras Hindu University (BHU). Dr. Sen has authored many research papers and articles in many reputed journals of national and international stature. He has also been participating and presenting research papers in many national and international seminars and conferences. He is a life member of Indian Accounting Association and the founder member of its Varanasi chamber. He also qualified UGC-NET for eligibility for Assistant Professorship in June, 2012 in the subject Commerce.

G. K. Sinha

G. K. Sinha shares his knowledge as Assistant Professor in Research & PG Department of Economics, Arya Mahila Post Graduate College (BHU), Varanasi, with MA (IRPM), MA (Economics), BJMC (Gold Medalist), MJMC, DEM (Bangalore), and PhD. He has authored five books: *Changing Scenario of Indian Financial System, Human Resource Management in Globalised Era, Skill Development and Entrepreneurship, Economic Growth in Modern Era,* and *Inclusive Growth of the Economy*. Many of his research papers and articles have published in various reputed journals of national and international stature. Dr. Sinha has presented more than 50 research papers in various national and international conferences and seminars. He is also as managing editor of bi-annual refereed *International Journal of Commerce & Social Science* and General Secretary of Uttar Pradesh Commerce Association. Dr. Sinha is life member of Indian Commerce Association, New Delhi, Institute of Rail Transport (Ministry of Railways), New Delhi, and Institute of Public Enterprise Research, Allahabad.

Mehmet Cem Şahin

Mehmet Cem Şahin, was born in Ankara in 1976; He graduated from Hacettepe University, Faculty of Literature, Department of Sociology in 1999. Şahin, who was appointed as Research Assistant to Gazi University, Gazi Education Faculty in 2000, graduated from Gazi University Institute of Educational Sciences in 2002 master's program with the title of "Turkish Education in the Republican Period in terms of Secular Education Practices (1923–1960)" and completed his PhD in 2008 with his thesis entitled "A Sociological Research on the Approaches of the Faculty of Theology on the Popular Culture Elements" in the Department of Philosophy and Religious Sociology, Department of Philosophy and Religious Studies (Religious Sociology), Ankara University Social Sciences Institute. In 2010, he was assistant at the Department of Religious Culture and Ethics Education at Ankara University, Faculty of Educational Sciences. Assoc. Dr. Sahin's staff was transferred to the Faculty of Theology of Ankara University in the Department of Primary Religious Culture and Moral Education in 2012. He received Associate Professor title in 2013; Şahin still holds the position of department chair in the same department. Since June 2017, he has been a Vice Director of the Institute of Social Sciences at the University of Ankara. He has published many researches related to his field.

Gönül Şayir

Gönül Sayir was born on June 6, 1989 in Konya-Selçuklu. After primary and secondary school education, she graduated from Selcuk University Ahmet Keleşoğlu, Faculty of Education Department of Computer and Instructional Technologies in 2011. In 2015, she completed her master's degree in Atatürk University in the field of education management inspection planning and economics. In the same year, she started a doctorate in the field of education management at Hacettepe University. The author is a computer teacher at the Zile Public Education Center of the Ministry of National Education.

Her national and international publications include several printed articles, notifications, and book chapter authorship presented at the symposium.

İlknur Şentürk

İlknur Şentürk graduated from Anadolu University, Faculty of Science, Department of Biology in 1997. She completed her master's degree in Eskişehir Osmangazi University, Institute of Educational Sciences, Department of Educational Administration and Supervision in 2000. In 2007, she graduated from Hacettepe University, Institute of Educational Sciences, Department of Educational Administration, Supervision, Planning and Economics, and she assumed the title of "PhD." She started to work as Assistant Professor in Eskişehir Osmangazi University in 2009. The author still works in Eskişehir Osmangazi University, Department of Educational Sciences. She has carried out scientific studies about "globalization and educational politics," "democratic classroom management," "neoliberalism, and educational administration," "gender and education," "organizational behavior," and "social justice in education."

Emel Terzioğlu Barış

Emel Terzioğlu Barış is an Assistant Professor at Educational Sciences Department of Karabuk University in Turkey. Her research interests are "adult education and psychology," "lifelong learning and guidance," "new teaching/learning methods, styles, and strategies," "visioner teacher training," "educational psychology," and "physical educational environments."

Belgin Tura

In 1999, she completed her undergraduate education at Gazi University, Business Education Department. In 2006, she received her MA degree in Gazi University, Department of Business Administration. She is currently a PhD student at Hacettepe University Educational Administration, Inspection Planning and Economics Department. In 2001, she started to teach as a teacher in the Ministry of National Education. She is still working at Ankara Sincan Melikşah Secondary School as a Technology and Design Teacher.

Kamil Demirberk Ünlü

Kamil Demirberk Ünlü graduated from the Business Administration Department in 2009 and International Trade Department in 2010 from Cankaya University. He earned his MSc from Graduate School of Applied Mathematics, Middle East Technical University in 2012. Currently, he is a

PhD candidate in the Graduate School of Applied Mathematics, Middle East Technical University, and a specialist at the Department of Statistics, Ankara University. His research focuses on applied probability and multivariate time series analysis.

Nursel Yardibi

Nursel Yardibi worked in the Central Bank of Turkey as an Educational Manager between the years of 1974 and 2003. She received Master's degree in the field of educational management, supervision, planning and economics at the Hacettepe University. Afterwards, she completed her PhD in the same field and at the same university in 1991. She worked as an Assistant Professor in the Faculty of Educational Sciences at Lefke Europian Üniversity in Cyprus. Until now, she has contributed numerous articles in leading journals and internationally published books. Additionally, she has given lectures at the Ufuk and Hacettepe University in the fields of educational sciences since 1995.

Mehmet Yılmaz

Mehmet Yılmaz is a Professor in the Department of Statistics at the Ankara University in Ankara, Turkey. As an author or coauthor, he has published more than 20 research papers in national and international journals. He has contributed two book chapters related to chaos and complexity published by leading international publishers. He has conducted undergraduate and graduate courses on "time series analysis" and "econometrics." He is also a reviewer of many international and national journals.

He graduated from the Faculty of Science, Statistics Department at Ankara University in 1997. In 1998, he started to work at Turkish Statistical Institute, immediately after he started to work as a Research Assistant at the Statistics Department in Ankara University. In 2004, he received a doctorate in Division of Statistics of Graduate School of Natural and Applied Sciences of Ankara University. Since March 2017, he has been working as a Professor. His research interests include theory of statistics, distribution theory, time series modelling, data analysis, and various related aspects of probability and statistics.

Leyla Yılmaz Fındık

Leyla Yılmaz Fındık has been working at Hacettepe University as an Education and Training Planner since 2013. Prior to this job, she worked as a teacher of English at Ministry of National Education between 2005 and

2013. She completed PhD in the field of educational administration supervision planning and economics in the Department of Educational Sciences and MA in 2012 in the same field in Hacettepe University, Ankara, Turkey. She received BA in linguistics at Faculty of Letters from Hacettepe University in 2005. She is the author of numerous articles and chapters on topics related to leadership, autonomy, higher education, and educational planning. Her primary research interests include educational administration, leadership, educational planning, autonomy, and higher education. She has worked on different projects related to quality assurance, Bologna Process, ECTS, and students affairs in Hacettepe University.

Derya Yılmaz Kılıçoğlu

Derya Yılmaz Kılıçoğlu received her BA from Middle East Technical University and her PhD in educational sciences from Eskişehir Osmangazi University. With the scholarship retrieved from Council of Higher Education, she visited to the University of Cambridge in 2014 as a visiting scholar. Dr. Yılmaz Kılıçoğlu's teaching and research interests include leadership in education, educational policy, organizational change, and research methods in education. She has published a range of articles and book chapters about organizational change and leadership in education. Dr. Yılmaz Kılıçoğlu also presented papers at regional conferences and at national conferences conducted by the American Educational Research Association, European Educational Research Association, and International Society of Educational Planning. He joined Eskişehir Osmangazi University in 2010 and currently, is an Assistant Professor in the Department of Educational Sciences.

Serhat Yüksel

Dr. Serhat Yuksel is an Assistant Professor of business at Istanbul Medipol University, Faculty of Economics and Administrative Sciences, Istanbul, Turkey. He received his BA from Yeditepe University. He received his MBA from Bogazici University and his PhD in Banking from Marmara University. His major interests are business, finance, and banking.

Contributors

Şuay Nilhan Açıkalın
PhD Student, Department of International Relations, Middle East Technical University, Ankara, Turkey. E-mail: suaynilhan@gmail.com

Yener Akman
Ministry of National Education, Ankara, Turkey. E-mail: yenerakman26@gmail.com

Salih Aydemir
Assistant Professor, Department of Sociology of Religion, Faculty of Theology, Harran University, Şanlıurfa, Turkey. E-mail: saydemirs@gmail.com

Emel Terzioğlu Barış
Assistant Professor, Karabük University, Karabük, Turkey. E-mail: emelterzioglu@gmail.com

Debasish Batabyal
Assistant Professor, Amity Institute of Travel & Tourism, Amity University, Kolkata, West Bengal, India. E-mail: debasishbatabyal@gmail.com

Selda Çelik
PhD Student, Department of Educational Sciences, Faculty of Education, Division of Educational Administration, Hacettepe University, Ankara, Turkey. E-mail: seldaceliks@gmail.com

Mukund Deshpande
Professor, Independent researcher. E-mail: mvdeshpande49@yahoo.co.in

Hasan Dinçer
Associate Professor, School of Business and Management, Istanbul Medipol University, Beykoz 34810, Istanbul, Turkey. E-mail: hdincer@medipol.edu.tr

Şefika Şule Erçetin
Professor, Department of Educational Sciences, Faculty of Education, Division of Educational Administration, Hacettepe University, Ankara, Turkey. E-mail: sefikasule@gmail.com

Niyazi Erdoğan
Associate Professor, Faculty of Economics and Administrative Sciences, Ufuk University, Ankara, Turkey. E-mail: drnierdogan@gmail.com

Behiye Ertaş
PhD Student, Vocational School of Health Services, Bozok University, Yozgat, Turkey.
E-mail: behiyeertas5884@gmail.com

Manuel Alberto M. Ferreira
Lisbon University Institute ISCTE-IUL, BRU-IUL, ISTAR-IUL, Lisboa, Portugal.
E-mail: manuel.ferreira@iscte-iul.pt

José António Filipe
Lisbon University Institute ISCTE-IUL, BRU-IUL, ISTAR-IUL, Lisboa, Portugal.
E-mail: jose.filipe@irete.iul.pt

Saheli Guha Neogi Ghatak
Assistant Professor, Department of Sociology, School of Social Science, Adamas University, Kolkata, West Bengal, India. E-mail: saheli.guhaneogi@gmail.com

Halime Güngör
PhD Student, Department of Educational Sciences, Faculty of Education, Division of Educational Administration, Hacettepe University, Ankara, Turkey. E-mail: halime_gngr@yahoo.com

Ümit Hacıoğlu
Assocociate Professor, School of Business and Management, Istanbul Medipol University, Beykoz 34810, Istanbul, Turkey. E-mail: uhacioglu@medipol.edu.tr

Gülşah İmamoğlu Akman
Teacher, Ministry of National Education, Ankara, Turkey. E-mail: ankaragulsah@gmail.com

Gökhan Kılıçoğlu
Assistant Professor, Faculty of Education, Eskişehir Osmangazi University, Ankara, Turkey. E-mail:gkilicoglu@outlook.com

Sevda Kubilay
Vocational School of Social Sciences, Omer Halisdemir University, Niğde, Turkey

Duran Mavi
PhD Student, Department of Educational Sciences, Faculty of Education, Division of Educational Administration, Hacettepe University, Ankara, Turkey. E-mail: duranmavi@hotmail.com

İlknur Maya
Associate Professor, Çanakkale 18 Mart University, Çanakkale, Turkey

P. K. Mishra
Associate Professor, Central University of Punjab, Bathinda, India. E-mail: pkmishra1974@gmail.com

S. K. Mishra
Associate Professor, Siksha 'O' Anusandhan University, Bhubaneswar, Odisha. E-mail: skmtite@gmail.com

Daniels Aide Okun
Independent Researcher. E-mail: danielsokun@yahoo.com

Mustafa Özmusul
Assistant Professor, Harran University, Faculty of Education, Şanlıurfa, Turkey. E-mail: mustafaozmusul@gmail.com

Nihan Potas
Department of Health Care Management, Faculty of Economics and Administrative Sciences, Hacı Bayram Veli University, Ankara 06100, Turkey. E-mail: nihanp@gazi.edu.tr

Nilanjan Ray
Associate Professor of Marketing and Examination Coordinator School of Management, Adamas University, West Bengal, India. E-mail: drnilanjanray@gmail.com

Debojyoti Roy
Assistant Professor, Adamas University, Kolkata, West Bengal, India. E-mail: debojyotiroy3333@gmail.com

Saurabh Sen
Department of Commerce, Sunbeam College for Women, Varanasi, India. E-mail: saurabhsen25@gmail.com

G. K. Sinha
Assistant Professor, Research & PG Department of Economics, Arya Mahila Post Graduate College, Banaras Hindu University, Varanasi, India. E-mail: gksinha30@gmail.com

Mehmet Cem Şahin
Associate Professor, Department of Sociology of Religion, Faculty of Theology, Ankara University, Ankara, Turkey. E-mail: m.cem.sahin@ankara.edu.tr

Gönül Şayir
PhD Student, Department of Educational Sciences, Faculty of Education, Division of Educational Administration, Hacettepe University, Ankara, Turkey. E-mail: gonulsayir@gmail.com

İlknur Şentürk
Assistant Professor, Faculty of Education, Eskişehir Osmangazi University, Ankara, Turkey. E-mail: ilknurkokcu@gmail.com

Belgin Tura
Professor, Hacettepe University, Ankara, Turkey. E-mail: belgintura@gmail.com

Kamil Demirberk Ünlü
Specialist, Department of Statistics, Faculty of Science, Ankara University, Ankara 06100, Turkey. E-mail: kdunlu@ankara.edu.tr

Nursel Yardibi
Assistant Professor, Independent researcher. E-mail: dryardibi@gmail.com

Mehmet Yılmaz
Professor, Department of Statistics, Faculty of Science, Ankara University, Ankara 06100, Turkey. E-mail: mehmetyilmaz@ankara.edu.tr

Leyla Yılmaz Fındık
Student Affairs Department, Hacettepe University Rectorate, Ankara, Turkey. E-mail: leylayilmazfindik@gmail.com

Derya Yılmaz Kılıçoğlu
Assistant Professor, Faculty of Education, Eskişehir Osmangazi University, Ankara, Turkey. E-mail: dderyayilmaz@gmail.com

Serhat Yüksel
Assistant Professor, School of Business and Management, Istanbul Medipol University, Beykoz 34810, Istanbul, Turkey. E-mail: serhatyuksel@medipol.edu.tr

Abbreviations

AIC	Akaike information criteria
ADF	augmented Dickey–Fuller
ANP	analytical network process
CBR	crude birth rate
CDR	crude death rate
CRD	annual credit to private sector
CVM	contingent valuation method
DTM	demographic transition model
DTM	population transition model
FPE	final prediction error
ECTS	European Credit Transfer System
EDI	economic vulnerability index
EU	European Union
GDP	gross domestic product
GIE	gender inequality index
GII	gender inequality index
HDI	human development index
Kendall's W	Kendall's coefficient of concordance
MoNE	Ministry of National Education
NCD	noncommunicable disease
NGOs	nongovernmental organizations
ÖSYM	Universities' Student Selection and Placement Center
PIACC	Programme for the International Assessment of Adult Competencies
PPP	polluter pays principals
SUR	seemingly unrelated regression
TDV	Turkish Religious Foundation
TFR	total fertility rate
TIKA	Turkish Cooperation and Development Administration
TOPSIS	technique for order preference by similarity to an ideal solution
TRD	annual trade openness
TÜBİTAK	The Turkish Scientific and Technological Research Council

TYYÇ	Turkey Higher Education Qualifications Framework
UNDP	United Nations Development Program
USP	unique selling proposition
WEF	World Economic Forum
WFP	World Food Program
WTP	willing to pay for pollution and consumption
YAYKUR	Common Higher Education Institution
YÖK	Higher Education Council

Preface

The aim of this book is to present a new perspective on the discussions of demographic transition, economic growth, and national development via exploration of the third world economies with a close relationship between the said concepts with chaos and complexity theory as a well as a multidimensional approach. This unique book also brings into the fold discussions of demographic transition, economic growth, and national development, taking a deliberate glance at the plight of academicians in policy formulation for demographic transition, economic growth and development, and third world countries.

This book is a conglomeration of various aspects and issues related to the effect of demographic transition on socioeconomic development in third world countries, especially in the postglobalized era, with an applicability of chaos and complexity theory in order to elicit transformational policies. This book discusses and predicts future projections of the new world of economic growth policies with an interdisciplinary approach and new discussions on theories such as chaos and complexity theory. True development requires government action to improve significant factors of development and to remove barriers against marginalized sections of the society. However, economic growth in the sense of expanding gross national product and related variables is one of the most fundamental inputs to the overall development process. The basic objectives of development should focus on the expansion of human capabilities, which have been neglected for too long in developing economies.

Development is a multidimensional phenomenon. The most significant factors for the overall development process relate to basic needs such as education, availability of food, minimum purchasing power, and facilities such as safe drinking water, health care, infrastructure, etc. The socioeconomic development in third world countries is not uniform because of wide disparity in socioeconomic infrastructures. The main cause of this disparity is the utilization of demographic dividends by the society. If the society takes full advantage of the demographic dividend, its people become more valuable assets. Take education, for example, the positive correlation between education and earnings is well known. As life expectancy increases, parents are likely to choose to educate their children to more advanced levels. Healthier children, in turn, tend to experience greater cognitive development per year of schooling than their less healthy counterparts. The parents also know that there is a good chance that each child will benefit from schooling investments over a long working life and, with fewer children, can devote more time and money to each child. The result of this educational investment is that the labor force as a whole becomes more productive, promoting higher wages and a better standard of living. People, therefore, tend to enter the workforce later, partly because they are being educated for longer, but they are likely to be more productive once they start working.

The relation between education and development and skill and development is much more essential to increasing development levels. It is also a development indicator; it is mostly true that education increases the human skill both mentally and physically. So, people can participate and cooperate in development activities of their society or country due to their education and skills. All these aspects of demographic transition directly and indirectly influence the development process of a country. Development is very much affected by the demographic transition, especially in developing countries. Low population growth rate and increase in school enrollment lead to improved human capital. On the other hand, demographic dividend as a result of demographic transition is critical to the economy of a country and has a direct impact on a country's economic growth and development. Though demographic transition gives us a challenge of old-age population (above 60 years), it is clear that present developing countries, such as Bangladesh, India, and China, and their development are so much affected by the demographic transition.

Economies of third world countries are being considered as a potent mark on the world economy. On the eve of such recognition, academicians, researchers, and scholars feel it is their responsibility to frame future polices after contemplating the development in retrospect.

Relationships between societies, groups, or individuals and problems and chaotic situations that people and societies have to cope with have engaged in a different dimension with recent developments in technology, computer science, and physics. New developments and what transpires on the other side of the world can be learned in a few seconds via the internet. Premised on this, peoples', organizations', or societies' lives have been influenced more quickly. However, classic problem-solving theories of social sciences are not adequate to augment today's complex and interdependent relationships. This book presents unique chapters mostly composed from the branch of economic growth and demographic transition in third world nations. Developed and developing countries have made many innovations in order to improve their levels of economic growth in the last 30 years. In the rapidly changing and increasingly complex world, a crucial need exists for better knowledge about how third world nations can be organized and led most effectively.

Prof. Dr. Şefika Şule Erçetin
Ankara, 2019

Introduction

The major area of strength thus will be an infusion of chaos and complexity theory as well as a multidimensional approach to discussing the aspects of demographic transition, economic growth and development, as well as third world countries. So, this book carries peculiar features that significantly differentiate it from other conventional books in the same realm. Given that the demographic variable is always sidelined in economic growth and national development works especially in developing countries, it was imperative to come up with a book to address the said gap. Moreover, the emergence of chaos and complexity theory is sometimes overlooked. Conversely, the plight of third world countries is, in most cases, ignored. This book tasks itself with this by exploring the intricate nature of economic growth and national development in lieu of demographic transition. The claim is that, in a bid to realize economic growth and the subsequent national development, there is need to transform the demographic structure and make it competitive enough. Exploration of chaos and complexity especially from the perspective of third world countries is also a scared need. This can be from both government policies and private entities. Preparing governments and other entities for such an undertaking as well as offering a guiding tool is a role of academicians, researchers, and scholars whose responsibility is in framing future polices after contemplating the development in retrospect. Otherwise, the usually well-intentioned policies will fail to yield any dividends.

The objective of this book is to discuss the efficiency of good strategies and practices and its impact on business and economic growths, depending on the depth and diversity of infrastructure sector in particular and overall socioeconomic development in general. This book is important for:

- University students in the fields of education, development studies, management studies, political science studies, economics, project planning and management studies, etc.; the book will be a useful resource for their courses.
- Teachers at high school level handling economics and entrepreneurship; the book will act as a resource guide during the teaching process.
- Academic staff at university level in the fields of education, development studies, management studies, political science studies,

economics, project planning and management studies, etc.; the book will be a useful resource for their courses.

- Planning directorates at ministries of different countries; the book can guide policy-making and implementation.
- National and international agencies in the realm of developing countries; the book can guide them to refocus their targets in the process of aiding country development programs.
- Career personalities such as economists, planners, educationists, managers, etc. at any level; the book may help them sharpen their knowledge credentials and improve on practice.
- Researchers with a bias on economics, development, planning, education, etc.; the book may serve as a literature review base for their various studies.
- Politicians at different levels with a vision of designing a credible and election-worthy development plan or strategy (manifesto); the book can offer practical values and policies toward designing such a development plan.
- Random readers with a bias on economics, development, planning, education, etc.; the book may improve their awareness about the fields in question.

This book also brings readers' and researchers' attention to new developments in economic growth and demographic transition in third world nations as enshrined in chaos and complexity. Finally, it is extremely important that this book is an essential reference source to improve current literature in related fields and provide newer ideas for future developments. This book presents itself as serious and excellent discussion pertaining new approaches to economic growth and demographic transition in third world nations form the interdisciplinary point of view. The understanding of those issues, however, requires a broader perspective of analysis as well as invoking interdisciplinary and multidisciplinary perspectives of discussion. That is why this book is exceptional and unique in all senses.

Prof. Dr. Şefika Şule Erçetin
Ankara, 2019

PART 1
Demographic Transition

CHAPTER 1

Relationship Between Demographic Transition and Economic Growth

ŞEFİKA ŞULE ERÇETİN[1,*] and SEVDA KUBİLAY[2]

[1]*Department of Educational Sciences, Faculty of Education, Division of Educational Administration, Hacettepe University, Ankara, Turkey*

[2]*Vocational School of Social Sciences, Ömer Halisdemir University, Nigde, Turkey*

[*]*Corresponding author. E-mail: sefikasule@gmail.com*

ABSTRACT

The aim of this article is to examine the relationship between demographic transition and economic growth on the basis of the literature review. The demographic transition involves four phases in which fertility and mortaliy rates and life expectancy show differences according to the developmental features of each phase. There are different theories on the relationship between demographic transition and the economic growth. The pessimistic theory which has its root from Thomas Malthus claims that the population growth causes economic deterioration. The optimistic theory opposes the Malthusian approach and asserts that there is a positive relationship between the population and the economic growth. The neutralist theory supports the idea that there is no significant relationship between them. Although this discussion on population and economic growth continues, there is a current challenge for the world which is having ageing populations. The implications give the signal that both developed and developing countries are following this trend and it is a threat hindering to have effective labor force, savings, and investments.

1.1 INTRODUCTION

Economic growth is affected by several factors. Inner growth models address population growth and human capital accumulation. Those models take technological developments and public investments as an internal variable. Theories of external growth argue that capital accumulation is a significant determinant for economic growth and the long-term growth is influenced by technology (Pala and Teker, 2014). As it is seen, population and human capital are important factors to explain economic growth. The change in population growth through history has been interpreted differently by scholars. Since ancient times, the connection between the population and the economy has been the center of attention for politicians, philosophers, and scholars. In the past, populous kingdoms were popular and favorable because the crowd used to mean wellbeing and power. The wars were done to occupy territories and expand the borders so that the population would get larger and the kingdom will be more sound and powerful. Soon enough it was seen that populations were not influenced by policies easily and crowded populations caused lower living conditions (Reher, 2012).

The first theory in relation to population and economy was developed by Thomas Robert Malthus and the theory emphasized that population is the most significant economic variable and it asserted that population growth is determined by the economy. That is to say, they both affect each other reciprocally (Reher, 2012). The change in population growth is explained by the demographic transition theory. The theory gives an idea about the link between population and economy during a period of momentous change. According to the theory, fertility and mortality rates are high and fluctuating at the beginning and then the rates gradually get lower and stable. This change affects the population age distribution as well. However, the demographic transition is not experienced at the same time by all the countries. This period of change was experienced by some countries earlier in the past but some did not have such a change until the 1950s. Together with the demographic transition, the high ratio of children in the population gets lower and the ratio of elderly gets higher. The demographic transition has been affected by the industrial revolution, urbanization, and educational and technological developments (Lee and Reher, 2011). In this chapter, the relationship between population and economic development will be examined within the scope of the demographic transition. Firstly, the demographic transition theory will be explained and then the connection between population and economic growth and theories about them will be discussed.

1.2 DEMOGRAPHIC TRANSITION THEORY

The theory has different definitions and interpretations because scholars and writers stressed different dimensions of the theory. The theory gives an explanation about the change in birth and death rates in developed nations (Potts and Selman, 1979); it puts an emphasis on the predictive feature of the theory; if developing countries follow the developed ones, they will have the same transitions in their populations as well (Demeny, 1972). McNamara (1982) stresses the transformation from a traditional society to an industrialized modern society. All the definitions have a common ground: demographic transition theory focuses on birth and death rates and change in population size. The theory relates these focal spots with socioeconomic conditions. According to the theory, ratio of birth and death is affected by the alteration in socioeconomic conditions (Nielsen, 2016). Mortality rates decrease when the living standards and health conditions get better, this improvement makes fertility rate decreased afterward (Lutz and Sanderson, 2005). The theory includes some stages to explain the relationship mentioned: the preindustrial stage; the postindustrial stage of explosion; the stage of slowing down growth; and the stage of a stable size of population. However, some researchers claim that there are more stages in the frame of the theory (Nielsen, 2016).

Phase 1: In this stage, high birth and mortality rates are common and the state of growth can be described as stagnant. Due to the fact that this stage reflects preindustrial age, living standards are low, health care and hygiene are poor. There is a strong fluctuation in the ratio of birth and death (Warf, 2010). Infant mortality and premature deaths of adults are quite common in this phase so a high birthrate hardly makes up for losses (Rimashevskaia et al., 2013).

Phase 2: This stage is identified with the industrial revolution so it is quite different from the first stage and it is called as the Early Industrial Age. A rapid growth of population is witnessed during this stage, in other words, there is a population explosion. Due to industrialization, life conditions get better, as a result, the death rate declines. Also in this stage, there is a continual high birth rate; there is a change from stagnation to growth (Galor and Weil, 2000). Life expectancy goes up and early mortality rate declines due to improved sanitation and developments in the field of medicine, especially in the cities. A large percentage of the population lives in the countryside but moving to cities is popular due to industrial developments. This phase is the beginning of the transition from the traditional to the modern type of reproduction (Rimashevskaia et al., 2013).

Phase 3: During this phase, a gradual decrease in the population growth occurs. The number of urban families is getting increased and most of the families live in the cities. There is a trend of having fewer children among urbanites, which spreads to the poor population as well (Rimashevskaia et al., 2013). This stage is also called the Mature Industrial Age and is characterized with the slowing growth. During this stage, birth control comes to the fore and women join the business world, educational opportunities increase, and living standards get improved (Nielsen, 2016).

Phase 4: This stage represents the Postindustrial Age; there is a balance between the ratio of birth and death so the population size becomes stable. The influence of infectious diseases is low, instead smoking and obesity become prevalent in this stage (Warf, 2010). A longer life expectancy, lower birth and mortality rates, and modern type of reproduction represent this phase which is also called as "modern stability" period. With this phase, rapid population growth ends because the birthrate is controlled with self-regulation methods (Rimashevskaia et al., 2013).

Shortly, according to demographic transition theory, the population growth becomes rapid when the ratio of death declines and the ratio of fertility remains high. After a while, balances change due to socioeconomic and technological developments, then the fertility rate starts to decrease, and the population growth rate goes down or decelerates. In modern times, demographic transition is effected by improved health care and education, urbanization, improved nutrition, and women's personal development and their wish to have a small family (Yaakop and Nor, 2013). As we look at the transition phase of the countries, it is possible to say that 60 countries worldwide had made the transition to the modern type of reproduction. The rest of the nations are still in different phases of demographic transition (Rimashevskaia et al., 2013).

1.3 SOCIAL AND ECONOMIC IMPLICATIONS OF DEMOGRAPHIC TRANSITION

Starting with a moderate population growth rate in earlier history, the world has been witnessing an unabated population growth for over 250 years. The maximum increase has been experienced between the years of 1930–1990 and its ratio was 1.75% per year (Reher, 2012). However, there are certain historical events that are determinants on fertility and mortality rate. For instance, after World War II the world witnessed a significant decrease in the ratio of mortality because it was the end of a bloody war and there were improvements in the standard of living in terms of health, control of illnesses, and preventive medicine (Bhende and Kanitkar, 1978). Due to birth control methods, fertility rates showed a decline in developing countries after the 1960s except for most African countries where socioeconomic development was so poor (Caldwell et al., 1992). Modernization and industrialization in developed countries resulted in the demographic transition (Agyei, 1988). Although there has been a decrease in the population growth during the past 40 years, the growth continues to be strong. As we look back to history, it is apparent: the world population was 791 million in 1750 and it is 7 billion today. However, recently there are indications that the population growth rate will continue to decrease in the future because low fertility is getting more and more common in societies. This pattern of growth is an explicit indicator of demographic transition (Reher, 2012). The demographic transition has significant societal and economic outcomes that are essential for the emergence of modern societies. It is a period of change that is in favor of social and familial change and/or progress (Dyson, 2010).

Firstly, population age structure is definitely affected by the demogaphic transition; the fertility rate is the responsible factor for it. The young generation and working-age population are shaped by the demographic transition. Low fertility means older age structure and lower rate of labor (Bloom et al., 2002). The population age structure has important results for the future of the society and its implications will be clearly seen in the long term. Labor force may not be sufficient if the process of aging continues faster than expected.

As the roles of women change in the society, they have begun to take part in working life. The women have become more knowledgeable and conscious about birth control methods, raising healthy children, and personal development. Women's active participation in the society has changed the family structure, declined the importance of marriage, and the number of children (Reher, 2011). This development has also resulted in delayed marriages, as a result, the proportion of unmarried women whose ages are 30 or older has increased. The fertility rate has been affected by the trend (Yaakop and Nor, 2013).

The demographic transition also affects the living standards and life expectancy. Longer life expectancy makes it possible to invest in human capital. The number of educated people gets higher and a better generation is raised in terms of health and that generation joins to the business life so it is possible to reach a more productive working life. On the other hand, a longer life also means a longer period of retirement and diminish in labor supply (Keyfitz, 1987).

Demographic characteristics of the societies are affected by some events such as mass migration. During the demographic transition, it is possible to come across with high population growth in some areas due to migration phenomenon. The population has decreased in the geographical regions where immigration has been experienced and the sudden and rapid increase in the population in the immigrated countries has had an impact on the social and economic life (Hatton and Williamson, 2006). The world is currently witnessing a new flow of migration from the Middle East Countries where political turmoil has been going on.

1.4 THE DEBATE OVER THE EFFECTS OF DEMOGRAPHIC TRANSITION ON ECONOMIC GROWTH

"How does population growth affect economic growth and development?" is the question economists, population scientists, and politicians have been talking about for many years and they explain it from different directions. The relationship between them is still blurred, some studies have revealed there is negative correlation; some have claimed that economic development is influenced positively by the population growth and some have asserted that there is no relationship between them (Günsoy and Tekeli, 2015).

In the literature, some researchers suggest that population growth inhibits economic growth or slows it down. There are also studies that have found that such a relationship is not the case. In a study involving 84 countries, Barro (2001) concluded that economic growth and population growth affect each other negatively by conducting a panel data regression analysis. In the same way, Blanchet (1991) asserted that after the 1970s, the relationship between the population growth and economy of the developing countries was negative. In his study of data from between 1960 and 1985, including 71 countries, Brander and Dowrick (1994) suggested that economic growth is negatively affected by the high population growth because of its impacts on savings and investment. Boltvinik (1999), in his analysis of 1941–1995, stated that, unlike the Malthusian hypothesis, there is a positive link between Gross National Product and population growth rates in Mexico. Lucas (1988) asserted that human capital is also significant for economic development. Kelley and Schmidt (1994) have found out that economic growth is negatively affected by population growth. Dawson and Tiffin (1998) examined India's long-term relationship between population and economic development and concluded that there was not an explicit link between them. Using similar techniques, Thornton (2001) analyzed the connection between population and economic growth for seven Latin American countries and stated that there was not a link between two variables. As it is seen from the studies, there are three different ideas about the relationship between population growth and economic growth. The proponents of these ideas fall back on theoretical models to back up their claims and findings. There are three main approaches to explain the relationship: pessimistic, optimistic, and neutralist theories.

1.5 THE PESSIMISTIC THEORY

The notion underlying the pessimistic theory goes back to 1790s when Thomas Malthus questioned the future of the world in case of higher population growth. Malthus claimed that the rapid population growth causes poverty (Darrat and Al-Yousif, 1999). According to the Malthusian model, the link between population and economic growth is twofold. Higher incomes promote early marriages and higher fertility, reduce the mortality rate caused by malnutrition and other factors, and cause population increase. However, the increasing population reduces the per capita income. This dynamic interaction forms the basis for the Malthusian model in terms of determining population and revenue (Çetintaş and Başel, 2004). Also according to this model, a long-term disagreement is present between the agricultural structure and the population structure. If the population growth rate is not controlled, the future of mankind will be jeopardized because foodstuffs will not be able to feed the growing population (Deliktaş, 2001). In other words, Malthus has argued that the population increases geometrically, but that the things people need are arithmetically increased. According to him, population growth will ultimately make the needs unattainable and cause human catastrophe. Another negative effect is the reduction in capital per individual and lower living standards; as the number of people gets higher, the need for houses, employment, social, and health security gets increased, for this reason, income per capita decreases accordingly (Güneş, 2005).

Although the Malthusian model was predominant in the 18th century, it remained with us during the 20th century as well. For instance, the US National Academy of Sciences (NAS) carried out studies at the beginning of 1970s and revealed that the population growth would have a negative effect on the economic development. The reason that the world's population had been getting larger, rapid population growth continued to bother the minds.

However, at the beginning of 1980s economists started to question the pessimistic theory and empirical studies weakened it. The more emphasis was given to technology and human capital accumulation and instead of short-term effects, intermediate or long-term effects were investigated (Bloom et al., 2001).

1.6 THE OPTIMISTIC THEORY

The second approach is Revisionist which opposes the Malthusian approach and claims that the accumulation of human capital goes up with the higher birth rate and economic growth is promoted. Both Malthusian and Revisionist approaches focus on the income per capita (Darrat and Al-Yousif, 1999). However, the latter has emerged, because in many countries the fertility did not increase with the increase in the income per capita as Malthus suggested, on the contrary, it showed a rapid decline (Becker et al., 1999). After the 1950s, together with population growth technological developments occured, capital and standards of living increased. Technological progresses occured both in agriculture and industry.

Significant innovations occurred in daily life of people and working conditions. Crucial institutional innovations were witnessed. There were improvements in the fields of education, health, and political participation (Bloom et al., 2001). These developments showed that Malthus's theory ignores the effect of technology on growth. Neoclassical growth models are largely based on technological improvements. Rapid and continuous technological improvements influence the individual earnings and the welfare level. In economic theory, with the advent of technological progress and the accumulation of human capital instead of physical capital, the idea that long-term economic growth could be possible with the accumulation of human capital has become widespread (Bloom et al., 2002). As technological development primarily increases the return of human capital, families start to have fewer children and increase the life quality of their kids. According to this, there is a two-way effect of technological development on population growth. The first effect is the increase in the resources to raise the children well and the increase in children's qualifications, and the second effect is the increase in the accumulation of children's capital by increasing the quality of the child. In this context, the rise in human capital accumulation, in the long term, has positive effects on technological development and economic growth (Galor and Weil, 1999).

Simon has an empirically pioneering work on population growth and economic growth. Simon (1989) stressed that population growth positively affects living standards, depending on economic variables and circumstances, but noted that this effect will emerge in the long run. Simon also points out that the level of technological development and population growth of the country is very important in relation to the growth of population and economy. The length of the period is important in determining the connection between population and growth. When the population growth rate and GNP per capita are observed for a long time, it could be seen that a rise in the birth rate has a negative influence on economy in the short term; however, in the long run, this effect tends to be positive (Simon, 1989). As Simon, Kuznets claimed that as population increases, the stock of human ingenuity indicates a rise as well. Although the growth of population causes a pressure, especially on resources, human beings are resourceful, creative, and innovative (Bloom et al., 2001).

Keynesian and post-Keynesian models also indicate that growth is influenced positively by the population growth. Keynesian model acknowledges that population growth will stimulate investments through effective demand. Population growth will increase demand for existing industries and encourage the establishment of new industries so the market will expand. In addition, the young population resulting from population growth will be more open to growth, as well as more susceptible to new ideas and resources (Telatar and Terzi, 2010).

1.7 THE NEUTRALIST THEORY

In the mid-1980s, another theory called as Neutralism became widespread among scholars. This theory explains that there is no positive or negative effect of population growth on economic growth. Empirical studies carried out on population and economic growth displayed that there was little significant connection between the two. Some countries having rapidly growing populations were prone to have worse economies. This fact was investigated and it was revealed that when the factors of country size, openness to trade, education level of the community, and the public institutions were considered, the negative correlation between the population growth and the economic development melts away. It implies that there are other variables influencing the correlation. The neutralist theory is the dominant view today (Bloom et al., 2001).

According to Allen Kelley, who is a supporter of Neutralist theory, three very important factors are affected by population growth: natural resources, savings, and diversification of resources (Kelley, 1988):

Natural resources: Neutralist theory does not support the idea that natural resources will be consumed by the population growth, which is the underlying theme of the Pessimistic theory. Natural resources can be saved with technology, conservation, and efficient market allocation of resources. Also, supply and demand of the sources are based on per capita income.

Saving: The so-called idea of population growth causes reduction in savings and that shows economic deterioration is not supported with scientific studies. It is difficult to talk about an apparent relationship.

Diversification of resources: According to the Pessimistic theory, the increase in population leads to a decrease in the natural resources and the investments, the direction of the investments turns from capital accumulation to the investments of nutrition, health, education, and social welfare programs which are seen as economically less productive areas. However, cross-national studies displayed that this did not happen to any great extent.

Population neutralism has motivated economists to disregard demography when conceiving the future welfare and improvement of the world's countries (Bloom et al., 2002).

1.8 NEW CHALLENGE FOR ECONOMIC GROWTH: POPULATION AGE STRUCTURE

The optimistic and pessimistic theories neglect the population age structure; however, this is a dynamic that is as important as the population growth. The reasons lying behind having an aging population are life expectancy, reduced fertility rates, and immigration. The societies having an aging population experience a strain on their economy particularly to supply support systems for the elderly (Anderson and Hussey, 2000). Health and educational investments for younger generation, healthcare, and retirement income for older generation cover a large part of the country income, which has a negative effect on labor supply and savings (Bloom et al., 2001). When the population gets older, the labor market has a challenge to get enough labor force, in the same way, aging of the existing workforce affects labor supply. In such a situation, it is easy to imagine that the effective labor supply will be shrinking (Du and Yang, 2014). Demographic aging leads changes in social life as well, especially in family structure, relations between generations, the education system, social insurance, and the entire set of values that guide society (Rimashevskaia et al., 2013).

As for the age of workforce, there can be stereotypes about older workers in terms of their performance, potential, productivity, and motivation. From the point of employers, it will be difficult to train older workforce, whereas we live in such an age characterized by technology and lifelong learning. According to a study conducted by Becker et al. (2006) on German companies, it was revealed that many companies are reluctant to deal with the training of older employees. Recruiting older workers and arranging trainings for them is a challenge for companies. On the other hand, with an older population, consumption behavior and product ranges will be affected. Namely, there will be an increase in older consumers' market segment with significant portion of pensioner households. Expenditures in education, health, and insurance will change accordingly (Perlitz et al., 2010).

As a result, the share of working-age people in a population significantly affects the country's economy. It is crucial for policymakers to consider the population age structure of their country because it affects their country's future. The aging population is pressing particularly in the developing world, which can cause unemployment and instability (Bloom et al., 2001). Although some countries have experienced this demographic shift before than others, the implications show that there is such a trend throughout the world. The governments should follow effective policies to handle this challenge.

1.9 CONCLUSION

The relationship between demographic transition and economic growth was examined in this study. The demographic transition theory claims there are four phases of population growth and all these phases are affected by social, technological, and industrial developments. The fertility and mortality rates show fluctuations as the societies get through these phases, and at the last phase, they reach a balance between the rates of fertility and mortality. There are different approaches explaining the connection between the demographic transition and economic growth. In fact, it can be said that the influence of the population on the economic growth will vary from country to country and time. Especially in some of the underdeveloped countries, growth in population may be an equal contribution to economic development, while in some cases it may be an obstacle to economic development. Again, the net effect may be negligible in some other countries (Kelley, 1988). The increase in population can create adverse effects on per capita income and lower living standards. On the other side, it provides a rise in population and labor, creating large markets and high production volume. Thus, as specialization and learning increase, life standard increases (Çetintaş and Başel, 2004). As it is seen there are contrasting claims about the relationship between economic and population growth. In literature, there are three theories to explain the effect of population growth on economy, these are optimistic, pessimistic, and neutralist theories. As the optimistic theory emphasizes the positive correlation between population and economic growth, the pessimistic theory asserts that population growth hinders the economic development. The neutralist theory, quite the contrary to optimistic and pessimistic theories, claims that there is no significant relationship between population and economic growth. Although this dispute has long been preceded, recent studies on economic growth have shown that a form of human capital focuses on the talent and experience of the workforce is more important. According to these studies, human investment has a more lasting influence on economic growth (Pala and Teker, 2014). The population policies to be implemented should not only be limited to increasing per capita income but also create the appropriate physical and social environment and aim at providing better education, health, work, and productive potential and equipment to the individuals (Farah, 2004), because there is another challenge for the world: aging population. The countries completed demographic transition tend to have aging population which is another threat to economic growth. A remarkable number of indicators give the signal that there will more and more elderly people living alone in both developed and developing countries, which influences the healthcare expenditures, issues about social support, pension programs, and finance (Yaakop and Nor, 2013). In such a situation, the countries will have older workforce and they will probably have difficulty in providing effective labor supply, which will bring strain on their economy. As a result, counties are supposed to design population and economic growth planning, which is specific to their needs and expectations.

KEYWORDS

- **ageing population**
- **demographic transition**
- **economic growth**
- **fertility and mortality rate**
- **life expectancy**

REFERENCES

Agyei, W. K. A. *Fertility and Family Planning in the Third World: A Case Study of Papua New Guinea*; Croom Helm: London, 1988.

Anderson, G. F.; Hussey, P. S. Population Aging: A Comparison Among Industrialized Countries. *Health Aff.* **2000,** *19* (3), 191–203.

Barro, R. J. Human Capital and Growth. *Am. Econ. Rev.* **2001,** *91* (2), 2–17.

Becker, G. S.; Glaser, E. L.; Murphy, K. M. Population and Economic Growth. Papers and Proceedings of the One Hundred Eleventh Annual Meeting of the American Economic Association. *Am. Econ. Rev.* **1999,** *89*, 145–149.

Bhende, A. A.; Kanitkar, T. *Principles of Population Studies*; Himalaya Publishing House: Bombay, 1978.

Blanchet, D. Estimating the Relationship Between Population Growth and Aggregate Economic Growth in Developing Countries: Methodological Problems. In *Consequences of Rapid Population Growth in Developing Countries;* Taylor & Francis: New York, 1991; pp 67–98.

Bloom, E. D.; Canning, D.; Sevilla, J. Economic Growth and the Demographic Transition. *National Bureau of Economic Research*, Working Paper 8685, 2001.

Bloom, E. D.; Canning, D.; Sevilla, J. *The Demographic Dividend, a New Perspective on the Economic Consequences of Population Change*; Paper 8808; RAND Corporation: Santa Monica, CA, 2002.

Boltvinik, J. Population and Economic Growth: Does Population Control have a Scientific Basis? *DEMOS* **1999,** *12*, 4–5.

Brander, J. A.; Dowrick, S. The Role of Fertility and Population in Economic Growth: New Results from Aggregate Cross-national Data. *J. Popul. Econ.* **1994,** *7* (1), 1–25.

Caldwell, J. C.; Orubuloye, I. O.; Caldwell, P. Fertility Decline in Africa: A New Type of Transition? *Popul. Dev. Rev.* **1992,** *18* (2), 211–242.

Çetintaş, H.; Başel, H. Türkiye›de Nüfus ve Hayat Standartları Arasındaki Uzun Dönemli Ilişkinin Ekonometrik Analizi. *Sosyal Siyaset Konferansları Dergisi* **2004,** *47*, 95–106.

Darrat, A. F.; Al-Yousif, Y. K. On the Long-run Relationship Between Population and Economic Growth: Some Time Series Evidence for Developing Countries. *East. Econ. J.* **1999,** *25* (3), 301–312.

Dawson, P. S.; Tiffin, R. Is There A Long-run Relationship Between Population Growth and Living Standars? The Case of India. *J. Dev. Stud.* **1998,** *34*, 149–156.

Deliktaş, E. Malthusgil Yaklaşımdan Modern Büyümeye. *Ege Akademik Bakış, İktisadi ve Sosyal Bilimler Dergisi* **2001,** *1* (1), 92–114.

Demeny, P. Early Fertility Decline in Austria–Hungary: A Lesson in Dsemographic Transition. In *Population and Social Change*; Glass; D. V., Revelle, R., Eds.; Edward Arnold: London, 1972; pp 153–172.

Du, Y.; Yang, C. Demographic Transition and Labour Market Changes: Implications for Economic Development in China. *J. Econ. Surv.* **2014,** *28* (4), 617–635.

Dyson, T. *Population and Development. The Demographic Transition*; Zed Books: London; New York, 2010.

Farah, F. M. Population and Social Developmental Policies. *Health Millions* **2004,** *29* (6), 5–6.

Galor, O.; Weil, D. N. From Malthusian Stagnation to Modern Growth. *AEA Papers and Proceedings* **1999,** *89* (2), 150–154.

Galor, O.; Weil, D. N. Population, Technology, and Growth: From Malthusian Stagnation to the Demographic Transition and Beyond. *Am. Econ. Rev.* **2000,** *90* (4), 806–828.

Güneş, Ş. Türkiye'de Nüfus Artışının Ekonomik Büyümeyle Ilişkisi Üzerine Ekonometrik bir Analiz. *Ankara Üniversitesi SBF Dergisi* **2005,** *60* (3), 123–136.

Günsoy, G.; Tekeli, S. Nüfusun Yaşlanması ve Ekonomik Büyüme Ilişkisi: Türkiye Üzerine bir Analiz. *Amme İdaresi Dergisi* **2015,** *48* (1), 35–87.

Hatton, T. J.; Williamson, J. G. *Global Migration and the World Economy: Two Centuries of Policy and Performance*; MIT Press: Cambridge, MA, 2006.

Kelley, C. A. Economic Consequences of Population Change in the Third World. *J. Econ. Liter.* **1988,** *27*, 1685–1728.

Kelley, C. A.; Schmidt, R. M. Population and Income Change: Recent Evidence. *World Bank Discussion Paper 249*, 1994.

Keyfitz, N. Form and Substance in Family Demography. In *Family Demography: Methods and Their Application*; Bongaarts, J., Burch, T. K., Wachter, K. W., Eds.; Clarendon Press: Oxford, 1987; pp 4–16.

Lee, R.; Reher, D. S. Introduction: The Landscape of the Demographic Transition and Its Aftermath. In *Population and Development Review*; Demographic Transition and Its Consequences; Lee, R., Reher, D. S., Eds.; The Population Council Inc.; New York, 2011; Vol. 37, pp 1–7.

Lucas, R. E. On the Mechanism of Economic Development. *J. Monet. Econ.* **1988,** *22*, 3–42.

Lutz, W.; Sanderson, W. C. Introduction. In *The End of World Population Growth in the 21st Century*; Lutz, W., Sanderson, W. C., Ŝerbov, S., Eds.; Earthscan: London, 2005; pp 1–16.

McNamara, R. Demographic Transition Theory. In *International Encyclopedia of Population*; Ross, J. A., Ed.; The Free Press: New York, 1982; Vol. 1, pp 146–147.

Nielsen, R. W. Demographic Transition Theory and Its Link to the Historical Economic Growth. *J. Econ. Polit. Econ.* **2016,** *3* (1), 32–49.

Pala, A.; Teker, D. AB-27 Ülkeleri ve Türkiye'de Ekonomik Büyümeyi Etkileyen Faktörlerin Belirlenmesi: Statik Panel veri Modeli Uygulaması. *İşletme Araştırmaları Dergisi* **2014,** *6* (1), 151–162.

Perlitz, M.; Schulze, L.; Wilke, C. B. The Demographic and Economic Transition in Central and Eastern Europe: Management Implications. *J. East Eur. Manag. Stud.* **2010,** *15* (2), 149–176.

Potts, M.; Selman, P. *Society and Fertility*; Macdonald and Evans: London, 1979.

Reher, D. S. Economic and Social Implications of the Demographic Transition. In *Demographic Transition and Its Consequences*; Population and Development Review; Lee, R., Reher, D. S., Eds.; the Population Council Inc.: New York, 2011; Vol. 37, pp 11–33.
Reher, D. Population and the Economy During the Demographic Transition. *Economic Aff.* **2012,** *32* (1), 10–16 https://doi.org/10.1111/j.1468-0270.2011.02122.x
Rimashevskaia, N. M.; Dobrokhleb, V. G.; Medvedeva, E. I.; Kroshilin, S. V. Demographic Transition: The Specificity of the Russian Model. *Rus. Soc. Sci. Rev.* **2013,** *54* (4), 63–75.
Simon, J. L. On the Aggregate Empirical Studies Relating Population Variables to Economic Development. *Pop. Dev. Rev.* **1989,** *15*, 323–332.
Telatar, O. M.; Terzi, H. Nüfus ve Eğitimin Ekonomik Büyümeye Etkisi: Türkiye Üzerine bir Inceleme. *Atatürk Üniversitesi İktisadi ve İdari Bilimler Dergisi* **2010,** *24* (2), 197–214.
Thornton, J. Population Growth and Economic Growth: Long-run Evidence from Latin America. *South. Econ. J.* **2001,** *68*, 464–468.
Warf, B. Demographic Transition. In *Encyclopedia of Geography*; Warf, B., Ed.; SAGE: Thousand Oaks, CA, 2010; pp 708–711.
Yaakob, U. H.; Nor, N. N. F. The Process and Effects of Demographic Transition in Penang Malaysia. *Kajian Malays.* **2013,** *31* (2), 37–64.

CHAPTER 2

Demographic Transition and Economic Growth of India

G. K. SINHA*

Research and PG Department of Economics, Arya Mahila Post Graduate College, Banaras Hindu University, Varanasi, India

E-mail: gksinha30@gmail.com

ABSTRACT

Demographic transition along with the growth in number there are changes in the population age structure. While the importance of the population growth as a determinant of quality of life is universally understood, the profoundly serious consequences of changing age structure especially if it occurs too rapidly is not understood by many.

Urbanization—has been another key development which, almost certainly, has facilitated the reductions in mortality and fertility that are integral components of the demographic transition. Among other things, urban areas offer economies of scale in terms of providing both health and family planning amenities. It is much harder to raise a large number of children if you live in a town. Also, modern forms of water supply, sewage, and access to education are usually somewhat better in the urban sector.

Thus, India's future demographic evolution will also have significant implications for the economy, education, and the environment. Economic growth may be enhanced by the demographic bonus. Benefits will arise if there are consequential increases in savings and investment. But such increases cannot be taken for granted.

2.1 INTRODUCTION

Demography is the systematic study of population. The origin of the word "demography" is Greek and it is a combination of the two words, namely, *demos* (people) and *graphein* (describe), which connotes the description of people. Demographic studies are associated with the trends and processes of population. These include studies on the changes in population size; patterns of births, deaths, and migration; and the structure and composition of population, such as the relative proportions of women, men, and different age groups. For convenience of the study, demography is usually divided into two categories: first, formal demography, which is a largely quantitative field; and second, which focuses on the social, economic, and political aspects of the population of a country. Demographic studies are based on the processes of counting or reckoning, namely, census or survey. A demographic study involves the systematic collection of data on the people residing within a specified geographical territory.

Toward the end of the 19th century, demographic transition was discussed globally. During the post-Malthusian era, the exceptional increase in population growth was ultimately reversed, bringing a significant decline in fertility rates and population growth in various regions of the world. Economies gained from technological progress, and this led to the growth of countries' per capita income by demographic transition. It increased labor productivity and growth process in three steps. First, the fall in population growth reduced the retardation of the growing stocks of capital and infrastructure, increasing the amount of resources per capita. Second, the decrease in fertility rates allowed the reallocation of resources from the quantity of children toward their quality and enhancing human capital formation and labor productivity. Third, due to the effect of declined fertility on the age distribution of the population, a fraction of labor force temporarily increased in the population and, therefore, mechanically increased the per capita productivity.

2.2 DEMOGRAPHIC TRANSITION

The term "demographic transition" refers to the secular shift in fertility and mortality from high and sharply fluctuating levels to low and relatively stable ones. Due to the emergence of democratic governments, industrialization, increase in urbanization, and progressive growth in educational levels of human populations during the past half millennium, the human society has remarkably changed. These are the most significant changes that have widely affected the human society.

The demographic transition theory is a well-known discussion topic on the human population phenomenon around the world. During a transition period, mortality declines, and, after some decades, fertility declines. This changes population growth rates, size, and age distribution, and the trend continues for many decades. Thus, demographic transition, along with growth, changes the structure of population.

2.3 DEMOGRAPHIC TRANSITION THEORY

The demographic transition theory is curiously paradoxical. While there is no general agreement on its explanatory frameworks, it is constantly invoked as if there were. This is due, in part, to demography's post-World War II desire to be a science and to establish a "grand theory" of population and fertility decline. Yet, even in this quantitatively based discipline, wider intellectual currents have intruded: postmodernism and cultural and anthropological explanations have recently entered the field. Short-term theory is now the order of the day.

In 1937, Kingsley Davis (1908–1997), an early exponent of the theory, suggested that ultimately the reproduction of species is incompatible with an advanced industrial society. Thus, he established the major framework of the theory, which held that socioeconomic development and modernization are major causal forces. This element of the theory remains robust. In a 2006 volume titled *Demographic Transition Theory*, one of the most influential 20th-century exponents, John C. Caldwell, argued that in industrial societies, the old values essential to agricultural societies—strong family, virginity, loyalty, and legitimate births—are no longer necessary except as "social pacifiers." Industrial societies and, even more, postindustrial knowledge economies need mobile, educated individuals unencumbered with babies and family ties.

2.4 GLOBAL POPULATION SCENARIO

The global population scenario is an important feature to appraise a country's demographic transition scenario as well its role in the world population. The world population in 1901 was 1.6 billion. It was 3 billion in 1960 and 5 billion in 1987, while in 1999, it was 6 billion. Every 12–13 years, 1 billion people are added. The birth rate decreased substantially during the 1990s. Increasing employment among women, urbanization, and rising literacy, as well as decline in infant mortality rate have been some major causes responsible for a growing desire for smaller families. Apart from this, the increasing awareness and easy availability of contraceptive devices have made it possible for the majority of the couples to attain a desired family size.

However, in some countries, the rate of population growth has slowed down because of increase in mortality, for example, HIV-related mortality in the Sub-Saharan Africa. The impact of these downfalls on the growth of global population was steeper than the earlier estimates in decades of the 1990s. Presently, the annual population increment is about 80 million, but it is expected to decrease to 64 million by the years 2020–2025 and 33 million by the years 2045–2050. Developing countries play a very vital role in this movement. Many demographers think that the current trends of rapid decrease in population growth will continue for the next few decades. The Population Division of the United Nations stated that the global population will likely grow to 8.9 billion by the year 2050.

2.5 DEMOGRAPHIC GROWTH IN INDIA

In the 20th century, the population of India was only around 238.4 million. In a period of 110 years, the population of India increased more than fourfold, that is, 1210 million in 2011. In the first half of the 20th century, the population growth of India was one and a half times. However, in the latter half, it registered a phenomenal three-time increase. Table 2.1 shows the population statistics of India since 1901. Apart from this, the table also divulges some other indicators of growth rate, such as decadal growth, average annual exponential growth rate, and progressive growth rate since 1901 over each decade.

2.6 THEORY OF DEMOGRAPHIC TRANSITION

The demographic transition theory implies that population growth is coherent with the overall levels of economic growth. It also implies that every society follows a distinctive pattern of population growth related to development. The Indian demographic history suggests four basic stages of population growth: the first stage was stagnant population (1901–1921); the second was steady growth (1921–1951); the third stage was rapid and high growth (1951–1981); and the fourth stage was high growth with definite signs of slowing down (1981–2011).

The demographic cycle has been described in three stages. In the first stage, the growth of population was low in underdeveloped and technologically backward societies. The death and birth rates were very high. Thus, the difference between the two (or the net growth rate) was low. In the third stage in a developed society where both the death rate and the birth rate have reduced considerably, a low growth rate persists and the difference between them is again small. Between these two stages is a transitional stage of movement from a backward to an advanced stage, and this stage is characterized by very high rates of growth of population.

Currently, India is the second most populous country after China and will soon become the most populous country in the world. Studies reveal that the four principal elements have work behind the demographic transition and change of population size in a society, that is, mortality rate. Similarly, decreasing infant mortality rate in the early stages of high fertility also accelerates the growth of population.

For critical analysis of the change in the trends of the demographic structure of India, it is necessary to analyze the historical trends and pattern of population size. In the year 1800, the first census was promulgated in England, which was also the dawn of census in the demographic history of world. In India, the estimation of the population size mainly came from the decadal census by the Government of India. In 1871, the first census of India was done; however, modern census began in the year 1881.

Every 10 years, the Census of India provides continuous population statistics. Perhaps it is fairly chances to changes in population size of the country within the 10-year duration, because a 10-year period is not a small interval. Hence, to get a somewhat good picture of changes in population size, some experts suggest that every 5-year interval based census would be beneficial for underdeveloped economies.

2.7 ECONOMIC GROWTH POST-1991

The main intention of liberalization was eliminating the license-permit raj by decreasing government intervention in business and business processes, thereby pushing economic growth through reforms. The liberalization policy opened up the country to the global economy. This promoted competition in the market and discouraged the monopoly of the public sector.

The size of the economy often reflects the strength of a country. Likewise, gross domestic product (GDP) represents the total worth of the goods and services produced in a country in a particular year. In 1991, India's GDP stood at Rs. 586,212 crore. After 25 years, it rose to Rs. 13,576,086 crore, an increase of 2216%. In terms of US dollars, India's GDP crossed the \$2 trillion mark in 2015–2016. In terms of nominal GDP, India is currently ranked ninth in the world. It would be the second largest economy in the world by 2050.

After Chinese postreforms, India remained the second fastest growing economy in the world. Especially, during 2005–2008, the economy showed an annual 9% growth. After revision of the GDP growth figures by the NDA government and beginning of China's slow down, India is now being billed as the fastest growing major economy in the world, with a growth rate of 7.6% during 2015–2016.

During the initial year of reform, India recorded 74 million USD as foreign direct investment. After that, it was steadily increasing, except between 1997 and 2000 and from 2008 to 2012 due to global economic slowdown. During March 31, 2016, the country received a total FDI of 371 billion USD. Since 1991, the highest FDI inflow of 43.40 billion USD was recorded during the year 2008. In the year 2005–2006, inflow was 175.54%. India attracted 10.55 billion USD worth of FDI in the year 2016. Apart from this, India received 63 billion USD (nearly Rs. 4.19 lakh crore) and replaced China as the top FDI destination, according to *The Financial Times* during the year 2015.

Structure of Indian economy has changed by the growth dynamics with a decline in the share of primary sector, that is, agriculture from 28.4% in the 1990s to about 15% in 2009–2011. Likewise, the contribution of service sector, including construction, has changed from 52% to 65% during this period. Although, the share of industry remained unchanged, that is, 20% of GDP. It reveals that the service sector dominated India's growth acceleration during the two decades. Before being interrupted by the global financial crisis, India's pace of average annual industrial growth was picked up from 5.7% during the 1990s to 9% during 2004–2008.

TABLE 2.1 India—Economic Growth Scenario.

Year/indicators	India's real GDP growth rates (factor cost)	Agriculture growth (%)	Industry growth (%)	Services growth (%)	Forex assets (exc. gold) (US$ billions)
1999–2000	8.00	2.67	5.96	12.05	35.1
2001–2002	4.15	−0.01	6.03	5.07	39.6
2002–2003	5.39	6.01	2.61	6.61	51.0
2003–2004	3.88	−6.60	7.21	6.74	100.6
2004–2005	7.97	9.05	7.32	7.89	131.2

TABLE 2.1 *(Continued)*

Year/indicators	India's real GDP growth rates (factor cost)	Agriculture growth (%)	Industry growth (%)	Services growth (%)	Forex assets (exc. gold) (US$ billions)
2005–2006	7.05	0.18	9.81	8.28	137.2
2006–2007	9.48	5.14	9.72	10.91	177.3
2007–2008	9.57	4.16	12.17	10.06	275.6
2008–2009	9.32	5.80	9.67	10.27	254.6
2009–2010	6.72	0.09	4.44	9.98	283.5
2010–2011	8.59	0.81	9.16	10.50	297.3
2011–2012	8.91	8.60	7.55	9.67	296.7
2012–2013	6.69	5.02	7.81	6.57	296.6
2013–2014	4.47	1.42	0.96	6.96	295.7
2014–2015	4.47	4.71	0.35	6.78	303.7

Source: Planning Commission, DataBook—2014–15.

Table 2.1 reveals that the India's GDP growth was fluctuating during 1999–2000; it was highest during 2007–2008, that is, 9.57 and was lowest 4.15 during 2001–2002. Similarly, agricultural growth rate was shown fluctuating during this period, it was highest, that is, 9.05% in the year 2004–2005 whereas it was negative during 2001–2002 and 2003–2004, that is, −0.01 and −6.60%. Likewise, highest growth rate of industry was 12.17% during 2007–2008 while lowest, that is, 0.35% overall trend was shown fluctuating. Apart from this, growth rate of services was also shown fluctuating during this period; it was highest 12.05% during 1999–2000, whereas lowest was 5.07% in the year 2001–2002.

However, trend of forex asset was shown to be increasing during this period; it was 303.7 billion USD during 2014–2015 while lowest was 35.1 billion USD in the year 1999–2000.

2.8 POPULATION GROWTH IN INDIA

Underdeveloped countries such as India have always struggled with balance of their population size and economic growth. Perhaps it is not easy for any underdeveloped states. In the year 2010, the estimated global population was 6908.7 million. Currently, 10 countries account for nearly 60% of the world population. China, India, and the United States were the three most populous country, it together accounts for four of every 10 people in the world. Apart from this, one out of every six persons in the world is from India. The population of India is almost equal to the combined population of six countries, that is, the United States, Indonesia, Brazil, Pakistan, Bangladesh, and Japan put together—the population of these countries totals 1214.3 million.

TABLE 2.2 Population and Its Growth 1901–2011.

Census year	Population	Decadal growth		Change in decadal growth		Average exponential growth rate (%)	Progressive growth rate over 1901 (%)
		Absolute	Percent	Absolute	Percent		
1901	238,396,327	–	–	–	–	–	–
1911	25,209,339	13,697,063	5.75	–	–	0.56	5.75
1921	251,321,213	−772,177	0.31	−14,469,240	−6.05	−0.03	5.42
1931	278,977,238	27,656,025	11.00	28,428,202	11.31	1.04	17.02
1941	318,660,580	39,683,342	14.22	12,027,317	3.22	1.33	33.67
1951	361,088,090	42,427,510	13.31	2,744,168	−0.91	1.25	51.47
1961	439,234,771	78,146,681	21.64	35,719,171	8.33	1.96	84.25
1971	548,159,652	108,924,881	24.80	30,778,200	3.16	2.20	129.94
1981	683,329,097	135,169,445	24.66	26,244,564	−0.14	2.22	186.64
1991	846,421,039	163,091,942	23.87	27,922,497	17.12	2.16	255.05
2001	1,028,737,436	182,316,397	21.54	19,224,455	10.54	1.97	331.52
2011	1,210,193,422	181,455,986	17.64	-860,411	−0.47	1.64	407.64

Source: Census—2011.

It is evident from the Table 2.2 that the population and decadal growth rate has increased during census year 1901–1971. It was highest in 1971, that is, 24.80% while lowest was 5.75% in 1901. Similarly, average exponential growth rate was lowest, that is, −0.03% in 1921, whereas it was highest in 1981, that is, 2.22%. From 1971 to 1991, it was ranging between 2.20% and 2.16%. While progressive growth rate was highest during 2011, it was 407.64%, and lowest was 5.42% in 1921.

During the 2001–2011 decade, the population increment was lesser when compared to the previous decade, while 1911–1921 was exception in this sense. This connotes that because of the combination of population momentum and somewhat impeded fertility, although India continues to grow to size, in pace of net addition it is on the decrease.

Similarly, during the decade of 2001–2011 in the absolute terms the population of India increased by about 181 million. Even so, during this period the net addition in population increased consistently. In the decade of 1960s, the net addition showed a steadily declining trend. While during 1981–1991 and 1971–1991 decade, 27.9 million more people were added. However, this number declined to 19.2 million for the decades between 1981–1991 and 1991–2001.

2.9 GENDER COMPOSITION IN INDIA

Census of India has provided various aspects of information of the population statistics of country. Gender composition is vital component of the census and it reflects socioeconomic condition of both genders. It provides logical and meaningful demographic analysis of a country. Any changes occurring in gender statistics mostly reflect the cultural, social, and economic patterns of the society in different ways. At a given point of time in a society, sex ratio has provided a vital social indicator to measure the extent of prevailing equity between males and females.

In India, sex ratio is inflected in favor of males and has continued to rise and expand in various forms. The policy-makers should take this issue seriously to check and control this trend to bring it back to parity. Demographers argued that some factors have influenced changes in sex ratio such as sex-selective out-migration, differential in mortality, and skewed sex ratio at birth are the major contributory factors.

Apart from this equality between man and women is vital issue of human rights and a condition for social justice. It is also a necessary and fundamental condition for equality development and peace for a country's overall growth.

Gender composition presents migration, mortality, and nasality as a character of a given population. For cross-sectional analysis to measure gender balance and equality, sex ratio is a widely used tool. Distribution pattern of men and women in a population affects relative roles and economic relationship. The gender composition is an important demographic characteristic of human population around which logical and meaning analysis is prepared.

TABLE 2.3 Sex Ratio in India 1901–2011.

Census year	Sex ratio (females per 1000 male)
1901	972
1911	964
1921	955
1931	950
1941	945
1951	946
1961	941
1971	930
1981	934
1991	927
2001	933
2011	940

Source: Census—2011.

Table 2.3 discusses the sex ratio of India from 1901 to 2011. Sex ratio was not in favor of females. After independence, it became more pathetic. In the pre-independence period, the sex ratio declined continuously up to 1951, when it rose marginally. Postindependence it continued and the sex ratio slipped down for two consecutive decades after 1951 to reach 930 in 1971. A steep fall of 11 points was seen in the sex ratio during the decade of 1961–1971. Highest sex ratio was 972 during 1901, which is pre-independence period while lowest in 1991 Census, that is, 927.

2.10 CRUDE BIRTH RATE AND TOTAL FERTILITY RATE IN INDIA

Crude birth rate and total fertility rate are the vital components for analyzing of demographic transition in a society. Crude birth rate gives a scientific and logical way to find the change in population. Birth rate and total fertility rate provide a demographic change in a society in a specific period. Crude birth rate reflects the number of live births during the year per 1000 populations estimated at mid-year. The total fertility rate is the most effective and useful measure of fertility in understanding population. It represents the number of children the typical women in that population will have over their child-bearing years. To find the rate of natural increase we subtract the crude death rate from the crude birth rate which is equal to the rate of population change in the absence of migration.

Mortality rate provides detail and essential statistics in analyzing population change and demographic studies and public health administration in a society. It is the vital component for projection of life tables and population in a country like India.

TABLE 2.4 Crude Birth Rate and Total Fertility Rate in India—1971–2013.

Year	CBR	TFR
1971	36.9	5.2
1981	33.9	4.5
1991	29.5	3.6
2001	25.4	3.1
2002	25.0	3.0
2003	24.8	3.0
2004	24.1	2.9
2005	23.8	2.9
2006	23.5	2.8
2007	23.1	2.7
2008	22.8	2.6
2009	22.5	2.6
2010	22.1	2.5
2011	21.8	2.4
2012	21.6	2.4
2013	21.4	2.3

TFR, total fertility rate.

Source: Health & Family Welfare Statistics in India—2015.

It is evident from the Table 2.4 that the crude birth rate at national level was 36.9 in the year 1971; it declined to 33.9 in 1981, about 8% fall during this period. This trend was persisting during 1991–2013; the decline was about 27.5%, from 29.5 to 21.4. The rural–urban differential was also meager during this period. Whereas, during the last three decades, crude birth rate in rural areas was continuously higher in comparison to urban areas. Likewise, the trend of total fertility rate (TFR) declined during 1971–1981 from 5.2 to 4.5 and from 3.6 to 2.3 during 1991–2013. From 1971 to 2013, the TFR in rural areas declined from 5.4 to 2.5. While in the urban areas, it declined from 4.1 to 1.8 during the same period. During 2013, around 74% of the deliveries were institutional which includes Government hospitals as well as private nursing home and hospitals. The percentage share of institutional deliveries performed in urban areas and rural areas was significantly different, that is,

92.1% delivery in urban areas while in rural areas it was recorded 69.7% during this period.

2.11 CRUDE DEATH RATE AND INFANT MORTALITY RATE IN INDIA

Infant mortality rate is considered a primary and important indicator of a geographic territory (country, state, or province) over health status or quality of life. It is the most sensitive and extensively used device to examine and analyze the overall development and change in mortality decline and improvement in population health.

Various public health research studies have found that the development and availability of modern health care have great impact on decline in infant mortality rate.

In the Change Response Model, Davis suggested that it is the first component of demographic transition. In the recent bracing persuasion of demographic transition, Dyson believed "mortality decline is the crucial catalyst; once it occurs, the other four components necessarily follow, and most often sequentially because one component generates the next. Mortality decline leads to an increased rate of natural increase, which produces the conditions that cause fertility decline, which in turn leads to increase in life expectancy and population aging."

TABLE 2.5 Crude Death Rate and Infant Mortality Rate in India 1980–2013.

Year	Crude death rate	Infant mortality rate
1971	14.9	129
1981	12.5	110
1991	9.8	80
2001	8.4	66
2002	8.1	63
2003	8.0	60
2004	7.5	58
2005	7.6	58
2006	7.5	57
2007	7.4	55
2008	7.4	53
2009	7.3	50
2010	7.2	47
2011	7.1	44
2012	7.0	42
2013	7.0	40

Source: Health & Family Welfare Statistics in India—2015.

Table 2.5 reveals that during year 1971–1981, the crude death rate in India declined significantly from 14.9 to 12.5, and thereafter from 9.8 to 7.0 during 1991–2013. The decline was steeper in rural areas as compared to urban areas. The infant mortality rate is very important component of health planning in India. During the decade of 1971–1981, it declined from 129 per 1000 live births to 110 and after that, it declined in the year 1991 and 2013, that is, 80 and 40, respectively. Likewise, the child mortality rate also portrayed a noticeable decline, that is, 51.9, 41.2, 26.5, and 11.0 in the year 1971, 1981, 1991, and 2013, respectively. In 2013, about 39.2% of the deaths were institutional and 60.8% received medical attention other than in institutions.

2.12 CONCLUSION

The present study divulges that the economic growth has been playing very important role to change demographic scenario in developing countries such as India. Therefore, demographic transition gives an excellent framework for analyzing various aspects of growth and development in an economy. Furthermore, in wide sense, it forecasts what will happen. Hence, India's population would continue to grow in next few decades, but in a slowing rate.

It is aftereffect of growth with demographic transition, the changes in the age structure of population. It is universally understood that the importance of the population growth is a determinant of quality of life. Their deeply serious consequence of changing age structure, especially if it is happening too apace, is not understood by many.

Another important component is urbanization, which has been playing a key role in development, which, almost certainly, has facilitated the reductions in mortality and fertility that are integral elements of the demographic transition. Apart from this, in terms of providing both health and family planning amenities urbanization offers economies of scale. It is very difficult to raise a size of family if you live in a town or urban areas. Also, public amenities such as water supply, sewage, and educational institutions, are usually somewhat better in the urban sector.

Therefore, economic growth may be increased by the demographic dividend. Benefits will ascend if there is consequential growth in savings and investment, but this increase cannot be taken for granted. Apart from this, India's future demographic evolution would also have significant implication on the economy.

KEYWORDS

- **demographic**
- **fertility**
- **mortality**
- **industrialization**
- **transition**
- **economic growth**

REFERENCES

Census–2011 Report; http://www.censusindia.gov.in/2011Census/pes/Pesreport.pdf

Dyson, T. *India's Demographic Transition and its Consequences for Development*; Institute of Economic Growth: Delhi, 2008.

Govt. of India. *Economic Survey—2015–2016*.

Govt. of India. *Health & Family Welfare Statistics in India*—2015.

Planning Commission. *DataBook 2014–2015*.

RBI. *Hand Book of Statistics on Indian Economy: 2015–2016*.

SRS. *Statistical Report*; 2013.

Teitelbaum, M. S. Relevance of Demographic Transition Theory for Developing Countries. *Sci. New Ser*. **1975,** *188* (4187), 420–425.

The Economic Times; https://economictimes.indiatimes.com/

The Times of India. https://timesofindia.indiatimes.com/blogs/toi-editorials/(accessed Mar 13, 2017)

www.encyclopeadia.com.

www.firstpost.com.

www.unfpa.org.

CHAPTER 3

A Sociological Approach to the Phenomenon of Forced Mass Migration: The Case of Syrian Asylum Seekers in Turkey

SALİH AYDEMİR[1,*] and MEHMET CEM ŞAHİN[2]

[1]*Department of Sociology of Religion, Faculty of Theology, Harran University, Şanlıurfa, Turkey*

[2]*Department of Sociology of Religion, Faculty of Theology, Ankara University, Ankara, Turkey*

Corresponding author. E-mail: saydemirs@gmail.com

ABSTRACT

Mass and forced migration can cause very serious crises in terms of societies and states due to many social, economic, and cultural problems it has in its nature. As a result of the internal conflicts that started in April 2011 in Syria, mass migration movement from Syria toward neighboring countries, particularly to Turkey, is a social incident, which is very unique due to its numerical size and its realization in a short period, and which has many problems that can be described as a social trauma in terms of the people of the source country and the state and society of the target country. However, it is also a fact that this troubled process has been successfully managed by Turkey. When the period before the arrival of Syrian asylum seekers to the country is compared to the period after their arrival, it is seen that there is no statistically significant difference or negative process with regards to Turkey's current problems related to health, education, security, and unemployment.

3.1 INTRODUCTION

The phenomenon of migration is a social reality, which mankind often encountered in the historical process. When the subject of migration is addressed, the approach that assumes that settled life is an essential output of human nature and migration is an accidental situation is often dominant. In fact, at the beginning of the adventure of life, even primitive human beings who had to meet their basic needs such as food and accommodation had been in search and as a natural result of this search, forced migration movements constituted an essential aspect of human existence.

Today migration is not described as a simple relocation of an individual or a community. Migration is described as a whole process that starts with a change in perception, continues with displacement, and terminates with adaptation to the place migrated (Çakır, 2011, p. 131). In this sense, the "*actor*" of migration is "*human*," who is acting in line with change, experiencing the processes from beginning to the end, and facing its consequences for years including the negative ones (Güvenç, 1972, p. 78). Migration begins with the transformation of voluntary/intentional or involuntary motives, which are brought about by various factors in the human mind, to actions with the objective of displacement. Displacement is followed by the stages of accommodation, settlement, adaptation, and integration. In case no adaptation is achieved in the place of migration and settlement, migration is not regarded as a successful process. That is because individuals, families, or groups who have not been able to adapt to the place where they migrated are considered as potential migrants and it is highly probable that they will tend to relocate at the earliest opportunity (Çakır, 2011, p. 131).

Social scientists working on migration issues have formed an important literature accumulating on the phenomenon of migration taking into consideration many variables ranging from the geographical conditions of the migrated place to the reasons for migration, from quantitative and qualitative characteristics of the migrant population to the social environment conditions of the migrated place. In this context, taking into account the migrated regions and borders, classifying migration types as internal migration and external migration, in general, constitutes the starting point. On the other hand, considering the common reasons and common results of migration, migrations that take place due to driving forces created by natural (geographical) environment may be studied under various titles such as "primitive," "forced," "impelled," "free," and "mass" migrations (Petersen, 1958). However, an important aspect that almost every social scientist working on migration has to confess is that although migration movements have common reasons, forms and consequences, each migration movement is at the same time unique and distinctive. The most important point that should not be overlooked by those who are subjects of migration is to be prepared for the possible consequences of the specific, distinctive, and unique dimensions of the migration they are dealing with. In the context of these developments and changes that take place in the social structure, every society undoubtedly needs to develop a new scientific approach for its migration practice or adapt scientific perspectives and theoretical approaches put forth to date.

Numerous push and pull factors can be listed as reasons for migration. Push factors include: (1) environmental, demographic, (2) economic, and (3) political repressions. Environmental and demographic factors that drive people to migrate include droughts, floods, and earthquakes. The economic conditions that force people to leave their places of residence are poverty, unemployment, and economic crises. Another factor that forces people to leave their places of residence is political repression, among which the most important factors are wars, uprisings, and application of martial law and totalitarian-authoritarian regimes on the people. At this point, it can be said that the best factor describing the situation of Syrian refugees in Turkey, which constitutes the subject of this study, is political reasons.

For whatever reasons people migrate, they do not only bring themselves to the place of migration but at the same time bring the values of the culture they come from (Kirman, 2016, p. 119). Migrants, who realize a displacement with many dimensions between two places, try to maintain the cultural structures and values they were born into and internalized on the one hand, while trying to adapt to the sociocultural conditions of the place of displacement on the other hand. This process brings with it a variety of adaptation problems depending on qualitative and quantitative parameters of the migration that took place. From this point of view, migration should not be regarded as just a change of place that an individual or a group has made. With migration, a series of essential interaction processes and relationships are developed between migrants and those who receive the migrants, such as influence, compromise, amalgamation, exclusion, competition, and conflict resulting from cultural similarities and differences (Fichter, 1994, pp. 110–116). These processes are undoubtedly no momentary but they are social problems that are spreading over time, emerging, and progressing gradually.

The social, economic and cultural problems that take place due to migration also vary according to migration types. For example, in voluntary migration, migration problems are less encountered because the migrant or migrants migrate by analyzing possible problems that may arise in the place of migration. In forced migration, since people often have to abandon their homes because of economic, social or political sanctions of the states, they are classified as asylum seekers or refugees, and therefore they have to deal with various problems resulting from that situation. As a result, asylum seeking and migration are nowadays regarded as an important human rights problem (Aker et al., 2002).

Although migration is as old as human history, studying migration as a scientific subject, treating it as a social problem, and trying to produce scientific solutions is not dated back to old times. Although many institutions and organizations at the national/international level dealing with migration movements and their consequences have formed a considerable literature in order to understand its consequences, the first studies about the theory of migration appeared in the last quarter of the 19th century. Various theories began to be developed from the end of the 19th century in order to reveal the general laws of migration as an important social reality with the beginning of development in social sciences.

3.2 THEORETICAL AND CONCEPTUAL FOUNDATIONS OF THE PHENOMENON OF MIGRATION

Migration, which can be defined as a human movement from one geography to another, should not be undoubtedly evaluated as a momentary event, but rather as a process with effects that will emerge from the point of view of the target country and the source country and on the basis of the interests and expectations of migrants. With its current interpretation, migration means the change of place of residence within a nation state or on an international scale. The aspects that differentiate migrations from each other are the reasons of migration, the content, the form of realization and the way of living (Kanbir, 2016). When we look at the general nature of the scientific paradigms used in the 20th century international migration studies, we see that various theoretical approaches developed at the macro and micro level have become prominent. Macro-level research is more concerned with the causes of migration and migration trends, while micro-level research focuses more on migration results and migrant experiences (İçduygu et al., 2014, p. 34).

Migration studies have been an interdisciplinary subject, in which many disciplines such as sociology, psychology, anthropology, political science, history, geography, economics, demography, and law have been involved as it has a very complex domain of influence affecting all dimensions of social life (Brettell and Hollified, 2000).

Many migration theories have been developed from the end of the 19th century to the present day in order to explain the types of migration, their general characteristics, and the functioning of the process. Since this study is not a complete study of migration theories, within the scope of this study, not all migration theories but those that will help explain the intensive migration influx from Syria to Turkey between the year 2011 and the present date (August 2017), and also some studies that have an important place in the history of migration theories will be mentioned in summary.

Massey et al. (1993) have summarized migration theories in two categories in their study titled "Theories of International Migration." They form a binary classification as theories developed by taking into account the initiation of migration and the ones developed by taking into account the perpetuation and persistence of migration. These are (Massey et al., 1993):

Initiation of migration	*Perpetuation of migration*
Neoclassical macro-migration theory	Migration as a system
Migration as a system	World systems theory
Dual labor market theory	Social capital theory
World systems theory	Institutional theory
Mobility transition	Network theory
Lee's push/pull factors	Cumulative causation
Neoclassical micro-migration theory	Behavioral models
Theory of social systems	

The first academic studies on migration, however, are the two articles titled "The Laws of Migration" that were published in 1885 and 1889 by E.G. Ravenstein. Ravenstein's first article on 1885 is based on the British population census statistics of 1871 and 1881. In the second edition, he published with the same title in 1889, he tried to put forward the general laws of migration with the data collected from more than 20 countries (Lee, 1966, p. 47). Ravenstein's migration laws are as follows:

1. We have already proved that the great body of our migrants only proceeds a short distance and that there takes place consequently a universal shifting or displacement of the population, which produces "currents of migration" setting in the direction of the great centers of commerce and industry which absorb the migrants.

In forming an estimate of this displacement we must take into account the number of natives of each county which furnishes the migrants, as also the population of the towns or districts which absorb them.

2. It is the natural outcome of this movement of migration, limited in range, but universal throughout the country, that the process of absorption would go on in the following manner:
The inhabitants of the country immediately surrounding a town of rapid growth flock to it. The gaps thus left in the rural population are filled up by migrants from more remote districts, until the attractive force of one of our rapidly growing cities makes its influence felt, step by step, to the most remote corner of the kingdom. Migrants enumerated in a certain center of absorption will consequently grow less with the distance proportionately to the native population which furnishes them, and a map exhibiting by tints the recruiting process of any town ought clearly to demonstrate this fact. That this is actually the case will be found by referring to maps 3, 4, 8, and 9. These maps show at the same time that facilities of communication may frequently countervail the disadvantages of distance.
3. The process of dispersion is the inverse of that of absorption and exhibits similar features.
4. Each main current of migration produces a compensating countercurrent.
5. Migrants proceeding long distances generally go by preference to one of the great centers of commerce or industry.
6. The natives of towns are less migratory than those of the rural parts of the country.
7. Females are more migratory than males.

These propositions either have been considered or supported by facts, in the preceding portion of this paper, or they will be considered in connection with the towns (Ravenstein, 1885, pp. 188–189).

As it can be seen, Ravenstein's laws concerning migration are based on economic, social and cultural indicators of the period he lived in. Therefore, it is incomplete and inadequate to understand the current multidimensional migration processes that take place in a more complex manner (Çağlayan, 2006). However, this situation undoubtedly does not make him worthless. Because he prepared a very rich theoretical ground for the migration theories and models developed after him. Castles and Miller, for example, put forth some evaluations and criticisms when they mention Ravenstein's migration laws. In this context, they basically emphasized that this theory was a historical and was based on individual preferences. Because, according to them, Ravenstein's theory is based on the assumption that migrants make rational choices about where they will go, and that they behave according to the cost–benefit calculation. Ravenstein's theory on the basis of individual benefit–cost calculation laid the groundwork for neoclassical economic-based migration studies. Castles and Miller argue that migration movements from low-income regions to high-income regions can also be evaluated within the context of the classical push–pull theories because they are based on economic factors. Because, socioeconomic variables that lead to migration such as low income, low standard of living, lack of economic opportunities, and political pressures are push factors. Variables such as high income, labor demand, economic opportunities, and political freedom are considered as pull factors (Çağlayan quoted from Castles and Miller, 2006, p. 71).

The study of Ravenstein, which is the first example of scientific studies on migration, does not cover all types or different forms of migration such as subtypes. For example, it is insufficient to explain the direction and course of displacements that cannot be expressed as migration such as forced migration, which is the source of situations such as asylum seeking and refugee status that are often encountered nowadays. In particular, evaluating forced, mass population movements from Syria toward its neighbors, which constitute the subject of this research, in line with Ravenstein's laws of migration will lead to certain deficiencies.

Another theory of migration is Stouffer's "The Theory of Intervening Opportunities," which focuses on the migrant as a social actor, an agent rather than the subject of migration itself as a phenomenon. For the first time in 1940, Stouffer mentions intervening opportunities (Yalçın, 2004, p. 28). According to this theory, the number of people migrating to a particular distance is directly proportional to the multiplicity of opportunities in those places. In other words, the more opportunities there are in the place of migration, the more migrants will go there.

This theory tends to explain the phenomenon of migration with a one-dimensional variant and explains migrations to developed industrial cities, metropolis, or megacities for mainly economic purposes. This theory, which is functional in terms of its basic principles, is inadequate in explaining mass migrations due to political reasons such as war.

Migration theories are usually identified with sociologists who have developed that theory. Many theories, however, have been developed by those who contributed to it later than the first theorist who initially put it forward. These contributions sometimes take place as structural changes, quite different from the original form of the theory, and sometimes do not go beyond small additions made to the original structure set forth.

The "pull and push factors" (İçduygu, et al. 2014, p. 36), a paradigm commonly known as the theory of push–pull between migration theories, which are often referred to in understanding and explaining the reasons and course of international migration movements, were first described by Everett Lee in a trim manner, in his article published with the title of "A Theory of Migration" in 1966. The push–pull theory is one of the most emphasized theories among migration theories. With such attribute, many researchers contributed to this theory and led to its development. However, these contributions did not disrupt the original structure of the theory, and Lee's original formulation was preserved.

Like many theorists studying on migration, Lee begins by making a general evaluation of the migration laws of Ravenstein and criticizes the subsequent theorists' studies by expressing that they are no more than a general tendency of the demographic structure of migrants (Todaro, 1980, p. 16). In his theory, Lee focused more on migration than migrant, however, emphasized the need for migrants not to be ignored.

Lee has identified four basic pull and push factors that form the basis of the sociological analysis of migration. These are: (1) factors related to the place of residence, (2) factors related to the place of relocation, (3) encountered obstacles, and (4) individual factors (Lee, 1966, p. 50).

These factors, which Lee identified within his theory, constitute the basic functioning and components of the pull–push theory. According to him, there are pull and push factors both in the place of residence and the place to be migrated. Lee uses the plus (+) sign for pull factors, minus (−) sign for push factors, and zero (0) sign for neutral factors of equal value to all. Neutral values of Lee's theory correspond to factors that are the same for all and have no positive or negative contribution to migration. The coexistence of both pull and push factors creates an integrity. The integrity that this association reveals clearly shows that there are positive and negative variables, namely push and pull factors, which are present both in the place where they live and the place where they will migrate.

Although Lee's theory is one of the important theories among the migration theories, it is more functional in explaining the nature of free migration rather than the processes that are controversial in evaluation with respect to its nature and described by the concepts of asylum seeking and refugee status in migration literature.

Migration theories generally focus more on migration than on forced mass displacements, such as exile, asylum, or refugee. However, since some theories are more abstract and comprehensive, they have the functionality to cover all sorts of individual and mass displacements. In this context, it is regarded that it will be a right choice to evaluate Petersen's migration theory in this category and state that it is more functional to explain the situation of the Syrian refugees, who are subjects of our study.

In his article titled, A General Typology of Migration published in 1958, Petersen created a five-level typology of migration. These are: (1) primitive migrations, (2) forced, (3) impelled migrations, (4) free migrations, and (5) mass migrations.

Primitive migrations: Migrations that are caused by push factors created by natural conditions. It includes migrations that originate from natural conditions such as climate, drought, etc., and mostly carried out collectively.

Forced migrations: It refers to migrations where migration does not occur as a result of the conscious choices of the migrants but occurs in the wake of a decision not taken by themselves but taken by a social or a political mechanism.

It is a term used to describe the migration movement that is carried out because of the constraining elements that include all kinds of threats to human life. Movements made by refugees, internally displaced persons or migrations made due to reasons such as natural, environmental, chemical, nuclear catastrophes, hunger, development projects, etc. For example, the laws that the Nazis put into force in order to enforce the Jewish migration, their anti-Semitic actions, and their policies of prisoning the Jewish people to the ghettos and moving them to the concentration camps is an example of forced migration (Holocaust) (Petersen, 1958, p. 261). Similarly, the migration wave from Syria to other neighboring countries, mainly to Turkey, can be evaluated under this heading.

Impelled migrations: In this type of migration, similar to forced migration, natural or social conditions play a role as the cause of migration. However, in this type of migration, migrants can make their own choices about migration (Petersen, 1958, p. 261).

Free migrations: It refers to the type of migration, where migration decisions are made by migrants as a result of their own conscious choices. This type of migration does not describe mass migration, but personal and small group/family migrations made with individual preferences. In free migration type, there is no coercive state or pull force on people, communities and societies (Petersen, 1958, p. 263).

Mass migrations: The basic motivation of mass migration is technological innovations and progresses. Petersen emphasizes that as transportation opportunities have developed in the world, alternatives have increased, and migration has become a massive situation. The most significant and distinctive characteristic that separates this type of migration from others is that migration has come to be a collective phenomenon (Petersen, 1958, pp. 263–264). Although Petersen did not express, the forced and impelled migrations brought about by causes such as war, civil war, uprising, and political oppression are also mass migrations in terms of quantity. It would be appropriate to describe the migration of the Syrians, which is the subject of this study, as mass migration (considering that about 10 million Syrians have been displaced and about 3 million of them have migrated to Turkey).

This typology of migration, which Petersen has put forth, can be considered as one of the most successful theories of migration in terms of its comprehensiveness and ability to explain all migration movements made in the historical process. Petersen's study allows us to analyze and explain various international migration movements, particularly those of our time (such as the people who fled from war in Africa, the migrant problem between some African countries and France during the colonial period, asylum seekers who escaped from civil war in Syria, etc.) as it includes the effects leading to migration such as push–pull factors, and also individual, social, natural, and political motivations (Çağlayan, 2006).

After the migration typology developed by Petersen, theories emerged that set forth an approach based on the economic dimension and the colonial relations such as the center-periphery theory (dependency school). The dependency school thinkers have divided the world into two parts, center and periphery. When they adopt this approach to migration, they state that migration is directed from undeveloped or underdeveloped countries (periphery) toward developed capitalist Western countries (center). The center-periphery theory emphasizes the dependence of the periphery to the center and the characteristics of the colonial period, and the modern day equivalent of this point of view lies on the basis of capitalism and nation-state (Yalçın, quoted from Marwska, 2014, p. 35).

The evaluation of the dependency school that the underdeveloped countries are one-sidedly dependent on developed countries constitutes a very critical dimension of this theory. Because the dependence of developed countries on underdeveloped countries in terms of raw materials and cheap labor force is undeniable. As a matter of fact, the advocates of this theory, which have studied the phenomenon of migration in the framework of colonialism, argue that the colonialist expansion of the past continues today through neocolonial governments and multinational corporations (Massey et al., 1993, p. 445).

Another theory that studies migration in a scientific way is "Migration Systems Theory." This theory is based on economic and political grounds on the exchange of migrants of two or more countries. According to this theory, two or more countries form a migration system and a relationship chain through mutual exchange of the migrants. There is no doubt that geographical distance, mutual economic interests as well as historical, social, and political processes and relations are influential in the formation of relations (Çağlayan, 2006, p. 82).

Thomas Faist considers "Migration Systems Theory" as a theory of transition between "World Systems Theory" and "Network Theory." According to Faist, world systems theory is an approach with macro perspective. However, Faist, in his study, evaluated the migrant as a basic research object of the situation of migration and focused on the migrant network through emphasizing a micro approach (Çağlayan, 2006, p. 83).

Another theory of migration is Castles and Miller's theory according to which the phenomenon of migration has four basic tendencies. These tendencies are related to (1) migration globalization, (2) acceleration of migration, (3) migration differentiation, and (4) womanization of migration (Castles and Miller, 2003).

The globalization of migration means that the convergence of different cultures, which is the main feature of globalization, is also reflected in the migration movements, and that the migration regions are diversifying in this process. For this reason, it is emphasized that the individuals who come from very different geographical regions are bringing about multicultural structures in the migrant-receiving regions.

The acceleration of migratory movements means that people are more likely to take the risks of migration, and as a result, the volume of migration is increasing. For that case, the migrant-receiving countries have to revise their migration policies and update their policies against this acceleration.

Diversification and differentiation of the outcomes of the current migration movements, not only relate to labor or labor market, but also bring up different status such as asylum seekers or refugees, who deal with many legal, social, cultural, and psychological dimensions that migrants and states may face.

Womanization of migration means that women are now frequently involved in migratory movements due to their increasing participation in the labor force market.

3.3 BASIC CONCEPTS AND DEFINITIONS OF MIGRATION AND MIGRANT STATUS AND LEGAL SITUATIONS OF SYRIANS IN TURKEY

In this study, the basic concepts used to describe the sociological, economic, legal, cultural, and psychological conditions of the Syrians who had to leave their countries are as follows: (1) "migrant," (2) "asylum-seeker," and (3) "refugee." These concepts are often confused with each other in everyday use. These concepts are described in the dictionary published by International Organization for Migration (IOM) as follows:

a) **Migrant:** There is no internationally recognized definition of a migrant. It is accepted that the term migrant covers situations in which the person concerned decides to migrate freely, for "personal comfort" and without any external coercion. The term, therefore, includes people and family members who migrate to another country or region to improve their economic and social status and to increase their future expectations for themselves or their families (IOM, 2009, p. 22).
b) **Asylum seekers:** Persons seeking refugee status in a country within the framework of relevant national or international documents and waiting for the result of their application for a refugee status. As a result of a negative decision these people have to leave the country and if they are not allowed to stay in the country due to humanitarian or other reasons they can be deported like any foreigner in an irregular situation in the country (IOM, 2009, p. 49).
   Asylum is also classified as mass and individual asylum considering the numerical size of asylum seekers and reasons for asylum. While mass asylum is defined as crossing of the boundaries by a large number of refugees as a result of mass human rights violations and widespread violence, individual asylum refers to cases where people who have been subjected to persecution leave their homeland and demand asylum from another country (Kirişçi, 2014, p. 14).
c) **Refugee:** The definition of the refugee concept is taken directly from the 1951 Geneva Convention. According to this, it is defined as a person who is outside the country of citizenship due to a well-founded

fear of persecution because of his race, religion, nationality, certain social group membership, and political opinions and is unwilling to avail himself of the protection of that country owing to such fear (IOM, 2009, p. 43).

d) **International protection:** Legal protection provided by a covenant-based organization in order to ensure that states respect the rights set forth in the legal documents such as the 1951 Refugee Convention, 1949 Geneva Conventions, and 1977 Protocols, the right of initiative of the International Committee of the Red Cross (ICRC), ILO conventions and human rights documents (IOM, 2009).

International protection is regarded as a migration movement but it is a displacement that is handled separately from other migration types due to its specific reasons. The concept of international protection, which emerges through the influence of social and political events and has social, economic, humanitarian, religious, and moral dimensions, maintains its importance for the world and for humanity as it has been in the past. As a consequence of this importance, the international protection area has gained a separate ground in main international human rights instruments. Article 14 of the Universal Declaration of Human Rights states that "Everyone has the right to seek asylum and to benefit from asylum in other countries in the face of persecution."[1]

The concepts in the literature on migration are undoubtedly not limited to the aforementioned ones. However, the terms such as irregular migrant, irregular migration, and mass/collective migration, which defines the status of the forced and mass migrating community from Syria to Turkey since 2011 and which can be added to the concepts, meanings of which are quoted from the IOM dictionary, can also be incidentally mentioned here. These concepts may also be referred to from time to time while explaining the statutes of the Syrian migrants.

3.4 SITUATION OF THE SYRIANS IN TURKEY IN TERMS OF IMMIGRATION LAW

The most important international legal document on migration is the "Convention Relating to the Status of Refugees" signed on July 28, 1951 in Geneva. According to this convention, as a result of events occurring

[1]http://www.goc.gov.tr/files/files/2016_yiik_goc_raporu_haziran.pdf (accessed Aug 1, 2017)

before January 1, 1951 in Europe, every person who has a well-founded fear of being persecuted for reasons of (1) race, (2) religion, (3) nationality, (4) membership of a particular social group, and (5) political opinion; who is outside the country of his nationality and is unable, or owing to such fear, is unwilling to avail himself of the protection of that country; or who, not having a nationality and being outside the country of his former habitual residence as a result of such events, is unable or, owing to such fear, is unwilling to return to it shall benefit from this convention. For the recognition of a foreigner's refugee status, it is necessary that he faces danger in relation to these five elements, is afraid of being persecuted, has a well-founded fear of being persecuted, and is outside the country (Kaya and Eren, 2014, p. 18). On the other hand, Turkey has abolished the time limit with the Additional Protocol signed in 1967, but no regulations on geographical limitation have been made. As the refugee status in Turkey is only given to European origins, the highest status that non-European citizens can receive is the "asylum seeker" status. For these reasons, the Syrians who escaped from the civil war in their countries and took refuge in Turkey can access the "asylum seeker" status legally or become "illegal migrants," as opposed to widespread use (refugees, migrants). As in the aforementioned definition of IOM, asylum seekers are those who want to be accepted as refugees in the country, who are waiting for the result of the refugee application they have made and can be deported if their application is negative.

Another important regulation concerning the legal status of foreigners coming to Turkey from foreign countries is the directive of 1994. According to this directive, asylum seekers are granted a temporary residence permit in Turkey until a safe third country is encountered for asylum seekers.

Some status conceptualizations of the legal and social status of these persons have been mentioned, namely,"

> "Conceptually conceivable status in this regard is a "forced migrant" or "displaced person" status. In general, the phenomenon of forced migration, which can be defined as the removal of people from their natural habitat for reasons other than their will and wishes, is a manifestation of the structural deterioration of national or international systems or natural conditions. When we look at the reasons behind the forced immigration processes that have taken place today, in some cases the conditions (internal causes) in the countries, and in others the developments (external causes) experienced at the international system level have led to forced migrations. Internal causes leading to forced migration can be stated as weak and unsuccessful governments, weak social structures, overpopulation, mass unemployment,

conflicts over sharing and control of resources, poor management of economy/poverty, natural and reasons related to humans (Saraçlı, 2010, p. 41).

The Syrians in Turkey have entered the country with asylum seeker status, which was granted to them upon their first arrival on April 28, 2011, and made them legally prone to deportation at any given time. However, the State of Turkish Republic responded to the basic needs of the Syrians, which were defined as "guests" by the authorities, such as accommodation, food, and hygiene by taking the necessary immediate measures for this temporary situation which was expected to be temporary in the first place. At the first stage, certain solution-oriented practical resolutions were implemented, which were realized through a humanitarian situation assessment regardless of the legal situation of the people or their rights and responsibilities. Currently, considering that it has been quite a long time since the first refugee influx, we have come to the point where solution efforts must go beyond mere efforts of saving the day (Kaya and Eren, 2014, p. 28).

The most important legal arrangement made after entry of the Syrians to Turkey is the immigration regulation issued in accordance with Article 91 of the "Law on Foreigners and International Protection" of the Council of Ministers published on October 22, 2014. According to it, due to the events in Syria, the aforesaid migrants coming to Turkey as of April 28, 2011 were taken under "temporary protection" and the concept of "conditional refugee" was started to be used in relation to the people from outside of Europe, who had been accepted as asylum seekers in the Regulation of 1994 and have the criteria to be refugees. In addition, this regulation includes stateless persons and refugees from Syria beside Syrian citizens (Kaya and Eren, 2014, p. 28; Kula, 2015, p. 46).

Although concepts are perceived as words that are abstract and that reflect reality through independent generalizations regardless of space and time, it is necessary to evaluate them in their historical context, therefore in the contexts of space and time (İçduygu et al., 2014, p. 27). In this context, while we evaluate the concepts used in this study within their general usage limits on the one hand, on the other hand, it is important to consider them within their specific contexts of time and space. For example, although the Syrians who came to Turkey because of the civil war in their country cannot be legally defined as "refugees" but as "guests" in the official discourse, they are referred to as "refugees," "asylum seekers," and "migrants" every time they are mentioned from daily conversations to news media coverage. In this study, the concepts of "asylum seeker," "refugee," and "migrant" are used for the Syrians in Turkey. Despite the fact that their legal statuses are "asylum seekers," even international primary-level institutions, such as the United Nations High Commissioner for Refugees, prefer to use the concept of "refugee" irrespective of their legal status. Although the definition of a migrant includes the phrase "... the situations in which the person decides to migrate freely without any external coercion ..." the term "migrant" is also used in this study for the Syrians in Turkey in order to describe their situations, as it has a more comprehensive content including all types of displacement.

3.5 METHOD OF RESEARCH

In this research, participant observation and in-depth interviewing techniques were used as data collection tools from the field as well as literature review. The reason why Ravenstein, who set forth the first systematic work on migration, focused on migration was to reveal the incorrectness of the idea of William Farr, who said "migration does not depend on any definite law" (Yalçın, 2004, p. 22). After Ravenstein, many other theories were developed that demonstrated that migration, like other social events, operates within the framework of certain laws and rules. In this study, since we believed that it would help the sociological analysis of the outcomes of the migration movement from Syria to Turkey since 2011 and its current course, a broad review of the migration-related sources was made, and an overview of migration theories was summarized.

The literature review was not only limited to the sources involving migration theories, but also the studies on Syrian refugees in the world and in Turkey were also reviewed. Regarding the Syrian migrants, the data from the United Nations High Commissioner for Refugees, AFAD, General Directorate of Migration, some non-governmental organizations as well as outcomes of the recent scientific studies on Syrian immigrants were also included in the research process. Despite the fact that it is a very up-to-date subject, information was provided from panels, symposiums and workshops that were carried out in few and the findings obtained were evaluated. However, considering that each migration is unique and specific and that migrations that are faced today depend on various reasons and have complex relationships network, face-to-face, and focused interviews were held with the official authorities of the Republic of Turkey, which were thought to contribute to the understanding and explanation of migration from Syria to Turkey, and with some Syrian migrants, who accepted our interview requests. Persons interviewed: an officer coordinating the asylum seeker camps on behalf of the AFAD, two translators making translation between the residents of the camps and national and international officials who visit the camps, a former provincial deputy director and a civil servant of the migration management, a department director of a provincial national education directorate responsible for issues related to education of the Syrians, and a school principal were conducted in-depth interviews. In addition, approximately one-hour interviews were conducted with a teacher who had been teaching to the students we brought from Turkey before the conflicts started in Syria and has been staying as an asylum seeker in Şanlıurfa since the conflicted has started, a doctor, a journalist, and a Syrian shopkeeper who deals with trading. In addition, open-ended, in-depth interviews were conducted with the headmen of the neighborhoods where the Syrians lived intensively (two people), six shopkeepers, and two persons working in an aid organization for Syrian people.

In the study, qualitative research methods were used and semistructured, in-depth interview method, and focus group interview method were applied. Qualitative researches try to understand situations and events from the perspective of the participants. Therefore, in qualitative researches, researchers are directly involved in the research (Büyüköztürk et al., 2010, p. 24). In the context of the research, in-depth interviews were conducted with the public institutions and organizations, local government departments, and NGO representatives, who continuously interact with the Syrian asylum seekers due to their official duties, who witnessed various problems experienced by asylum seekers, and who are better aware of their particular views.

3.6 DEVELOPMENT OF THE ARAB SPRING AND THE SYRIAN CIVIL WAR

On December 18, 2010, a protest started in Tunisia when Mohammad Bouazizi, a college graduate street vendor, burned himself and it rapidly spread to Libya and then to Egypt through quick organization via social media and in such a short period time of 1–2 months it turned into a social uprising overthrowing the governments. This process has been named as "Arab Spring," "Arab Awakening," and "Arab Uprising" by the public. This social movement had its effects in Syria a few months later, in March 2011; however, it did not terminate in a short period in Syria as in other countries. Conflicts have become permanent and the movement that had started against power has turned into a civil war with expanding fronts due to the ongoing conflicts between different opposition groups.[2]

Although the Arab Spring, which started in Tunisia and influenced many countries in a short period of time, seems to involve social and mass movements that have similar claims and are similar to each other in many respects, the ongoing process necessitates to consider, and evaluate each of them as a specific formation due to different sociological variables such as historical, political processes, religious, and ethnic structures of the relevant country.

As a matter of fact, President Zeynel Abidin Bin Ali, who remained in power for 23 years in Tunisia, announced that he put forward many democratic initiatives in his country, shortly after the demonstrations started with the Arab Spring in order to overthrow the government, but on January 14, 2011, nearly a month after than the uprising started, he was forced to leave power as well as the country (ORSAM, 2015, p. 11).

On January 25, 2011, the people gathered at the Tahrir Square in Egypt started to protest the government, and Hosni Mubarak's 30-year rule since 1981 ended on February 11, 2011.

In Libya, Muammer Gaddafi opposed the uprising that started on February 16, 2011, through very harsh statements, but he had to abandon his seat with the intervention of the international coalition started on March 19, 2011. After a few months, he was caught in a drainage pipe and could not escape from being lynched.

Following the protests in Yemen, President Ali Abdullah Saleh was overthrown from power following his 32 years of rule and he was replaced with Abd Rabbuh Mansur Hadi, his assistant. Leadership of Mansur Hadi did not bring spring, but rather caused internal conflicts to take place and Yemen was brought to a position open for all kinds of intervention by different countries. As a result, the process referred to as the Arab Spring directly or indirectly affected the entire Middle East and Turkey.

The country in which the process has been experienced most painfully is undoubtedly Syria. The uprising started in Syria with a demonstrator burning himself similar to that in Tunisia. On January 26, 2011, a demonstration was held in the city of Rakka to protest the killing of two Kurdish-born soldiers two days after the event, which had taken place in the region of Haseke where mainly Kurdish population lived. Following this demonstration that included ethnic demands, events took place in Damascus, the capital city, on February 4–5, 2011.

---

[2]https://www.ihh.org.tr/haber/18-aralik-2010-arap-baharinin-baslangici-1975 (accessed June 6, 2017)

Following the Friday prayers on March 18, 2011, demonstrations of a size not seen in the history of Syria for 30 years took place in the cities of Damascus, Aleppo, Daraa, Qamishli, Humus, Banias, and Deir Zor of Syria. The event that started the demonstrations in the city of Daraa, where the bloodiest incidents took place, was arresting of a group of young people, who wrote the slogan "people want to overthrow the regime" on the walls of the city on March 6. The police harshly intervened in the crowd who gathered to protest this and as a result, four demonstrators lost their lives and dozens were injured. One day after this event, conflicts took place once again in the city of Daraa during the funerals of the killed protesters. On the third day of the demonstrations in Daraa, the building of the Ba'ath Party, the court of justice, the office of the "Syria Tel" company belonging to Rami Maluf, the wealthiest person in the country, were burnt down. Upon the spreading of the events in Daraa, the army entered the city. Meanwhile, demonstrations continued in Hama, Humus, Aleppo, Banias, and Deir Zor.[3]

While the number of protests in Syria was increasing, the Turkish government made a series of diplomatic initiatives in order to persuade the Syrian government to respond positively and effectively to the protesters' demands for democratization. However, since the government of Syria was using delaying tactics, and the interventions of the government were very harsh, Turkey gave up its efforts to persuade. The refugee influx from Syria to Turkey started with the asylum requests of 250–300 Syrian citizens on April 29, 2011.[4] In a very short time, a crowded community piled up at the border gates and the process reached the figures expressed in millions today. The forced and mass migration from Syria to Turkey has brought many problems in terms of asylum seekers as well as host country governments and citizens beside reaching high figures in numerical terms. The process of evolution of the social status of those who came from Syria to Turkey with forced and mass migration can be followed in the course of time through the transformation of concepts that characterize the social status of migrants in the migration literature such as "our beloved guests," "our migrant brothers," "asylum seeker," and "refugee." The Government of the Republic of Turkey transformed its initial attitude of defining the refugees as "guests" without any legality into a "temporary protection regime" on October 2011, which allowed refugees to have much clearer legal status. This status means a protection regime far beyond what is presented in the

[3]http://www.orsam.org.tr/index.php/Content/Analiz/1478?s=turkmen%7Cturkish (accessed June 8, 2017)

[4]http://www.orsam.org.tr/files/OA/61/11sunaoner.pdf (accessed June 8, 2017)

Regulation of 1994 and the government automatically accepts that it will provide any kind of basic humanitarian service such as health and accommodation for the Syrians and also the "non-refoulement" principle defined within the framework of international law (Kirişçi, 2014).

As of April 2011, the migration wave that started with the arrival of 250–300 asylum seekers to the border gate of Turkey from Syria, spread rapidly in a short period of time and caused the number of refugees outside the established camps to reach to much higher figures than of those in the camps. The government had not predicted that the asylum seekers who had reached considerable numbers would spread out of the camps at such a rate.

In the second half of 2012, there was a significant increase in the number of refugees staying out of the camp. This date also points to a turning point for Turkey calling for international aid. However, in the previous period, the Turkish government had thought that it could manage the refugee crisis alone.

The list of the difficulties both Turkey and the international community have faced in trying to produce solutions to the Syrian refugee crisis is quite extensive. The difficulties discussed in this chapter are ranked neither according to their importance nor the need for more effort. Nevertheless, the most difficult issue facing Turkey is the question of what will happen to the Syrian refugees in the long term. In this context, the way of dealing with other difficulties related to the problem depends on the answer to this question. When the crisis started, there was a widespread conviction that the Syrians would temporarily stay in Turkey and they could return to their homeland within a few months. It was initially accepted as a humanitarian and moral requirement to welcome these people, who had escaped from a brutal war, generously, and with open arms. It was also expected that this policy would provide Turkey with an international reputation and interests after the war ended in Syria, and the country was reconstructed. At the moment, most people are aware of the fact that the Syrians will continue to stay in Turkey for a longer period of time and bring economic burdens. Given the situation of civil war in Syria, these observations are not so surprising. This is because of the enormous physical destruction that has accompanied the collapse of the economy and the basic public services throughout the country (Kirişçi, 2014, p. 22–23).

In the first period, when approximately 3 million Syrians under the temporary protection status started to come to Turkey from Syria during the forced and mass migration process, the migrants presented a timid, loser, shy, and embarrassed image. In fact, this attitude reflects the widespread psychology of all mass migrations. On the other hand, Turkish landlords, feeling pity for them, also acted with the desire to help within the range of their power as well as with compassion to these people, who had to leave their homes, hardly saved their lives, and had to travel without even taking their most important belongings with them. As the stay of the Syrian people got longer, the Syrians initiated certain businesses in order to ensure their livelihood and not to burden more on the other side. In neighborhoods where the Syrians live intensively, shops such as grocery stores, hairdressers, restaurants, clothing stores, stores such as small home appliances and repair workshops were opened by Syrian merchants. On the other hand, they started to compete in the business fields of the local people as a cheap labor force. Local people, who had tried to help the migrants voluntarily in the early days of migration, started to realize that the Syrians turned into competitors after a while, and the "guests" or "Ansar-migrant fellowship" began to evolve into the competition, mutual dissatisfaction, complaints, and some conflicts although not serious. Many activities are carried out in order to keep this large mass forced migration out of mutual exclusion, alienation, and conflict, to live together at the lowest possible level of conflict, and to keep two communities in harmony. These activities are executed at national and international level by official or semiofficial institutions as well as non-governmental organizations.

3.7 RESEARCH FINDINGS

3.7.1 DEMOGRAPHIC DATA ON THE SYRIAN ASYLUM SEEKERS IN TURKEY

The demographic data of the Syrian asylum seekers in Turkey were provided from the periodically published reports by the United Nations High Commissioner for Refugees (UNHCR), Ministry of Interior General Directorate of Migration Management, Prime Ministry Disaster and Emergency Management Presidency, known as the AFAD as well as briefings and reports prepared in order to inform the authorities and public.

According to United Nations data, as of 2016, 244 million people, in other words, 3.2% of the world's population are international migrants. Of these, 65.6 million were displaced by force, 22.5 million are refugees, and 2.8 million are asylum seekers. The most striking aspect of this report is that the relatively poor countries are hosting people displaced by force. It is stated in the report that in the first half of 2016, 3.2 million people displaced by force mostly took refuge in low- or middle-income countries.[5] Some of them are listed according to the number of refugees they have. By the middle of 2016, Turkey became the host country that had the highest number of refugee population of 2.8 million among all countries. Turkey was followed by Pakistan (1.6 million), Lebanon (1 million), Iran (978,000), Ethiopia (742,700), Jordan (691,800), Kenya (523,500), Uganda (512,600), Germany (478,600), and Chad (386,100).[6]

UNHCR stated that as of May 2017, the total number of Syrian asylum seekers in Turkey reached 2,969,669,[7] of which 8.60% were located in campsites such as container towns and tent cities, and 91.40% were residing in off-camp areas.

According to the distribution per cities, the highest rate was in Istanbul (553.453), which was followed by Şanlıurfa (477.202), Hatay (447.541), Gaziantep (375.633), Adana (191.564), Mersin (205.240), Kilis (130.119) Bursa (142.791), İzmir (135.548), and Konya (104.360).

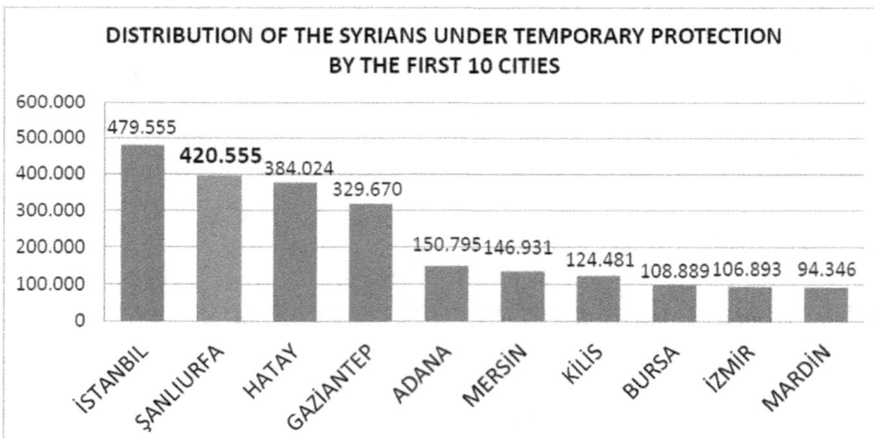

FIGURE 3.1 (See color insert.) Distribution of the Syrians under temporary protection by the first 10 cities.
Source: http://www.goc.gov.tr/icerik6/gecici-koruma_363_378_4713_icerik

[5]http://www.unhcr.org/turkey/home.php?content=733 (accessed July 13, 2017)
[6]http://www.unhcr.org/turkey/home.php?content=733 (accessed July 13, 2017)
[7]The main source of the differences in the numerical data on Syrian asylum seekers in Turkey is that the reports published with respect to that data have been published once every few months and that the asylum seeker entry to the country is ongoing or that unrecorded asylum seekers are shown as an estimate.

As seen in Figure 3.1, the Syrians living in Istanbul constitute the greatest proportion of Syrians in Turkey who obtained temporary protection status. Considering the populations of provinces and Syrian asylum-seekers, it is seen that the distribution of asylum seekers is much higher in the provinces bordering Syria. This is because, the proximity of destination is an advantage in terms of migration, geographical conditions, presence of kinship relations, similarity of ethnicity, and cultural characteristics are also effective. The most noteworthy aspect of the Syrian population's concentration in the border provinces is Kilis, where Syrian population exceeds local population. Asylum seekers are spread throughout Turkey, mainly in big and border cities, although they are not homogeneous.

3.7.2 SERVICES AND ISSUES REGARDING ACCOMMODATION, FOOD, AND EMPLOYMENT PROVIDED TO SYRIAN ASYLUM SEEKERS

The main problems that the Syrian asylum seekers face from the moment they arrived to Turkey have been the provision of accommodation, food, and alimony (livelihood). The problem of accommodation, which started with the entry of Syrian asylum seekers into Turkey, was tried to be met with the camps established in the border cities. According to data from migration administration, Turkey has hosted 256,971 Syrian foreigners in 26 temporary accommodation centers established in 10 cities since April 2011.[8] In addition, there are 2,957,454 Syrians who are recorded with biometric data outside temporary shelters. Considering those not recorded, it can be said that Turkey has provided more than 3 million Syrians with accommodation, health, education, and food aid. In the statement made from the official twitter account of the General Directorate of Migration Management, 1,580,866 of the Syrians living in Turkey are male and 1,376,588 are female.[9]

The government has a highly successful food aid program in cooperation with the World Food Program (WFP). Within this scope, an electronic card is given to each camp resident and a total of 100 TL is charged to these cards, 50 TL of which is supplied by AFAD, and 50 TL by the WFP and Turkish Red Crescent. Asylum seekers have the opportunity to buy both food and other necessities with this card from the grocery stores in the camp. This amount of money loaded on cards may seem insufficient at first glance,

[8]http://www.goc.gov.tr/icerik3/gecici-korumamiz-altindaki-suriyeliler_409_558_560 (accessed June 20, 2017)

[9]http://www.haberturk.com/gundem/haber/1427703-yaklasik-3-milyon-suriyeli-kayit-altinda# (accessed Aug 1, 2017)

however, considering that each family member is provided with a card and big expenses such as accommodation, heating, electricity, water, health, and education are covered by the Republic of Turkey, it is seen that this aid is enough to cover basic human needs. Moreover, in case anyone living in the camp finds a job in the camp or outside the camp, there is no deduction from the subsistence benefits he receives.

A Syrian asylum seeker, who was interviewed on TRT News channel on the "Seeking Our Compassion" program published on August 21, 2010 at 21:00, described the camp conditions as "five-star hotel comfort." In the satisfaction and expectation survey conducted with the asylum seekers staying in the camps in Turkey in 2012, 64.1% of the participants stated that they were satisfied, while 16.5% stated that they were not satisfied. It is noteworthy that the results were quite high despite the fact that the camps were still in the stage of establishment and were not fully functional yet.

It cannot be said that the same conditions apply to Syrian asylum seekers who stay outside of the camps. They face serious difficulties, particularly in order to fulfill their needs for accommodation. The increase in the need for rental housing has already caused rents to rise in cities and neighborhoods where Syrians live. Moreover, most refugees had to live in very low standard houses, and those who did not have enough financial means, begun to live in depraved buildings or even worse in scratchy tents. The efforts of local governments and some NGOs to meet these needs avoid potentially big problems that this process will bring about. The Syrian asylum seekers who are economically in higher status can meet their accommodation needs much more easily in the districts appropriate for their status whether they buy or rent houses; however, increases in the sale and rental prices lead to dissatisfaction among local people as well as some conflicts although not physical.

Another area of integration and conflict related to Syrian asylum seekers is their entry into the labor market and business life. In a study on the Syrians living in Istanbul, it was stated that residences can only meet their basic needs (Esin et al., 2014). In addition, studies show that the number of working Syrians in Turkey is very low. This is due to the legal requirements governing worker–employer relations regarding their permission for work. Syrians must have residence permits before they can work in Turkey; however, the rate of the Syrians with residence permits is only 3% of total asylum seekers, which corresponds to approximately 80,000 people. Otherwise, those who do not have the right to work will have to face sanctions (Erdoğan and Ünver, 2015).

However, it was stated in the interviews conducted by us that the number of working asylum seekers is well above this number, and no sanctions are applied in order to prevent them from working. It was even identified that Syrian people who had temporary protection identity were working without work permits at various workplaces. When interviews were held on the issue in Şanlıurfa, it was stated that in almost every workplace in the industry there was a working Syrian, and it could be observed this number could easily increase according to the capacity of the workplace. As a matter of fact, in many research works on Syrian asylum seekers in Turkey, it is stated that every two Syrians, who are able to work, are working or Syrians working without work permits are much higher than the registered employees (TTB, 2014; Erdoğan and Ünver, 2015). Despite the serious problem of unemployment in Turkey, the reason why Syrians are able to find a job easily is that they are cheap labor force. As a matter of fact, it is stated that the average monthly income of the Syrians working in any job is 230.9 dollars (AFAD, 2014).

In a research carried out on the Şanlıurfa scale regarding Syrian asylum seeker women workers, it is stated that although many women were not working in Syria except for housework or family farming, they started to work after they came to Turkey (Kaya, 2016, p. 162). However, it is also stated that the absence of work permit leads to various problems. It is stated that, for example, a Syrian asylum seeker who opens a hairdresser's shop on the name of a Turkish citizen without a work permit, gives a significant portion of his earnings to that Turkish partner who has no contribution other than to permit his name used (Kaya, 2016, p. 165).

The employment problem that the Syrians are facing is one of the most important problems. In this context, lack of work, unemployment, lack of livelihood, poverty or need for others' help, and support for a long time, is a very serious problem humiliating the person and injuring human dignity. Here, the contradictory situation with regard to work and business life is also the problem of conflict arising from labor force competition. According to the interviews conducted, the biggest complaint of local people is that asylum seekers are cheap labor force. The complaints in this regard are that Syrian asylum seekers who open medium and small-scale business places are exempted from tax, resulting in unfair competition among local traders.

3.7.3 EDUCATION AND HEALTH SERVICES PROVIDED TO SYRIAN ASYLUM AND PROBLEMS

As the conflicts in Syria are expected to come to end in a short period and the asylum seekers are expected to return to their countries, at this stage, it is discussed how many of the asylum seekers will return to their countries and how many will stay in Turkey in case the crises is resolved in Syria. The current status of Syrian asylum seekers in Turkey will have significant effects on the future peace and prosperity of Turkish society. For this reason, execution of education policies for Syrian children and youth, and development of short, medium and long-term programs and projects with a solution-oriented stance are crucial.

More than 40% of the Syrian refugee children in school age in Turkey are unable to continue their education due to various obstacles. This situation is a loss as well as a threat both in terms of future of Turkey and of Syria. The Turkish government that has been aware of the threat and the non-governmental organizations dealing with the future of the Syrians has been seeking various solutions to this problem since the first-day asylum seekers came to Turkey.

Considering the size of the Syrian refugee population living in urban areas, their demands continue to increase day after day in several areas, particularly in health, education, and employment. The Turkish Government served as a model and adopted a legislation in January 2016 that allowed the refugees to create self-sufficient and additional opportunities in a larger scale and enabled access of the Syrian refugees under Temporary Protection to registered employment.

It was announced by the UNHCR through an urgent funding call more than US$ 3.5 billion that additional support was necessary in order to strengthen the existing capacity in the implementation of the relevant legislation as well as to promote employment creation and recruitment.[10]

One of the most important obstacles to the schooling of Syrian refugee children in Turkey is a language barrier. Although there were some difficulties in the countries outside of Turkey, such as Jordan and Lebanon, where the Syrians migrated heavily, there was no language problem. Since the language of education in Turkey is not Arabic, the most important obstacle that Syrian immigrant children faced in terms of education has been the language. For this reason, Syrian refugees who came to Turkey in the first period came together in the border provinces where they lived and started

[10]http://www.unhcr.org/turkey/home.php?content=726 (accessed Aug 1, 2017)

to educate their children with Syrian curriculum in houses, associations, camps, and various educational settings in their own languages and in their curriculums (Duman, 2016, p. 49).

As it was believed in the early years that the conflicts in Syria would end shortly, and the Syrians would return to their countries, in 2012, the Ministry of National Education preferred to support their education in Arabic curriculum instead of teaching them Turkish (SETA, 2016, p. 13). However, due to the increase in the number of Syrians who came to Turkey as well as the severity of the civil war in Syria, the Ministry of National Education (MoNE) tried to solve the problem of Syrian education with various activities and circulars published on April 26, 2013 and September 26, 2013. It is seen that the first circular was prepared to determine the places that gave or might give education to these children, rather than a policy focused on solving the educational problems of Syrian children living outside the camp (SETA, 2016, p. 13). As a matter of fact, in the second general circular issued on September 26, 2013, titled "Educational Services for Syrian Citizens under Temporary Protection in Our Country," following provisions were agreed (MoNE, 2013b):

- Syrian children in the camp, which had to pause their training, will be trained in order to prevent them from losing year.
- Coordination of educational services to be carried out in this context will be under the responsibility of MoNE.
- The need for teachers will be fulfilled by teachers in excess in the relevant city, in case they are not sufficient, Arabic-speaking people who meet the conditions will be assigned and paid through tuition by MoNE.
- In a similar way, those among the Syrian citizens who want to educate the children or those who are eligible, will be assessed by the Ministry of Education and assigned provided that they will not claim any payment and work on totally voluntary basis.
- The content of the training will be prepared by the Higher Education Commission of the National Coalition of Syria under the control of the Ministry of National Education and in case the Syrians with Turkish origin request, they may have education in Turkish curriculum.
- Turkish and vocational training courses will be opened for Syrians who want to learn Turkish in accordance with the facility capabilities.
- Syrians who reside outside of the camps and have residence permits will be enrolled with respect to the circular on "Foreign Students" numbered 2011/48 published on August 16, 2010.

As it is seen above, the circular is an important step toward the educational needs of Syrian children. However, it is understood that the education of those who reside outside of the camps and who don't have temporary residence permits, which constitute the majority of the children, is left to local administrations, voluntary national and international organizations.

One of the institutions that initiated a project for lack of education of Syrian children in Turkey is the "Turkish Religious Foundation" (TDV). In 2012, the women's branch of TDV KAGEM (Turkish Religious Foundation Women Family and Youth Center), which acknowledged a lack in this area, allocated Quran courses as schools by preparing a comprehensive project for school education for Syrian children. In this process, "Syrian Schools" started to be established through the leadership of TDV, where the Syrians live intensively, and the Directorate of Religious Affairs appointed a mutiassistant on every province to deal with the project. Then in 2014, MoNE issued a circular numbered 2014/21 and moved aforementioned "Syrian Schools" to the buildings of MoNE, stipulated that they should study here and called them "Temporary Training Center" (Duman, 2016, p. 49).

With the uncertainty of how long the Syrians will remain in Turkey, MoNE offers two types of education for Syrian students at primary and elementary school age in the border provinces where the Syrians live intensely. The first is the practice that the Syrian children, who have learned Turkish in various ways, are educated according to the curriculum of Turkish National Education together with the Turkish students. The intensity of the students studying in this program is, for example, in Şanlıurfa, a total of 9400 children of the Syrian asylum seekers at primary school age with 4256 female and 5144 male, a total of 1388 children at elementary school age with 657 female and 731 male, and a total of 379 children at high school age with 275 female and 104 male. Those who enter this group are educated according to Turkish National Education curriculum in Turkish schools. Additional Turkish courses are also available on request. The other is allocation of the schools affiliated to the MoNE to the Syrian students after the completion of the education of the Turkish citizens and establishment of GEM temporary training centers (Geçici Eğitim Merkezi in Turkish) and also an education practice formed by the Turkish academics who know Syrian and Arabic through the assistance of the Syrian teachers and carried out with a specific curriculum created specifically for children by re-examining the Syrian curriculum before the civil war. The curriculum applied is the one that was applied in pre-war Syria and with the addition of Turkish language support. For example, there are 45 schools in Şanlıurfa that provide education within the scope of this practice.

The Ministry of National Education provides an online service similar to the MoNE e-school data system offered to Turkish citizens, for Syrian refugee students, titled Foreign Student Information Management System (YÖBİS). Through this system, records and written statements about all kinds of student activities of Syrian students can be received in Arabic and Turkish. All the processes required by the educational activities of the students attending the Temporary Training Centers are followed. In addition, those who succeed in the "High School Proficiency and Equivalence Examination for Foreign Students (YÖLYDS)," which is a centralized exam held by the MoNE for students who have successfully completed their education in these schools (and those who have previously completed high school or equivalent in Syria) receive a "High School Equivalency Certificate." Asylum seeker students with secondary school equivalency certificates can obtain the right to study at higher education institutions if they fulfill the conditions laid down in the legislation. Apart from Temporary Training Centers, asylum seeker students can also be admitted to Turkish schools where the children of Turkish citizens are educated. The asylum seeker students who are educated in the schools affiliated to the MoNE are subject to same educational and managerial processes and procedures that Turkish citizens are subject to (Erdönmez, 2016, pp. 28–30).

On July 23–24, 2017, a coordination meeting was held in Istanbul on the pedagogical development of Syrian children under temporary protection. In this meeting, it was stated that 169,120 students in 14,742 schools were educated in Turkish Curriculum with the exception of "Temporary Training Centers (GEM)" affiliated to MoNE as of July 9, 2017 as a result of the efforts to register the Syrian population at the age of education gradually to official schools affiliated to MoNE. On the other hand, it is stated that 291,039 students, all of whom are Syrians, are studying in 404 units of Temporary Training Centers in 20 provinces, based on intensive Turkish teaching programs, and that there are 8002 students registered in open schools and a total of 492,544 students who arrived with mass migration to Turkey are provided education service (MoNE, 2017).

The education services provided to Syrian asylum seekers in Turkey should not be limited only to the activities of the formal education institutions. We also have institutions that provide nonformal education such as provincial public education centers. For example, a deputy director of a provincial public education center, with whom we interviewed in Şanlıurfa, stated that there were two public education centers in the province and they solely organized 116 tri-monthly courses for Syrian refugees so far. A total of 80,742 children were trained in the centers of refugees organized by AFAD. On the other hand, as a result of the steps taken by the cooperation of the Ministry of National Education, a total of 508,846 children were provided school training from kindergarten to the end of high school (AFAD, 2017). The point reached today is that the educational problems of the Syrian children have been solved to a large extent, and the Syrian children under temporary protection, trained in the Turkish education system, will contribute to the adaptation rather than conflict in the long term.

Modern medicine defines the state of being healthy not only as a condition of illness or disability but also as a state of full well-being in the physical, social, and mental respect. Today, contrary to traditional theories, a holistic health approach is adopted, in which health is addressed through a bio-psycho-social approach. Migration, which means to leave the environment in which a person has been born and grew up and to be involved in a new social and cultural environment, is a phenomenon in which human life is affected socially, culturally, physically, etc., therefore undoubtedly affects health and health-related variables (Topçu and Beşer, 2006, p. 30).

Since mass and forced migration due to reasons such as war or natural disasters do not occur willingly and in selective, healthy travel conditions, migration itself involves health problems. The main health problems faced by the migrants are not limited to those faced during migration, but they continue to exist after they reach where they migrated (Beşer, 2012, p. 22). As a matter of fact, smallpox illness that has not been seen in the World and Turkey since 1977 has started to be seen both in asylum seeker children and in the children of Turkish citizens after the Syrian asylum seekers arrived. According to the records of national and international related institutions and organizations, the most common health problems that asylum seekers face are; vitamin deficiencies, anemia, unintended pregnancies, risky pregnancies, miscarriages, birth complications, retarded growth and development of children, chronic illnesses and complications, diarrhea, infectious diseases such as meningitis, typhoid fever, dysentery, etc., and vaccine-preventable diseases (measles, tuberculosis, hepatitis, etc.), sexually transmitted infections including HIV/AIDS, physical violence and related injuries, several mental problems such as depression, anxiety disorders, and posttraumatic stress disorder as well as dental diseases, (War, Migration and Health, TMA Report 2016, p. 23).

An important part of the diseases that occur during migration are health problems that can be prevented by pre-disease vaccination system. For this reason, Turkey took precaution against this priority issue regarding asylum seekers from Syria and the masses that entered Turkey from the official border entrance gates and that needed to be vaccinated were vaccinated. As of the end of 2014, the total number of vaccinations made to Syrian asylum seekers in and outside the camps was 888,000. In 2015, while the number of vaccinations made to Turkish citizens residing in Kilis was 25,965, the number of vaccines made to Syrians was 26,671. This data is an example of the sensitivity of the vaccination process with Syrian asylum seekers (Akçacı, 2016, p. 394).

In our country, health and care services given to the Syrian citizens are carried out under the best conditions (Doğan, 2016). Health services and medicines for all Syrians registered in Turkey are provided in the same conditions provided to the Turkish citizens in cooperation with the Ministry of Health. Until today, 953,466 thousand operations of Syrian patients were performed, 1,143,393 patients were hospitalized and 25,919,750 patients were provided ambulatory care services.[11]

Registered Syrian refugees are able to benefit from the health services by the identity cards provided to them and their identification numbers. Moreover, they can buy medicines by paying 20% of the cost of the drug exactly like the host Turkish citizens. Due to the amplitude of ill and wounded refugees in the state hospital of Kilis, where the Syrian population has reached the same level as the local population, there are disruptions and delays in the service of the Turkish citizens, causing local people to be uncomfortable with the Syrians' presence and to react from time to time (IGAM, 2013). However, another aspect regarding migration and health problems that needs to be stated is that difficulties of adaptation to the cultural, climate and geographical conditions of the migrated society can be determining factors in the emergence of health problems. If the sociocultural and economic conditions in which migrants enter are similar to the conditions of their environment where they came from, they may face less health problems, otherwise they may face more problems. However, violence during migration, bad weather conditions, and nutritional difficulties negatively affect the health of migrants. Communication problems in the locations where they are migrated, inability to reach health care services, and exclusionary policies applied to immigrants lead to increase in faced health problems (Ertem, 2015).

Syrians under temporary protection in Turkey are able to access health care of almost any kind. When assessed in terms adaptation and conflict, Turkish citizens sometimes consider health services offered to Syrians as

[11]https://www.afad.gov.tr/upload/Node/2373/files/Suriyeli_Siginmacilara_Yapilan_Yardimlar+7.pdf (accessed Aug 20, 2017)

excessive, and this causes conflict or discomfort. For example, the interviews we conducted in Şanlıurfa in order to determine the problems of adaptation and conflict of the Syrians, an officer, indicating that the health service provided to Syrians was of better quality than that of provided to them and showing the hearing device in his ear, said: "For this device that costs 4000 TL, a state contribution of only 400 TL is provided to me, while a Syrian person is provided at no charge." It is seen that the most serious complaints of the Syrians about the health services provided to them were not about the inability to access health services but rather about the referral chain, which is considered to be very necessary for the health system.

In terms of adaptation and conflict, it can be said that health problems will not cause serious conflict between Syrian asylum seekers and host Turkish citizens in the long run.

3.7.4 NONGOVERNMENTAL ORGANIZATIONS SUPPORTING THE INTEGRATION OF ASYLUM SEEKERS

It was improbable to host about 3 million asylum seekers since April 2011, employing them without serious problems, and offering the conditions to live in harmony with minimum problems. However, since more than 6 years, we have been witnessing this experience throughout Turkey, especially in border provinces. There are many reasons for this. Having geographical proximity, common historical background, common beliefs and cultural values, many relationships such as kinship, and the hospitality and tolerance of the Turkish nation and state enable such harmony. However, it is necessary to emphasize the contribution of charitable organizations such as foundations and associations, which we can generally call as nongovernmental organizations, in order to maintain this harmony at a level that can be called almost seamless.

The term civil society (NGO) is used to describe the economic and social area, which is outside of public resources in general and operates by its own means. This concept, which emphasizes the initiatives and autonomy of the individuals and groups that carry out their own organizations, also has a meaning including the control of state–society relations and audit of the functioning of public organizations by society (Bilge, 2009, p. 105). Non-governmental organizations (NGOs) are, in a general sense, networks of intermediary institutions operating as a third area between the state and society. As a communication network, civil society has many functions such as social solidarity, social trust, social responsibility, stability, solutions to problems, and serves on the basis of civil rights and freedoms. While civil society is viewed as an integral part of a virtuous society in terms of politics and ethics, it is also regarded as an important actor and area in transition to democracy from military administration or authoritarian and totalitarian political regimes (Keyman, 2006). NGOs are organizations that rush to the aid of people in situations where public authorities are limited in extraordinary circumstances, such as war, natural disaster, and mass migration. In ordinary terms, they are organizations that mobilize, transform society and contribute to social activities in issues such as social consciousness, emancipation, democratization, and human rights. It would not be a faulty evaluation to explain the drama that the Syrians are facing today is due to lack of adequate civil society organizations in precivil war Syria.

There are several national and international charity organizations in Turkey that existed before the advent of Syrian asylum seekers and helped people in need. However, there are also NGOs with intensive activities for the Syrians, especially in the border provinces. Some of these charities have been established by the Syrians, some by Turkish citizens and some by the citizens of both countries. In addition, some of them can serve all kinds of people in need, while others can only help those who are Syrians and even those among Syrians who have specific characteristics. General information on main NGOs is provided below:

Şanlıurfa Platform: Representatives of 60 unions and other organizations in Şanlıurfa came together in 2012 and formed "Şanlıurfa NGO Humanitarian Aid Platform." This platform includes charities such as Deniz Feneri, IHH, Mazlum Der, as well as some trade unions and foundations. The platform carried out its activities in close cooperation with the governorate and some metropolitan municipalities and provided aid especially during the first years of the intensive advent of asylum seekers, and contributed significantly to the overcoming of the crisis. The platform provided 30 tons of material aid from the Nuremberg Medina Foundation in Germany and delivered it to Syria (IGAM, 2013, p. 13).

The IHH is another NGO that continues to provide aid to Syrian refugees independently and effectively from the platform although it is a member of the platform. The IHH is the NGO that has the widest area of activity with regard to the Syrians and continues its efforts effectively since the outbreak of the crisis.

In addition, many national and international aid organizations such as MAZLUMDER, Human Rights Association (İHD), International Blue Crescent Foundation, Support to Life Foundation, Helsinki Citizens' Association, Humanitarian Aid Platform, and Hayrat Humanitarian Aid Association are conducting activities for Syrian asylum seekers in Turkey (IGAM, 2013, p. 15).

The NGOs listed above are the NGOs that are actively involved in providing assistance to Syrian asylum seekers, although they were not generally established for Syrian asylum seekers. There are also associations founded by the Syrian asylum seekers or host Turkish citizens, especially for Syrian asylum seekers. They play an intermediary role between the state and the migrant community and assume important functions such as creating social and economic opportunities, expressing problems about living conditions, political lobbying and propaganda, improving legal rights, human rights and democracy education (Kaya, 2015, p. 266). In other words, they act as a means for adaptation and integration between the asylum seekers and the host community bringing solutions to the problems and avoiding them to evolve into conflicts.

The legal procedures in Turkey for refugees and asylum seekers do not allow Syrian temporary asylum seekers to establish associations. In order to overcome this obstacle in the short-term, the Republic of Turkey has made arrangements to regulations and circulars. Some of the associations were established with official status according to the laws of the Republic of Turkey, while some were established with informal status due to reasons such as passport, residence, lack of an adequate number of members (Kaya, 2015, p. 267). The most active ones established by the Syrians are as follows: Namaa Solidarity and Assistance Association, Damascus Education Assistance and Solidarity Association, Bunyan Assistance and Development Association, Compassionate People Association, Human Development and Human Services Association, Syria Urfa Association of Ansar-migrants Fellowship, Association of Hope for the War-Disabled People in Syria, El Homadi Charity Association, Furat Social Assistance and Solidary Association, Rahmet Civil Society Association, Giyath Mattar Solidarity and Assistance Association, Yasmin Sham, Association of Syrian Turkmen, Rakkahouse Association, Center for Civil Society and Democracy In Syria, and Syrian Women Marched for Peace Association.

While some of these associations are general and comprehensive, some are associations established for more specific areas, such as the "Association of Hope for the War-Disabled People." However, in general, refugee associations are structures that provide all kinds of assistance to its members and the people they address. The associations are functional spaces where education, information, accommodation, rehabilitation, and health-related activities are held alongside meetings, recreational, artistic, and cultural activities. For example, a RakkaHouse manager stated that between 2013 and 2015, they provided hotel services for around 1000 people. Although the association is an Arabic Association, he stated that they hosted and assisted to approximately 30–40 Kurdish Syrians (Kaya, 2015, pp. 267, 268). A section of the RakkaHouse Association was decorated with photographs of people died in the war. In a room where photographs of women, men, children, and babies were found, the association manager showed a photograph of a baby and called him "our smallest martyr." This room serves to motivate individuals in terms of keeping alive the purpose of revival, social memory, and mourning. For example, the Yasmin Sham Association, which operates for children with disabilities, provides assistance and guidance to 90 physically and mentally challenged people for their needs such as medicine, physical therapy, medical support, baby formula, milk, food, and clothing (Kaya, 2015, p. 269).

Although these associations are mentioned in the category of associations founded by the Syrians, there are also citizens of the Republic of Turkey among their founders who often overcome legal barriers. The reasons for both informal activity and involvement of Turkish citizens in the management of the association are absence of passport of the asylum seekers who had to leave their countries in a war environment, loss of passports, expiry of passports, or absence of men and women who are eligible to become the manager of an association (Kaya, 2015, p. 268).

As the problem of refuge and asylum that emerged during the Syrian civil war is a multidimensional issue that concerns the whole world, international NGOs are also sensitive to the issue. Associations with activities regarding temporary asylum seekers in Turkey: IMC: International Medical Association, RET International: Association for the Protection of Refugees through Education, Concern Worldwide Turkey, AAR Japan: Japanese Assistance Association, CARE International: Association for Care and Protection, ASAM: Association for Asylum Seekers and Migrants, DRC: Danish Refugee Council Association, IBC: Association of Charity and Zakat, Association of RIZK Syria Employment Office, Doctors Without Borders (MSF).

In addition to these, there are UN agencies that help solve the problems of Syrian asylum seekers and help them to adapt to the society instead of having conflict. These are: UNHCR: United Nations High Commissioner for Refugees, UN WFP: World Food Program, UN UNICEF: United Nations Children's Fund, IOM: International Organization for Migration, UN UNIDO: United Nations Industrial Development Organization, UN UNFPA: United Nations Population Fund, UN UNDP: UN Development Program.

3.7.5 MUTUAL EXPECTATIONS AND COMPLAINTS OF SYRIAN ASYLUM SEEKERS AND TURKISH CITIZENS

After the civil war that broke out in Syria, the most important problem brought by the forced mass migrations from Syria to Turkey since the middle of 2011 are rivalries and conflicts between the Syrians and the local people that showed up soon after the arrival of asylum seekers to Turkey. In such a crisis environment, the state is expected to make plans and turn many situations that seem as disadvantages into advantages. The most important work that needs to be done in order to achieve this is to develop relations between the societies of the countries of origin and target countries, to reveal the sources of problems and conflict areas with scientific researches, and to carry out and direct the process by planning according to the research data. In this context, some researches are prepared about the situation of Syrian migrants in Turkey although they are not sufficient quantitatively or qualitatively. On the other hand, several speculative information about the Syrian asylum seekers is created by the public. Valid and reliable information can only be obtained by researches made about the asylum seekers from Syria as well as the local people who are in a one-on-one relationship with them. In this context, as mentioned in the method section of this study, the data of the research were gathered from the Syrians, with whom in-depth interviews were conducted, translators, migration management authorities, AFAD coordinator, some shopkeepers, one headman, and some residents living in the neighborhoods. According to these findings, opinions, expectations, and complaints of the Syrian asylum seekers about the Turkish official authorities and citizens can be summarized under the following titles:

3.7.5.1 EVALUATIONS OF SYRIAN ASYLUM SEEKERS ABOUT THE TURKISH AUTHORITIES AND CITIZENS

In the focus group interviews with a group of asylum seekers consisting of Syrian teachers, students and doctors, they stated that when they first arrived in Turkey, they assumed that they received abundant humanitarian assistance from different countries of the world, but the Turkish authorities passed them only a small part of it, however later they recognized that they were wrong; in fact they realized that the aid they received from the international organizations was almost nothing and all burden was bore by Turkey. Then they realized that the source of such negative opinions reflected the thoughts of the Syrian population in the lower socioeconomic status. In addition, the Syrians, who were interviewed, said, "The understanding that some of our demands are exaggerated complaints sometimes hurts us deeply. However, we believe that even brothers and sisters will not do to each other what Turkish people and state have done for us." They also expressed their satisfaction with education, health, security and cleaning services offered to them and their children and stated that the quality of these services is much better than the ones offered by their own states before the civil war in Syria. Similar opinions have also been confirmed in interviews with the translators making translations between the authorities and asylum seekers in Syrian asylum seeker camps. When the Syrians told that they were satisfied with the services provided by the Turkish State, it was questioned then why they complained about the minor issues. A teacher among the people we interviewed, for whom we thought his observations and evaluations were appropriate, replied that the main reason for their complaints was not the actual issue they complained about but rather these complaints were triggered by the frustration and despair they have been feeling since they have been living separately from their home country.

The Syrians and their translators, who were interviewed, expressed the following points as complaints: Problems they have been experiencing when renting housing, for example, uses of discriminatory expressions such as "no house to Syrians" and orientation to give rental houses to Syrians at exorbitant prices, generalization of negative behaviors of a limited number of Syrians such as panhandling, prostitution and theft for all Syrians, treatment of local people to them as if they were taking their business from their hands when they find a job or when they open a business themselves.

The most striking finding about the Syrian asylum seekers' expectations and complaints from Turkish citizens came from a refugee student. A student, who had been studying at Harran University and interviewed by us during the days there was an attack on the city of Aleppo in Syria by the Shia militiamen, said that the students who worked on behalf of some NGOs attached banners on the billboards of the faculties, on which there were statements such as "Aleppo children are cold," "Aleppo children are hungry," that those who hanged these banners were undoubtedly sensitive about the material needs of the Syrians. However, he also expressed that the same student who cared for the hunger and coldness of the children in Syria, lacked in greeting, communicating and socializing with the Syrian next to him, and this created a contradictory situation. He also said, "We expect to be approached as friends who share troubles, sorrows, and problems, not as the ones that need to be constantly assisted."

When the interviewed AFAD coordinator, migration management officer and translators were asked what issues the Syrians were complaining about, they responded by saying that asylum seekers exaggerated all kinds of situations they complained about. In addition, they stated that when a political, administrative, bureaucratic authority at national or international level came in, they made a habit of transforming their demands into complaints by exaggerating even the smallest requests and exhibited a case of solecism. In fact, an MoNE official, to whom we asked about a training service to the Syrians, did not want to give information about the service that was given very successfully and said, "This is private knowledge." After we insisted and received this information, we asked him "Why such a good service is private?" and we received the answer that "We are receiving reactions from the people when the service given to them is not the same as the service given to Syrians."

As it is understood from this dialogue, the services given to the Syrians from time to time are more specific and advanced than those given to the Turkish citizens, therefore local people may complain about it.

3.7.5.2 EXPECTATIONS AND COMPLAINTS OF TURKISH OFFICIAL AUTHORITIES AND CITIZENS FROM SYRIAN ASYLUM SEEKERS

In this framework, information about the Turkish citizens' view of the Syrian asylum seekers were obtained from the coordinator of the Şanlıurfa AFAD coordinator, migration management officer, and translators as well as from one headman of a neighborhood in Şanlıurfa, four shopkeepers, and some residents of the neighborhood who were near the shopkeepers during the interview. While Turkish citizens express their views on the Syrians, they begin with some wishes and wishes. They use expressions of compassion and prayer, such as "May God never hold anyone apart from his belongings," "We do not want anybody to experience what they have been through," "May God never set anybody away from their home, land, and country." It is understood that the interviewed Turkish officials and citizens made their evaluations by comparing the asylum seekers with the first period they arrived in Turkey. They expressed that at the first stage, the Syrians were embarrassed, shy, obedient, and compatible, but they changed over time and showed attitudes and behaviors that reached the level of immorality, greed, and impoliteness. It is understood that some of the interviewees based the issue on religious-moral behavior within the framework of Ansar-migrant fellowship, and they perceived their help and compassion as worship and a test of Allah. However, during the stage of complaints, a shopkeeper said, "In the neighborhood, shopkeepers close their shops and go home in order to avoid any noise that will annoy the neighbors, however they listen to music loudly until late at night in their workplaces and make noise, therefore, annoy people and also they unfairly compete with local shopkeepers through working after normal office hours.

Another shopkeeper stated that when a Syrian family rented a house, 3–4 families including 15–20 people stayed in houses where only 5–6 persons could live and they used houses very badly. Another interviewed shopkeeper said that a Syrian worker who did the same job with him (a hairdresser) opened a business place, and that no Syrian had ever been a customer to him since then. He also stated, "If we do the same, they will say it is discrimination; however, they showed the same attitude that they show in every case," and expressed that this minority solidarity was annoying and discriminatory. During the interview with the headman, when he was expressing minor faults of the asylum seekers, his wife interrupted and said that "they cause breakdown of families and corruption of the morality of our children," bringing subject to those marrying with Syrian women as the second and third wife in the neighborhood as well as those who prostitute. A third person who was present at the meeting said they were concerned that prostitute Syrians caused infectious diseases and that some Syrian youth living in their neighborhood would harm themselves and their families by drug use. Another interviewed shopkeeper and translators stated that a major complaint about the Syrians was that the Syrians were never satisfied with the services provided to them.

3.8 CONCLUSION

Mass and forced migration can cause very serious crises in terms of societies and states due to the many social, economic, and cultural problems it has in its nature. As a result of the internal conflicts that started in April 2011 in Syria, mass migration movement from Syria toward neighboring countries, particularly to Turkey, is a social incident, which is very unique due to its numerical size and its realization in a short period, and which has many problems that can be described as a social trauma in terms of the people of the source country and the state and society of the target country. However, it is also a fact that this troubled process has been successfully managed by Turkey. When the period before the arrival of Syrian asylum seekers to the country is compared to the period after their arrival, it is seen that there is no statistically significant difference or negative process with regards to Turkey's current problems related to health, education, security, and unemployment. For example, according to TUIK data, 16.9% of the population was below the poverty line in 2010,[12] whereas in 2015, 14.7% of the population was below the poverty line.[13] In the same way, the number of students per classroom at the secondary school level in education was 33 in 2010 and 25 in 2015. Despite these positive statements, when it is questioned why Syrian asylum seeker problem leads to expectations of constant major conflicts and social disorganization from the point of Turkey or why there is a perception that the asylum seekers are struggling with very serious problems, it is seen that the main problem arises from the socioeconomic levels and social status of Syrian asylum seekers in Turkey. Observations and examinations made within the scope of the research show that among the Syrians who reside outside the camps with local people, those living in the peripheral districts and suburbs are more prone to crimes, whereas those who live in places with higher in socioeconomic level and social status do not face any problems.

In the current stage, the evaluations of the two communities about each other are as follows: Turkish citizens view asylum seekers as "mass of people from abroad," "stayed longer than thought," "converted from guest to burden" and "a burden that needs to be get rid of." The asylum seekers are uncomfortable with the reduced interest and support toward them which they had received in the first period they arrived in Turkey and also with opinions such as "unwanted people," "we will be free of them when they leave," "asylum seeking," and "people who should go as soon as possible" that emerged due to the widespread generalization of very few Syrians who showed negative behaviors such as theft, prostitution, and disobedience.

There is no doubt that Syrian asylum seekers will be able to adapt to Turkey in the long term through the education of their children and their families. In this context, since it is observed that class-based reasons play role in problems and conflicts experienced with the asylum seekers beside them being foreigners in the country and asylum seekers coming from lower classes in terms of the sociocultural and economic statuses cause more problems and conflicts, it was concluded that various training courses should be organized in order to assist the process of integration and socialization. It is thought that it is important to organize courses to increase the knowledge and experience about the history and culture of the city where asylum seekers

[12]http://www.tuik.gov.tr/PreHaberBultenleri.do?id=21584 (accessed Aug 21, 2017)
[13]https://biruni.tuik.gov.tr/yayin/views/visitorPages/index.zul (accessed Aug 21, 2017)

live in order to improve their sense of belonging. It is concluded that these courses may be presented in their own language (Arabic) or in the language of the host country (Turkish), with a perception, understanding and mind appropriate to the sociocultural structure of the host country (Turkey).

It is evaluated that during the first years of migration, mutual trust between the peoples of the two communities was at a much higher level, and this trust was shaken over time by the generalization of a few and local undesirable events. Mutual trust must be built up again and more strongly through education with possible courses and seminars and through the inclusion of learning outcomes in the educational programs for asylum seeker children in order to establish mutual confidence of the two communities. Because it is observed that mutual insecurity has also been experienced among the authorities working in the field. For example, it is understood that cooperation between local NGOs and international NGOs is inadequate and this situation is caused by mutual insecurity. Therefore, it is aimed to increase the cooperation between the institutions and individuals in order to ensure the connection between the parties, to break mutual prejudices and to achieve the desired life satisfaction by providing complementary and effective cooperation.

KEYWORDS

- **forced mass migration**
- **Syrian asylum seekers**
- **sociological approach**
- **Turkish authorities**
- **mutual expectations**

REFERENCES

Akçacı, T.; Bulut, Ş. Göç Politikaları ve Sağlık Harcamaları: Kilis İli Örneği, Ortadoğu'daki Çatışmalar Bağlamında Göç Sorunu Sempozyum Bildiri Kitabı, 2016, s.387–408, Kilis.

Aker, T.; Ayata, B.; Özeren, M.; Buran, B.; Bay, A. Zorunlu iç göç: Ruhsal ve toplumsal sonuçları. *Anadolu Psikiyatri Dergisi* **2002,** *3*, 97–103.

Aron, R. *Sosyolojik Düşüncenin Evreleri* (Çev: Korkmaz Alemdar). İstanbul: İş Bankası yy, 1986.

Beşer, A. *Göç, Kültür ve Sağlık İlişkisi, Kültürlerarası Hemşirelik.* İstanbul: İstanbul Tıp Kitabevi, 2012.

Bilge, M. *Türk Modernleşmesi ve Sivil Toplum*. Ankara: Binyıl Yayınevi, 2009.

Brettel, C. B.; Hollifield, J. F. (Eds). Migration Theory: Talking Across Disciplines. Routledge: New York and London, 2000.

Büyüköztürk, Ş.; Kılıç Çakmak, E.; Akgün, Ö.; Karadeniz, Ş.; Castles, S.; Miller, M. J. The Age of Migration: International Population Movement In The Modern World. Third Edition; Palgrave Macmillan: New York, 2003.

Çağlayan, S. Göç Kuramları, Göç ve Göçmen İlişkisi. *Muğla Üniversitesi Sosyal Bilimler Enstitüsü Dergisi (İLKE)* **2006,** *17*, 67–91.

Çakır, S. Geleneksel Türk Kültüründe Göç ve Toplumsal Değişme. *SDÜ Fen Edebiyat Fakültesi Sosyal Bilimler Dergisi* **2011,** *24*, 129–142.

Demirel, F. Bilimsel Araştırma Yöntemleri. Pegem Akademi: Ankara, Turkey, 2010.

Doğan, U.; Atay, E. Göçün Sağlık Üzerine Etkileri: Türkiye Örneği, II. Uluslararası Ortadoğu Konferansları, Ortadoğu'da Çatışmalar Bağlamında Göç Sorunu, Kilis 7 Aralık Üniversitesi Yayını, 2016, pp 85–92.

Massey, D. S., et al. Theories of International Migration: A Review and Appraisal. *Population Develop. Rev.* **1993,** *19* (3), 431–466

Duman, T. Sosyal Uyumu Sağlamak İçin Suriyelilerin Eğitiminin Önemi: Türkiye Örneği, II. Uluslararası Ortadoğu Konferansları, Ortadoğu'da Çatışmalar Bağlamında Göç Sorunu, Kilis 7 Aralık Üniversitesi Yayını, 2016; pp 46–54.

Ekholm, K.; ve Friedman, J. Eski Dünya Sistemlerinde Sermaye Emperyalizmi ve Sömürü, Dünya Sistemi; Der. Frank, G. Gills, B. İmge: Ankara, Turkey, 2003.

Erdoğan, M.; Ünver, C. Türk İş Dünyasının Türkiye'deki Suriyeliler Konusundaki Görüş, Beklenti ve Önerileri, TİSK, Ankara, Türkiye, 2015.

Esin, N. M.; Ardıç, A.; Nar, Ş.; Yıldırım, B.; Sunal, N. İstanbul'un Bir Bölgesinde Yaşayan Suriyeli Göçmenlerin Yaşam Koşulları. 17. Ulusal Halk Sağlığı Kongresi, Edirne, 2014; pp 1390–1391.

Fichter, J. Sosyoloji Nedir? (Çev: Nilgün Çelebi). Atilla Kitabevi: Ankara, Turkey, 1994.

Gills, B. K.; Frank, A. G. Dünya Sisteminde Çevrimler, Krizler ve Hegemonik Değişiklikler: MÖ 1700-MS 1700, Dünya Sistemi, Der. Frank, G. Gills, B.; İmge: Ankara, Turkey, 2003.

Güvenç, B. *İnsan ve Kültür*. Ankara: Sosyal Bilimler Derneği Yayınları, 1972.

İçduygu, A.; Erder, S.; Gençkaya, Ö. F. Türkiye'nin Uluslararası Göç Politikaları, 1923-2023: Ulus-Devlet Oluşumundan Ulus-Ötesi Dönüşümlere, Koç Üniversitesi Göç Araştırma Merkezi, İstanbul, 2014.

Hagen-Zanker, J. Why Do People Migrate? A Review of the Theoretical Literature, 2008

Kaya, İ.; Yılmaz, E. E. Türkiye'deki Suriyelilerin Hukuki Durumu: Arada Kalanların Hakları ve Yükümlülükleri, SETA Raporu, 2014.

Kaya, M. Komşuda Misafirlik: Suriyeli Sığınmacılarca Kurulmuş Mülteci Derneklerinin Perspektifinden Türkiye'de Yaşamak. *J. Acad. Social Sci. Studies.* **2015,** *39*, 263–279.

Kaya, M.; Demirağ, E.; Orhan, H. Türkiye›deki Suriyeli Kadın Sığınmacıların İş Piyasasındaki Çalışma Koşullarına Sosyolojik Bir Bakış: Şanlıurfa Örneği, Ortadoğu'daki Çatışmalar Bağlamında Göç Sorunu Bildiri Kitabı, Kilis, 2016, pp 155–171.

Kanbir, Ö. Avrupa Göçmen Krizinin Uluslararası Ekonomi Politiği", Ortadoğu'daki Çatışmalar Bağlamında Göç Sorunu Bildiri Kitabı, Kilis, 2016, pp 102–115.

Keyman, E. F. Türkiye'de Sivil Toplumun Serüveni: İmkânsızlıklar İçinde Bir Vaha. Sivil Toplumu Geliştirme Merkezi. Ankara, 2006.

Kirman, M. A. Din Sosyolojisi Sözlüğü, Karahan Kitabevi, Adana, 2016.

Kula, C. Uluslararası Belgelerle İltica Hukuku ve Ülkemizdeki Suriyelilerin Hukuki Durumu, Hukuk Gündemi Ankara Barosu Dergileri: Ankara, Turkey, 2015, pp 42–54.

Lee, E. S. A Theory of Migration. *Demography* **1966,** *3* (1), 47–57.

Mcneill, W. H. Dünya Sistemi; Der: Frank, G. Gills B. İmge: Ankara, Turkey, 2013.

ORSAM (Ortadoğu Stratejik Araştırmalar Merkezi) Suriyeli Sığınmacıların Türkiye'ye Etkileri. Rapor No: 196, 2015.

Petersen, W. A General Typology of Migration. *Am. Sociol. Rev.* **1958,** *23* (3), 256–266.

Ravenstein, E. G. The Laws of Migration Authorls. *J. Stat. Soc. London* **1885,** *48* (2), 167–235. (Tun., 1885). Published by: Blackwell Publishing for the Royal Statistical Society Stable. URL: http://www.jstor.orglstable/2979181 (accessed June 24, 2008).

Saraçlı, M. Uluslararası Hukukta Yerinden Edilmiş Kişiler, Gazi Üniversitesi Sosyal Bilimler Enstitüsü Yayınlanmamış Doktora tezi Ankara, 2010.

SETA, (Siyaset, Ekonomi ve Toplum Araştırmaları Vakfı) Türkiye'deki Suriyeli Çocukların Eğitimi Temel Eğitim Politikaları, (Analiz Raporu), İstanbul, 2016.

Topçu, S. ve Beşer, A. Göç ve Sağlık. *C.Ü. Hemşirelik Yüksekokulu Dergisi.* **2006,** *10* (3), 37–42.

Yıldız, Ö. Türkiye Kamplarında Suriyeli Sığınmacılar: Sorunlar, Beklentiler, Türkiye ve Gelecek Algısı. *Sosyoloji Araştırmaları Dergisi.* **2013,** *16* (1), 140–169.

E-REFERENCES

Göç Terimleri Sözlüğü, (2009) Uluslararası Göç Örgütü (IOM) Yayını

http://www.goc.gov.tr/icerik6/goc-terimleri-sozlugu_0_382_727_icerik

http://www.goc.gov.tr/files/files/2016_yiik_goc_raporu_haziran.pdf (erişim 01.08.2017)

https://www.ihh.org.tr/haber/18-aralik-2010-arap-baharinin-baslangici-1975 (erişim tarihi 06.06.2017)

http://www.orsam.org.tr/index.php/Content/Analiz/1478?s=turkmen%7Cturkish (erişim, 08.06.2017)

http://www.orsam.org.tr/files/OA/61/11sunaoner.pdf (erişim 08.6.2017)

http://www.unhcr.org/tr/

http://www.unhcr.org/turkey/home.php?content=733 (erişim tarihi, 13.07.2017)

http://www.unhcr.org/turkey/home.php?content=733 (erişim tarihi, 13.07.2017)

http://www.goc.gov.tr/icerik3/gecici-korumamiz-altindaki-suriyeliler_409_558_560 (erişim 20.06.2017)

http://www.haberturk.com/gundem/haber/1427703-yaklasik-3-milyon-suriyeli-kayit-altinda# (Erişim 01.08.2017)

http://www.unhcr.org/turkey/home.php?content=726 (erişim 01.08.2017)

https://www.afad.gov.tr/upload/Node/2373/files/Suriyeli_Siginmacilara_Yapilan_Yardimlar+7.pdf (Erişim tarihi 20.08.2017)

http://www.tuik.gov.tr/PreHaberBultenleri.do?id=21584 (erişim tarihi 21.08.2017)

https://biruni.tuik.gov.tr/yayin/views/visitorPages/index.zul (erişim tarihi 21.08.2017)

CHAPTER 4

Demographic Transition: Human Capital Variable

İLKNUR MAYA*

Department of Educational Sciences, Faculty of Education, Çanakkale 18 Mart University, Çanakkale, Turkey

E-mail: mayailknur@gmail.com

ABSTRACT

Human capital is a key component of the economic development of a country. For this reason, the Human Development Index is used by the United Nations to measure the economic and social development of countries. In general, African countries are countries with low human capital. The aim of this research is to determine the level of human development of Sub-Saharan African countries and to reveal what is needed to be done in order to enable the human development indices in these countries to be able to reach from low levels to high in the long term through the new organizational principles of the theory of complexity. The focus of the research is the third world countries. The working group constitutes some selected Sub-Saharan African countries. These countries are Niger, Central African Republic, Chad, Burundi, Guinea, Sierra Leone, Mozambique, Mali, Liberia, Ethiopia. In this research, the Human Development Index was selected as an indicator. In the research, the following results were obtained: (1) Sub-Saharan African countries' Human Development Index values show imbalance characteristics over the years. Moreover, the values of human development of countries are chaotic by offering a fluctuating development. Sub-Saharan African countries have very low survival rates at birth, average schooling years and income per capita. (2) In Sub-Saharan Africa, Human Development Index values have increased in recent years, but they are still quite low. (3) In Sub-Saharan Africa, there are gender inequalities in terms of school enrollment, women's participation in parliament, and women's participation in the workforce.

4.1 INTRODUCTION

The global era, we live in, reveals the importance of education more clearly every day. An important fact known today is that education is a very important phenomenon both from the viewpoint of the individual and the countries in the global arena. From an individual's point of view, getting qualified and high-level education ensures that the individual has higher income and higher quality of living conditions as well as the development of personal, social, cognitive, psychomotor, and emotional development. Likewise, countries, as seen in the world, are seeking economic and social development by attaching importance to education. Through education, it is understood that many countries in the world have achieved high development and attained success in a sustainable competitive environment in the global arena. For this reason, many international organizations such as the Organisation for Economic Co-operation and Development (OECD), United Nations (UN), United Nations Educational, Scientific and Cultural Organization (UNESCO), United Nations International Children's Emergency Fund (UNICEF), and United Nations Development Programme (UNDP) report and document their work on education-individual relationship and education–country relationship.

Education offers very important contributions regarding reduction of infant mortality, diminishing inequalities in the longevity of lifespan, reduction in crime, establishing a society based on peace and security, and the formation of democracy consciousness and environmental awareness in a country as well as the economic development. All these benefits are expected to provide further development for the the country in the long term. Considering the benefits of education, investing in education in countries should be an urgent priority area. In this way, human capital may develop. It can be stated that qualified human capital is a necessity for the development of a country.

However, access to education and qualified education conditions in the world countries show significant differences. In this context, the "African countries," often mentioned as third world countries, are in a disadvantageous position. Sub-Saharan African countries are the most disadvantaged ones in African countries. Although there has been a significant growth in African countries, especially in sub-Saharan Africa over the last 50 years, it is still not at the expected level. In these countries, low levels of access to education and a low-quality educational environment cause low levels of human development.

In this research, it is attempted to determine what the level of human development of Sub-Saharan African countries is and to understand how human development indexes can evolve at a high level in the long term in these countries through new organizational principles of complexity theory. Because it can be said that countries, as complicated organizations in today's world, can achieve better living conditions by taking place at the edge of chaos and revealing their creative aspects.

4.2 DEMOGRAPHIC TRANSITION

4.2.1 WHAT IS HUMAN CAPITAL?

The 21st century—in which we live—is called the age of information. Economy is based on information in this century. Countries' or individuals' ability to benefit from this information-based economy depends on the qualities of manpower countries have and individuals' skills and abilities. As a consequence, it has become possible for countries today to raise high-quality individuals by attaching importance to human capital; and it will also be possible in this way in the future.

Education is the process of instilling in individuals the desired skills, abilities, and behaviors. Therefore, the most important way of making human capital is education. Education, on the other hand, can be considered as a three-dimensional group, as general education beginning with preschool and including primary school, secondary school, and high school; vocational education and lifelong education/learning. It is important from several aspects to consider the formation and development of human capital a learning process that continues lifelong.

The fact that the concept of human capital is the most important factor in nations' economic growth and social development makes the question of how individuals are to shape or program the phenomenon of education beginning with early childhood in the administration of countries important. The concept was first used by the Scottish economist Adam Smith, the author of the book "Wealth of Nations," causing the emergence of the discipline of economics in the 18th century. Then, in the 1950s–1960s, the concept was observed to be used as an important concept (OECD, 2007). Economists preferred using the concept of "capital" in those years to emphasize the importance of the phenomena of education and expertise in economic growth. In this context, human capital contains a combination of individuals' potentials, natural abilities and aptitudes, and the knowledge and skills they acquire later through general and vocational education. For all these reasons, education is important. Research has shown that raising 7-year primary education up to 8 years in Kenya—that is, to say, 1-year increase in the duration of education—results in great gains in the country (Agesa et al., 2016, s 131).

4.2.2 THE IMPORTANCE OF DEVELOPING HUMAN CAPITAL

Human capital is the stock of competence, knowledge, and personal traits embodied in the skills of the workforce to produce economic values. Human capital is sum of the properties individuals gain through education and experience. Human capital is a mean of production through which new areas of investment introduce new and qualified workforce. Human capital is the human resources containing knowledge, skills, attitudes, and motivations belonging to an institution or a community, and causing members to make efforts to raise their life standards. Consequently, human capital is the capital and productivity emerging through knowledge and skills acquired with education, teaching, and experiences (Chikwe et al., 2015).

Human capital is a nonmaterial capital when compared with tangible capital such as machinery, lands, mineral sources, and other raw materials. Therefore, it is internal and cannot be directly measured. The indicators of human capital are the welfare of nations, materialistic and social welfare of citizens, the effectiveness of institutions and systems, and the quality of assessable humanistic development indices. The human capital status of a nation influences economic and social indicators such as life expectancy, the rate of literacy, gross national product, income per person, the balance of trade, level of industrialization, and the quality of infrastructure precautions directly and is positively correlated. Human capital can have great impacts on political stability, national peace, and harmony and common moral values (Chikwe et al., 2015).

It may be said that human capital is one of the most important sources of a country. Moreover, human capital is now much more important than in the past in the development of countries. Therefore, we see that many countries invest in human capital, that they make policies for the development of this capital through applications and that they implement those policies.

The necessity of developing its human capital for a country can be considered in three aspects. First, it should be pointed out that this era is the age of knowledge. Thus, the knowledge economy is rising increasingly. Due to this, it is observed that production dynamics are based on services all over the world or on the production and management of knowledge. The second aspect is that the need for a qualified workforce has been increasing in every sector. The third is that the population is getting older, especially in developed countries. As a result of this, business life will also gradually be composed of old people. In this case, it will be possible for older individuals to adjust to the new economic and social order by developing their skills and by acquiring new skills through education.

Human capital is an indispensable component of national economic competitiveness, and it has been gaining more and more importance as a country develops, as is indicated in the analysis in World Economic Forum (WEF) index. More than 24% of the 2007–2008 WEF competitiveness scores of the most developed countries and more than 16% of the scores of the least developed countries are related to their human capital indicators. These simulations show that countries which are not ordered at the top regarding human development index can obtain great gains in competitiveness if they successfully implement reforms improving human capital. The countries which are ordered at higher positions, on the other hand, have to increase their human capital and maintain it if they wish to preserve their superiority. When scores for each indicator of human capital are analyzed in details, in which parts reform efforts will be effective most can be understood (Sabadie and Johansen, 2010).

4.2.3 HUMAN DEVELOPMENT INDEX AS AN INDICATOR OF THE VALUE OF HUMAN CAPITAL

Human development index is an index developed by the United Nations and aims to measure the economic and social status of a country. It is generally used as the indicator of basic development to show the average achievement of a country. Human development index has three basic components: The first is life expectancy. It is related to a lifetime in a country. The survival rate of babies at birth is used as a criterion in recent years. The second is education. The issue of education represents the primary school schooling rates, secondary school schooling rates, higher education schooling rates, and adult literacy rates in a country. The issue of education also demonstrates the current situation according to gender. The third is related to income status in a country. According to another source, human development index measures the average achievement of a country by three dimensions. They are as in the following (Bilbao-Ubillos, 2013): Lifetime, which reflects people's ability to lead a long and healthy life, population's level of reaching knowledge through the educational system, and average living standard of the population regarding purchasing power.

Human development index is an authoritative instrument providing support to improve the human development of poor nations in general. Thus, the index sets up physical and social connections between many aspects of the family and the surroundings of the house so that children and individuals are better off. Besides, human development index is based on a developmental paradigm focusing on human development and paying attention to the role of environment and the setting to develop and support human potential (OECD, 2007).

Human development index was chosen as a macro-level indicator functioning as the focus of economic and social development and as a reflection mechanism of a country in this study. Human development index changes between 0.00 and 1.00. Accordingly, the human development index is grouped into four levels. This fourfold grouping is used to describe very high, high, medium, and low-level countries. The levels used for the human development indices of counties are within the following intervals: (1) *Very high*: between 0.80 and 1.00, (2) *High*: between 0.70 and 0.79, (3) *Medium*: between 0.55 and 0.69, and (4) *Low:* between 0.00 and 0.54.

Human development reports presenting national, regional, and local level information have been published since 1992. The latest published report for human development was the one published on December 14, 2015 when this study was performed, and it was calculated for 188 countries.

4.3 THIRD WORLD COUNTRIES AND THEIR HUMAN DEVELOPMENT INDICES

African countries, with exceptions, are usually classified as "third world countries" because their human resources and natural resources are developed less and that most of their citizens have low living standards (Jogwu, 2010). Therefore, we see that mostly African countries are selected in studies conducted about human capital and economic development (Hadziahmetovic, 2016).

Since the 1970s, Sub-Saharan African countries have made negative economic growth measured in changes in gross national product per capita. It is known that many countries in Africa (Botswana, Egypt, Cameroon, Republic of Congo, Kenya, and Mauritius) achieved annual growth above 1% in gross national product per capita in the period between 1965 and 1998. For most of the countries in the region, on the other hand, it was difficult to consider their weak economic performance effectively in the 1980s and 1990s. It is commonly known that low economic performance in sub-Saharan Africa is due to such factors as drought, badly developed structures, lack of both human and physical capital, bad administration, political violence, ethnical wars, and diseases such as HIV/AIDS. Besides, economic policies of the countries in the north of Africa are another factor. Economy policies developed by the World Bank and by international finance systems of the International Monetary Fund (IMF) are sometimes held responsible for deteriorating the economic crisis in Africa (Oketch, 2005).

However, there is a general view that education is an instrument in economic, social, and political development. This claim is proven by the relationship between the level of educational development and living standards available in such developed countries as Singapore, the United States, the United Kingdom, Canada, China, South Korea, and Japan. Industrialized countries are also modern economies and democracies because such nations have well-educated, enlightened, and capable adult population. However, in African countries, the majority of the adult population is illiterate, and the rate of unqualified labor is high. For this reason, African countries are under-developed countries.

According to the UNESCO World Report (UNESCO 1991 and 1995), the continents of Africa and Asia have the highest percentage of adult illiterate people in the world. Almost all of the poor and under-developed countries are in these two continents and in South America. Undoubtedly, many African countries make huge investments to increase access to formal education and to raise the rates of registration in primary, secondary, and higher education. Despite the presence of praiseworthy policies and programs in Africa contributing to increasing the adult rate of literate population, there is an increasingly widening gap between theory and practice and between policy and implementation.

Africa is a continent having more than 52 countries. Africa is one of the countries making the best progress in human development index between the years 1990 and 2014. However, Africa also has the lowest level of human development when compared with other regions of the world. Therefore, there is a remarkable difference, for instance, between North Africa and sub-Saharan Africa and between Anglophone and Francophone. The differences between African countries emerge in educational, social, economic, political, cultural, and religious aspects. Thus, those countries are at different levels of development. Thirty-six of the African countries are in the low human development group while 12 of them are in the medium human development group and five are in the group of high human development. No African countries are belonging to the group of very high human development (UNDP, 2016).

Great differences are observed between African region and differing sub-African regions regarding values and ranking in the assessment made by UNDP by using different indicators of human development. North Africa is well above the average in African Human Development Index, and it is even above South Asia when compared globally (including Mauritania). Similarly, South Africa is the only subregion above the average African regional human development index. In general, Africa has the fastest development rate in human development in the last 20 years. Africa, when compared with other parts of the world, has the lowest level of human development. But this is not true for all African countries. Seventeen African countries—five of which are in the subregion—have attained medium or high levels of human development. Of African countries having medium or high levels of human development, five are in South Africa, five are in North Africa, four are in central Africa, two are in West Africa, and one is in East Africa. The highest levels of human development in Africa are in Algeria, Libya, Mauritius, Seychelles, and Tunisia. However, 36 African countries (out of 44 countries in the world) are classified into the group with the lowest level of human development (UNDP, 2016).

These countries, which were initially at low-levels of human development, obtain great gains. Some of the countries, in particular, have obtained the greatest gains since 2000. They are the United Republic of Tanzania, Burundi, Mali, Niger, Angola, Sierra Leone, Mozambique, Rwanda, and Ethiopia. The countries which are at low levels of human development at the beginning grew faster than the average, but their growth slowed down in 2010.

This study sets out from the assumption that underdevelopment of Africa, a third world country, is the result of its level of education and of having illiterate population. The level of literacy is important in reducing poverty and unemployment, in assuring participation in economic and political activities and in empowering women, who constitute approximately 50% of the population in some countries.

Evidence coming from research shows that, adults in many cities and some rural areas of Nigeria are enthusiastic about attending evening literary classes and thus overcoming the obstacle of literacy. But only a small part of them continue classes for a long time. According to the Federal Ministry of Education in the country (FME, 2004), dropout rates are high in schools. For example, only 814,133 out of 1,142,966 adults registering in the programmes continued the programs in 1996. This situation demonstrates that the rate of completing education is 71.2% but that the rate of dropout is 28.8% (UNDP, 2015).

The low-level quality of education in the Sub-Saharan African countries can have negative effects on the growth and human capital of those countries. It may be said that the economic growth of Sub-Saharan Africa and a great increase in school registration beginning with the year 2000 appear to solve the problem of lack in economic growth, but this is not the case indeed. The rapid increase in school registration results in crowded classrooms and makes existing sources inadequate, and this reduces the quality of education. High economic growth of the Sub-Saharan Africa in recent years stems from high prices it sets for its basic export commodities, and the prices can fall in the future. Sub-Saharan African countries can invest more in education and apply effective child health programmes so that they can raise the quality of their schools. In addition to that, they can build more schools, obtain more teaching materials, and they can apply methods for better school administration (Glewwe et al., 2014). Table 4.1 below shows the figures for the human development index in some selected African countries.

TABLE 4.1 Figures for Human Development Index in Some Selected African Countries (2015).

Countries	Ranking in human development (2014)	HDI (2014)	Survival rates at birth (2014)	Average schooling year (2014)	Per capita income annually (2011, $)
1. Niger	188	0.348	61.4	1.5	908
2. Central African Republic	187	0.350	50.7	4.2	581
3. Chad	185	0.392	51.6	1.9	2085
4. Burundi	184	0.400	56.7	2.7	758

TABLE 4.2 *(Continued)*

5. Guinea	182	0.411	58.8	2.4	1096
6. Sierra Leone	181	0.413	50.9	3.1	1780
7. Mozambique	180	0.416	55.1	3.2	1123
8. Mali	179	0.419	58.0	2.0	1583
9. Liberia	177	0.430	60.9	4.1	805
10. Ethiopia	174	0.442	64.1	2.4	1428

HDI, human development index.
Source: UNDP. Human development report 2016: Human development for everyone, 2016.

As is clear from Table 4.1, all of the selected African countries rank the lowest in 188 countries regarding human development, and are in the 0.00–0.54 interval and thus they are in the low-level group. On the other hand, the survival rate of babies at birth as an indicator of health is below 50% in African countries, and schooling rate as an indicator of education is below 5 years. In addition to that, gross income per capita in African countries is less than 2000$. Table 4.2 below shows the change of human development index in some selected African countries according to years.

TABLE 4.2 The Change of Human Development Index in Some Selected African Countries According to Years (1980–2014).

Countries	1980 HDI	1990 HDI	2000 HDI	2010 HDI	2014 HDI
1. Niger	0.190	0.214	0.257	0.326	0.348
2. Central African Republic	0.302	0.314	0.310	0.362	0.350
3. Chad	–	–	0.332	0.371	0.392
4. Burundi	0.230	0.295	0.301	0.390	0.400
5. Guinea	–	–	0.323	0.388	0.411
6. Sierra Leone	0.268	0.262	0.299	0.388	0.413
7. Mozambique	0.238	0.218	0.300	0.401	0.416
8. Mali	0.199	0.233	0.313	0.409	0.419
9. Liberia	–	–	0.359	0.405	0.430
10. Ethiopia	–	–	0.284	0.412	0.442

HDI, human development index.
Source: UNDP. Human development report 2016: Human development for everyone, 2016.

As is evident from Table 4.2, the values for human development in selected African countries between 1980 and 2014 improved. The Central African Republic, Sierra Leone, and Mozambique are exceptions to this. While the value for human development in the Central African Republic rose between 1980 and 1990, the value fell between 1990 and 2000, and then it rose again between 2000 and 2014. While human development index fell in Sierra Leone between 1980 and 1990, it rose between 1990 and 2014. In Mozambique, the human development index fell between 1980 and 1990, but it rose between 1990 and 2014. Therefore, it is apparent that the figures representing human development index in these three African countries fluctuate.

The majority of African countries is at low levels of human development and is remarkable with social gender differences. According to gender equality calculations made by UNDP, on the other hand, there is a considerable amount of gender inequality in almost all African countries. Women in Africa attain 87% of human development levels men attain on average. Many African countries have been extending the capabilities of their citizens in the fields of health, education, and other social services for the last 10 years. Thus, the citizens in Africa can be healthier, women can give birth to their babies in a safer way, and can extend their lifetime.

Nevertheless, many women in Africa face serious health problems due to such factors as marriage at early ages, sexual and physical violence, and high rates of maternity death. Women and girls in Africa can have access to all stages of education more now. About education, it is remarkable that gender equality has been attained in registration in primary school education. However, gender inequality has been continuing in secondary and higher education in Africa. The causes for children not to go to school vary in Africa. This is usually related to poverty, ethnic origin, social isolation, living in rural areas or slum areas, geographical remoteness, disasters, gun battles, impossibility, and low-quality education. Another important determiner of social gender equality is described as women's participation in business life and economic decision-making processes. There is gender inequality in Africa regarding access to economic assets, participation in the workplace, entrepreneurship opportunities and benefiting from natural resources and the environment. Due to inconsistencies between women's educational differences and qualities and the demands of the labor market, women are more likely to be employed without security in the form of weak regulations and limited social security (UNDP, 2016).

It was demonstrated that one of the most outstanding inequalities between men and women was in income differences. This situation is true for all subregions. North Africa has the greatest inequality in income. While the rate of women's participation in work force especially for the young population was 19.7% in North Africa in 2014, it was 52.1% in Sub-Saharan Africa (ILO, 2015), which stemmed from a high rate of gender inequality. The expected gender equality in school years was attained only in North Africa. The region also has the longest life expectancy for both genders at birth (UNDP, 2016).

The majority of African countries are remarkable with social and gender inequalities and low levels of human development. Even though many African countries—including Ethiopia, Rwanda, Mozambique, and Angola—are in the group of the low level of human development, they have made evident progress in increasing their values for human development index. Table 4.3 shows the human development indices according to gender in some selected African countries.

TABLE 4.3 The Human Development Indices According to Gender in Some Selected African Countries (2015).

Countries	Gender inequality index (GII)	Women's participation in the parliament	Minimum participation in secondary education (25 and +)		Participation in the workforce (25 and +)	
			W	M	W	M
1. Niger	0.713	13.3	2.4	7.8	40.0	89.7
2. Central African Republic	0.655	12.5	10.1	26.7	72.6	85.1
3. Chad	0.706	14.9	1.7	9.9	64.0	79.2
4. Burundi	0.492	34.9	5.3	8.3	83.3	82.0
5. Guinea	–	21.9	–	–	65.6	78.3
6. Sierra Leone	0.650	12.4	10.0	21.7	65.7	69.0
7. Mozambique	0.591	39.6	1.4	6.2	85.5	82.8
8. Mali	0.677	9.5	7.7	15.1	50.8	81.4
9. Liberia	0.651	10.7	15.4	39.3	58.2	64.8
10. Ethiopia	0.558	25.5	7.8	18.2	78.2	89.3

Source: UNDP. Human development report 2016: Human development for everyone, 2016.

Accordingly, the values for the human development index according to gender in some selected African countries are below 1.00. Therefore, it is clear that gender equality has not been attained in those countries. While gender inequality index (GII) is 0.492 in Burundi, it is 0.591 in Mozambique. These two countries have the lowest values regarding gender equality index. However, these two countries have the highest values regarding women's rate of participation in parliament. The rate of women's participation in the parliament is 34.9% in Burundi whereas it is 39.6% in Mozambique. This is not linear reasoning, and it is not an expected situation. Both countries are again in the best position regarding women's participation in the workforce. While women's participation in the workforce is 83.3% in Burundi, it is 85.5% in Mozambique. The two countries have better values than the others, and the rates of women's participation in the workforce are very close to those of men's in both countries. It is interesting that women's rates are higher than men's.

4.4 THIRD WORLD COUNTRIES AND THEIR HUMAN DEVELOPMENT INDICES IN THE LIGHT OF COMPLEXITY THEORY

Complexity theory has applications primarily in physical sciences and also in social sciences in recent years. There are a great number of studies researching the phenomenon of complexity in social sciences (Fenwick, 2008; Gilstrap, 2005; Marion and Uhl-Bien, 2001; McClellan, 2010; Osborn and Hunt, 2007; Schneider and Somers, 2006; Svyantek and Brown, 2000). In the simplest terms, complexity theory is against linear and mechanistic world view explaining physical and social phenomena in a simple cause-effect relationship. On the contrary, it argues for a nonlinear, organic, uncertain, and unpredictable worldview (Regine and Lewin, 2000).

While classical science searches for order and balance, familiarity and predictability in systems; complexity theorists expect even simple systems to be dynamic, unbalanced, unpredictable, and complex while describing them (Galbraith, 2004; Prigogine, 1997). Thus, complexity theory is against reductive logic which does not perceive a system from a holistic perspective and which ignores the effects of interactive dynamics. Instead, complexity theory assures that we see organizations as interacting with each other, composed of distinct units, and as a whole in a structure uncovering innovative behaviors (Marion and Uhl-Bien, 2001; Regine and Lewin, 2000). For this reason, complexity theory approaches problems more holistically.

According to complexity theory, an organization is seen as considering the formations emerging inside and outside the organization.

The traditional view of the organization is based on the idea of the familiarity of the world. From this perspective, organizations are composed of a set of strict rules, formalized control and a hierarchical structure. These organizations can give familiar and predictable responses to changes. Traditionally, organizations demand and search for order. Leaders, on the other hand, are the people who attain balance by reducing complexity through coding in traditional organizations, who use reductive logic rather than holistic thought, who understand the critical cause–effect relationship and who make complex plans for a predictable world (Ashmos et al., 2002).

The reductive approach is based on the idea that an individual can understand the whole if he/she understands the ports constituting the whole system. Therefore, the same reasoning has been accepted in studies for a long time. Either individuals in organizations or institutions as a whole are analyzed in the form of closed systems in those studies. Yet, several physical scientists like Gibbs, Einstein, Maxwell, and Boltzmann pointed out in the 20th century that the orbits of physical parts should also be analyzed and thus they carried out a transformation in science. In this approach, it is regarded that if a person knows the properties of a job; he/she can understand the behavior of the parts of a system by monitoring the direction of movement of many parts, the mutual interaction of diverse internal conditions and differing combinations of ties in the outer environment (Marion and Uhl-Bien, 2001: p 393).

Complex organizations are dynamic. They have many organizational actors inside and outside the organization. Those actors create a dynamic structure in the organization through continuous exchange of information. Decisions for the present day are influenced by actions in the past; whereas decisions for tomorrow are influenced by today's actions in complex organizations (Plowman et al., 2007: p 343).

Theorists began to conceptualize organizations as harmonic systems into which adaptation is possible (McKelvey, 2001). Complex organizations are organized by themselves. Self-organization happens with the help of roles. Systems organized by themselves establish order in the organization with independent units' sharing information, displaying action and their sharing feedback about the system rather than a central authority's imposing an upper-level planning. Thus, information is an energy-creating factor in complex organizations. Information can emerge in a multi-directional manner via dialogues and discussions (Dooley, 2004: p 360; Osborn and

Hunt, 2007: p 334). Creative and adaptive behavior appears when the actors in the system interact independently (Boal and Schultz, 2007: p 411).

Political systems are unbalanced just like other social systems. Imbalance can be the source of the new order (Galbraith, 2004: p 10). African countries considered in this study are the third world countries with low levels of human development index. On examining the human development index values of 10 African countries selected in this research according to years, the imbalance is observed. Besides, the values representing the human development of those countries are fluctuating and have a chaotic structure. Therefore, it is important for those selected African countries to determine the vision and use strange attractors so that their low human development indices can be changed into high. The reason for this is that complex organization are at the edge of chaos today, and they use strange attractors through vision by using the creative aspect (Gilstrap, 2005: p 55).

Moreover, according to complexity theory, it is not a good thing for a self-adaptive system to gain balance. A system in balance means that it is not effective and even it means that it is dead. The chaotic structure of the African countries selected explains their potential to function in the road between disorder and order.

Complex systems have internal unpredictability. African countries' human development indices have been fluctuating according to years. They are influenced by many factors and are thus shaped. Some of the African countries such as Central African Republic, Sierra Leone, and Mozambique have ups and downs with differing tendencies according to years. Although Burundi and Mozambique have low levels regarding GII—a factor influential in the formation of human development index values—the yield has more positive results regarding women' participation in the parliament and labor. These are the examples for internal unpredictability in themselves.

Complexity theory provides us with the new principles of organization of relations and patterns. It is difficult to present the best ways and privileged "objective" perspectives. Therefore, African countries should be addressed by considering that issues such as education, health, economic status, participation in politics, and freedom and equality as the determinants of human development levels are the components of human development apart from simple, familiar, and standard methods.

According to complexity theory, on the other hand, it is necessary to want to be a part of the solution in solving a problem. Successful leaders in complex organizations are observed to employ the following mechanisms while serving to the formation of the future: spoiling the existing patterns, encouraging innovations, making others sensitive to occurrences. The starting point is important in generating the new world reality in complex organizations. In this context, it is necessary for African countries to set high targets in such components as health, education, and income level by starting points so that they can attain the acceptable or desired levels of human development.

4.5 CONCLUSION, DISCUSSION, AND RECOMMENDATIONS

In conclusion, African countries implemented policies and programs to achieve socioeconomic development and to attain Millennium Development Targets until 2015. However, human capital is the determiner of economic growth and development in Africa. It should now be known that the way of developing human capital is education in African countries, as in all other countries. In this framework, it is evident that African countries can attain higher living standards by attaching importance to education. For this purpose, policies should be made for more individuals' access to and attendance in education, educational opportunities should be improved, and precautions should be taken to assure both teachers' and families' support. Therefore, African countries need to maintain their investment in human capital at present and in the future so that they can reach the industrial development levels. They should concentrate on investments in the development of human resources through education, and they should evaluate reforms which are appropriate for economic growth. Thus, it may be said that the need for educational reform is essential in the region (Seetanah, 2009). Research (Lam et al., 2016) conducted in 12 countries containing diverse cultures suggests that educational applications and teachers' and families' support play positive roles in individuals' participation in education.

It should not be forgotten that development is difficult if the majority of the population is illiterate in Africa or any place on earth. Ignorance and illiteracy restrict individuals' capacity and hinder their contributions to national development. Another requirement for African countries to go out of the category of third world countries and to become a developing country and gradually a developed country is that women's education should be considered important—because on examining the human development indices of developed countries, their schooling status and participation in labor in those countries according to gender; it is found that gender equality is established in those countries. To illustrate this, Norway can be used as an example (HDR, 2015). A study conducted by Turanlı et al. (2015)—with analyses of development indices of 44 countries between the years 1990 and 2010—found correlations between education and development and between education and participation in labor. Then, care should be taken especially with women and rural areas to raise the rate of literacy in Africa. The reason for this is that almost half of the African population is composed of women and most of the citizens in Africa live in rural areas. In consequence, governments in Africa should set up a partnership between nongovernmental organizations and private sector and cooperate with them to achieve national development and to make and implement policies and programs for attaining national targets in education.

The ties between human development and participation in business life are synergic. Participation in business life diminishes poverty, assures fair growth, provides income and sources of livelihood, and thus, improves human development. Human development increases human capital (by increasing information, skills, and awareness) and expands opportunities and options. Sustainable work reduces or eliminates undesired results and supports human development. Therefore, sustainable work has critical importance not only in maintaining the current situation but also in assuring working for next generations. A UNDP (2015) report states that the ties between human development and participation in business life are not automatic and that some activities such as making people work forcefully violate human rights and can give harm to human development. The report states that a job can have unequal opportunities and its rewards can be separatist, and thus it can maintain inequalities in society. Therefore, gender equality should be considered important also in the participation of the adult population in labor.

KEYWORDS

- **third world countries**
- **Sub-Saharan African countries**
- **human capital**
- **chaos theory**
- **complexity theory**

REFERENCES

Agesa, R.; Agesa, J.; Dabalen, A. Does an Extra Year of Primary Schooling Yeild Higher Earnings? Evidence from Kenya. *J. Dev. Areas* **2016,** *50* (3), 131–148.

Ashmos, D. P.; Duchon, D.; McDaniel, R. R.; Huonker, J. W. What a Mess! Participation as a Simple Managerial Rule to Complexify Organization. *J. Manage. Stud.* **2002,** *39* (2), 189–206.

Bilbao-Ubillos, J. Another Approach to Measuring Human Development: The Composite Dynamic Human Development İndex. *Soc. Indicators Res.* **2013,** *111* (2), 473–484.

Boal, K. B.; Schultz, P. L. Storytelling, Time, and Evolution: The Role of Strategic Leadership in Complex Adaptive Systems. *Leadersh. Q.* **2007,** *18*, 411–428.

Chikwe, C. K.; Ogidi, R. C.; Nwachuhwu, K. Challenges of Research and Human Capital Development in Nigeria. *J. Educ. Prac.* **2015,** *6* (28), 44–47.

Dooley, K. Complexity Science Models of Organizational Change. In *Handbook of Organizational Change and Development*; Poole, S., Van De Ven, A., Eds.; Oxford University Press: Oxford, 2004; pp 354–373.

Fenwick, T. Responsibility, Complexity Science and Education: Dilemmas and Uncertain Responses. *Stud. Philos. Educ.* **2008,** *28*, 101–118.

Galbraith, P. Chaos Mini-theme. Organisational Leadership and Chaos Theory. Let's be Careful. *J. Educ. Adm.* **2004,** *42* (1), 9–28.

Gilstrap, D. L. Strange Attractors and Human İnteraction: Leading Complex Organizations through the Use of Metaphors. *Complicity* **2005,** *2* (1), 55–69.

Glewwe, P.; Maiga, E.; Zheng, H. The Contribution of Education to Economic Growth: A Review of the Evidence, with Special Attention and an Application to sub-Saharan Africa. *World Dev.* **2014,** *59*, 379–393.

Hadziahmetovic, N. Human Capital and Economic Development Review of What Was Studied and Where Was Researched. *Eur. Res.* **2016,** *102* (1), 12–24.

Jogwu, C. N. O. Adult İlliteracy: The Root of African Underdevelopment. *Educ.* **2010,** *130* (3), 490–498.

Lam, S.; Jimerson, S.; Shin, H.; Cefai, C.; Veiga, F.; Hatzichristou, C.; Polychroni, F.; Kikas, E.; Wong, B. P. H.; Stanculescu, E.; Basnett, J.; Duck, R.; Farrell, P.; Liu, Y.; Negovan, V.; Nelson, B.; Yang, H.; Zollneritsch, J. Cultural Universality and Specificity of Student Engagement in School: The Results of an İnternational Study from 12 Countries. *Br. J. Educ. Psychol.* **2016,** *86* (1), 137–153.

Marion, R.; Uhl-Bien, M. Leadership in Complex Organizations. *Leadersh. Q.* **2001,** *12* (4), 389–418.

McClellan, J. L. Leadership and Complexity: Implications for Practice Within the Advisement Leadership Bodies at Colleges and Universities. *Complicity* **2010,** *7* (2), 32–51.

McKelvey, B. What is Complexity Science? It is Really Order-creation Science. *Emergence* **2001,** *3* (1), 137–157.

OECD. *OECD Insights. Human Capital: How What You Know Shapes Your Life*; OECD Publishing: Paris, 2007. ISBN-92-64-029095.

Oketch, M. O. Determinants of Human Capital Formation and Economic Growth of African Countries. *Econ. Educ. Rev.* **2005,** *25*, 554–564.

Osborn, R. N.; Hunt, J. G. Leadership and the Choice of Order: Complexity and Hierarchical Perspectives Near the Edge of Chaos. *Leadersh. Q.* **2007,** *18*, 319–340.

Plowman, D. A.; Solanosky, S.; Beck, T. E.; Baker, L.; Kulkarni, M.; Travis, D. V. The Role of Leadership in Emergent, Self-organization. *Leadersh. Q.* **2007,** *18*, 341–356.
Prigogine, I. *The End of Certainty*; Free Press: New York, 1997.
Regine, B.; Lewin, R. Leading at the Edge: How Leaders İnfluence Complex Systems. *Emergence* **2000,** *2* (2), 5–23.
Sabadie, J. A.; Johansen, J. How do National Economic Competitiveness İndices View Human Capital? *Eur. J. Educ.* **2010,** *45* (2), 236–258.
Schneider, M.; Somers, M. Organizations as Complex Adaptive Systems: Implications of Complexity Theory for Leadership Research. *Leadersh. Q.* **2006,** *17*, 351–365.
Seetanah, B. The Economic İmportance of Education: Evidence from Africa Using Dynamic Panel Analysis. *J. Appl. Econ.* **2009,** *12* (1), 137–157.
Svyantek, D. J.; Brown, L. L. A Complex Systems Approach to Organizations. *Curr. Dir. Psychol. Sci.* **2000,** *9* (2), 69–74.
Turanlı, M.; Taşpınar Cengiz, D.; Turanlı, R.; Akdal, S. Effect of Female Education and Labor Force Ratio on Economic Development. *Educ. Sci.* **2015,** *15* (2), 387–396.
UNDP. *Overview Human Development Report 2015: Work for Human Development*; The United Nations Development Programme: New York, USA, 2015.
UNDP. *Africa Human Development Report 2016: Accelerating Gender Equality and Women's Empowerment in Africa*. The United Nations Development Programme Regional Bureau for Africa: New York, USA, 2016.
UNDP, 2017. *Human Development Report 2016: Human Development for Everyone*, 2016. http://hdr.undp.org/sites/default/files/2016_human_development_report.pdf. (accessed June 20, 2017).

CHAPTER 5

Theoretical Assertions and Economic Frameworks on Demographic Transition

DANIELS OKUN*

Independent Researcher

**E-mail: danielsokun@yahoo.com*

ABSTRACT

Phenomenally, the term demographic transition has over time been perceived as an intellectual tool for policy and decision-making, although fluid and elusive the concept and theories of demographic transition has been a resource for policy makers over the years. From the late 18th century, the death rate dropped. With the development of new technologies in agriculture and production, as well as health and hygiene, more people have become healthy, increasing their average life expectancy and creating new trajectories for population growth. This sudden change has long created a shift in what people "understand" to be, the correlation between birth rate and mortality. Over the past 300 years, the population in most countries have continued to develop due to the relationship between birth and mortality in these countries.

This global phenomenon of observation and recording produces a pattern of demographic transition models that helps to explain and understand demographic changes. Using the demographic transition model, demographers can better understand the current population growth of a country based on its position between the four major stages and then transmit the data to address a country and its transnational economic and social policy. In the course of human history, there have been many people interested in the future of demographic characteristics and population growth. After analyzing the changes in the Western population over time, a pattern was found that shows that population growth is linked to the economic development of a country. It has been pointed out that in countries with high standard of living, population growth is slow and the population of countries with low living standards grows faster.

This finding leads to the concept of demographic transition, a series of stages in which a country has moved from nonindustrial to industrialized. The concept of "population change" involves four stages of population size and social behavioral change. As in the case of the United States, its history has undergone major demographic changes. Like many other developed countries in the world, traditional demographic changes related to declining mortality and declining fertility have also contributed to subsequent demographic changes. These changes include urban transformation, marital transition, family transformation, aging transformation, and ethnic transition. In these transformations, the most prominent at the beginning of the 21 century is the latter two: the aging transition and the racial and ethnic transition.

Pragmatically, if the number of baby boomers begins to reach 65 years between 2017 and 2031, the country will rapidly be aging. In addition, as the White population began to decline due to the low fertility and aging, the Hispanic population would dominate the national population because of its youthfulness. In fact, population projections show that the Latino population will double from 50.5 million in 2017 to 111.7 million by 2050, while the White population would decline from 196.8 billion to 186.3 million during this period. These major demographic transitions will have a significant impact on the demographic, social, and economic trends of the next few decades, which in many ways may lead to differences in experience between the United States and other developed countries. The discussions on the impact of these trends on the future of the United States have become a popular debate and lecture.

The world and most regions and countries at large are experiencing some unprecedented rapid population changes. The most obvious example of this change is the huge expansion of the number of manpower: since 1950, there has, been more than 40 billion growth. The forecast for the second half of the world is expected to develop rapidly in the developed countries, where parts of the region are stagnant or potentially declining and continuing to grow rapidly in the least developed regions. Other population movements have also changed dramatically: women's fertility has fallen sharply and life expectancy has risen to new heights. Past fertility and mortality trends have led to high fertility in developing countries. The growing population in most developing countries and developed countries is very young. Contemporary societies are now in a different stage from the demographic transition. This chapter summarizes the major trends in population size, fertility, and mortality, and aging structure during transitional periods. The focus will be from 1950 to 2050 in a century, the world in its fastest population changing period.

5.1 INTRODUCTION

After a researched and documented, hundreds of years of slow and statistically uneven growth, in 1800, the world's ever-growing population reached one billion. Human digital modernization then began with a slow but more stable pace from billions of years ago. However, the growth rate of the 20th century accelerated to an unprecedented level. As a result, in 2003, the world population doubled to 7.4 billion (United Nations 1965, 1983, and 2007). It was expected and envisioned that this population expansion would and could last for decades until the end of the 21st century. In 2080, the world's population will increase by more than 20 times in 1800 years. The near-term characteristics of drastic changes in the population of most of the world's population are arguably in the central stage of the secular process called the demographic transition. Right within, this transition period, the decline in the birth rate and the decline in mortality led to an era of rapid population growth. This transition was often accompanied by the gradual development of agricultural societies into the industrialization of the development process. Before this transition, population growth (which is equivalent to the indifference there is between birth and mortality without immigrants) which through an analysis is close to zero, since high mortality rates are more or less offset the typical high birth rate of agricultural societies before and after the preindustrial revolution age. After the transition, the population growth rate was close to zero, and most developed countries had lower birth rates and mortality rates.

At a national level, population growth has, however, been affected and influenced by migration, but for the average and the total number of regional populations used in this analysis and study, migration is often asserted as a secondary factor and is therefore and will not be discussed in details. It is important to notice that the annual birth rate and mortality of the population depend and have always depended and relied largely on individual fertility and mortality. The most widely used and recognized fertility indicator is the total fertility rate, and this ensures that the number of women born at the end of the childbearing age will be equal if fertility is and supposedly stays at a given age which is satisfied with a particular age.

5.2 DEMOGRAPHIC TRANSITION

Demographic transition is, theoretically, defined as the postindustrialization phase of a social change that occurs in any given society from the high birth rate and mortality rate to a low level, thus changing the aging population. The theory of population change is a widespread and broad analytical clarification of the patterns and diversions of changes in the mortality rate, fertility rate, and growth rates of societies, communities, and countries; as a whole from one growing demographic to another emerging demography (Lee, 2003).

The word was first created by an American demographer; Frank W. Notestein in the 1920s, but has since been elaborated and expanded by many other scholars, thinkers, and demographers alike.

The theory of demographic transition is an analytical study of the population of a country through a period of transition that adjusts with imbalances caused by the decline in mortality and the stable birth rate. As social prospects change, birth and mortality gradually fall to a lower recession, and then balance, leading to a decline in population growth. Demographic transition reveals a country's dynamic population through its formation process from high or low productive capacity, low mortality, and natural growth rates to low birth rates, low mortality rates, low natural growth rates, higher total population, and social demographic changes.

Demographic transition assertively emphasizes on the birth rate and the mortality rate of a population the way they affect, influence, and reduce a country's level of development and advancement from a preindustrialized phase to an industrialized economies. The intellectual assertion is that birth and mortality are associated with the progressive stages of industrial development in any given society (Cai, 2010).

5.3 ASSERTIONS AND THOUGHTS ON DEMOGRAPHIC TRANSITION

As a result of demographic trends, long-term poverty, economic inequality, environmental degradation, pandemic diseases, organized crime, repressive governance, and other developments that the state cannot control separately, it can cause serious security concerns. Weapons cannot solve these problems. *Ban Ki-moon: (Korean diplomat,*

January 2007–December 2016, the eighth Secretary-General of the United Nations).

India needs to work, and Germany needs talent, and cooperation is essential to meeting the needs, desires, and unarguably the demands of the population of the two countries. *Angela Merkel: (German politician and German prime minister since 2005).*

China and India and other countries of the population pressure on our wages and earnings put a lot of pressure. Technological advances have been advancing by leaps and bounds, and the information technology revolution has reduced transportation costs. *Lauren Fabius: (French politician).*

People, in general, are mostly, concerned about every detail of the population changes in every country, so, therefore, as a result, as people begin to will focus on the establishment of young people's likable things, they also focus on the establishment of other people's likable things. *Mark Zuckerberg: (American computer programmer and Internet entrepreneur, The co-founder of Facebook).*

Women, as a primitive demographic unit, run unprecedented forces on every detail and scope of the American life. *Rachel Sklar: (Canadian lawyer, CNN contributor, and media blog).*

Population aging and population reduction are serious problems in many of today's developed countries. In the case of Japan, these demographic changes are growing faster than any other country. *Fukui Toshihiko: (Japanese economist and central banker, 2003–2008, the Bank of Japan governor 29).*

The transition to the population is related to the improvement of the quality of healthcare and sanitation and the improvement of women's access to education. *Jane Rice: (Co-author of the American Scientist and textbook author, Campbell/Reece Biology).*

5.4 THE FOUR PHASES OF DEMOGRAPHIC TRANSITION

Demographic transition: The classical demographic transition through an ever-changing population transition model has four phases as follows:

Phase one: the pre-transitional phase—Classified as the preindustrial era—the first stage of the demographic transition model (DTM) has a low population growth rate due to high birth rates (annual births per thousand) and high mortality rates (per thousand deaths). In this first phase, the population as a whole is constantly changing due to the dynamic patterns of these variables. As the birth rate and mortality are relatively equal, the total population barely changes. But why are they both high? There are many factors that affect the high birth rate and high mortality rate in a country.

Historically, high birth rates are attributed to societies that rely heavily on agricultural productivity or unskilled physical labor because larger families mean greater labor. The high birth rate is usually a response to high mortality which is commonly seen when a country is trying to achieve or maintain and examine an alternative level (a total on average fertility rate of 4.1 births per mother). Through a more in-depth analysis of the situation leading to high mortality, demographers often point out that infant mortality and life expectancy is a criterion for determining high or low mortality. When the medical and maternal care is limited or inadequate, the infant mortality rate is high, when health and public health are low, life expectancy is low. Social, environmental, and political actions at this stage have serious consequences in terms of population, since birth and mortality are so fragile (Bocquet-Appel et al., 2008).

The first phase of the demographic transition model is considered to be prior to the preindustrial or transitional period, and no country is currently included in the first phase of the DTM. In view of human history until the 18th century, all countries were considered to be in the first stage. They are categorized and grouped according to high birth rates and high mortality rates.

This brought to attention the "prevention" (late marriage) and "positive" (famine, war, and plague) checks, population growth which remained at a low level. The total population is at a low; nonetheless because of the high birth rate and high mortality, it seemed to have been balanced (Ezeh et al., 2012).

Phase two: the early transitional phase—The categorization of the preindustrial age: The second stage of the population transition model (DTM) is categorically grouped by a rapid development while a country's supposed mortality rate declines and the birth rate is increasingly on a high. Thus, once, the number of births exceeds the number of deaths, as opposed; to the increase in the birth rate, the total population of a country in the second stage increases. The decrease in mortality rate is usually as a result, of a significant improvement in the overall health status of the population, especially in the pediatric care that affects the life expectancy rate and the ratio of the highest endangered population. However, in the scope and field of basic health care, the expansion of the education system, gender equality, inequality, and technological progress in the department of food production and health contributes a great deal to the reduction of mortality.

The transition to the second stage is notably argued to be a recent phenomenon in the human history. The first country moves from the first stage to the second stage until the first country attempts to begin the industrial revolution. Retrospectively, throughout the middle of the 20th century, most countries in the world have not only entered the second stage but have also continued all up to Stage 3 and Stage 4. But nonetheless, due to social and economic reasons, including sub-Saharan Africa, many countries in Guatemala, now additionally, Nauru, Palestine, Yemen, and Afghanistan, are still in the second phase of the population transformation. India is currently in the second phase of the population transformation, which can significantly reduce the number of deaths, but the birth rate is declining.

In the early stages of the transition, the mortality rate began to decrease. As the birth rate remained relatively high, the population grew rapidly. As a result of these improvements in health care and sanitation, the number of deaths then decreased and the overall population ultimately increased, and in turn the increase of a high birth rate was seen.

Phase three: the late transitional phase (categorized as a highly advanced industrialized phase)—In the third phase of the population transition model (DTM), low mortality and decreasing rates of birth were usually and supposedly impressive due to the improved economic conditions, improved women's status and education, and contraception. The decrease in the birth rate varied from country to country and in most cases, continents to continents; the experience is different everywhere. Some countries have, indeed, experienced and gone through a rapid transformation, the birth rate dropped and reduced by more than 60%, whereas in other countries, it has fallen more slowly. The rate of recession relied more on economic and social factors—declining in education and gender equality.

Education has repeatedly and uncontrollably delayed women's fertility, providing women with an opportunity to increase mother's understanding of her body and health—all of which have resulted in the reduction in the number of families and ultimately, a reduction in the birth rate. In addition, women seeking and demanding higher education was, however, more likely, to plan small families or decide not to give birth at all. The birth rate reduction of a country occurs by shortening the combination of child commitment and the capability to limit and reduce the number of families by delay or contraception. In most countries, gender and reproductive education are becoming the new form and features of public discourse, debates, and engagements, and as such, as women are increasingly affected, they have become knowledgeable. The transition to the third stage of the country is somewhat more stable—economically, socially, and politically.

But the very big question that then pops up in long discussions is: Will these factors affect the birth rate and mortality, or does the birth rate affect the development of the country?

Be that as it may, stable population growth provides a highly significant advantage for any country, providing an opportunity to strengthen the economy, as its citizenship strives through willingness and might to increase over its many years of work and engagement. Therefore, the third stage is often more or not seen as a sign of highly significant development. Examples of the third-stage countries are notably Botswana, Colombia, India, Jamaica, Kenya, Mexico, South Africa, and the United Arab Emirates, where the birth rates are impressively began to decline. Population growth is slowing down rapidly, even though the total population is still growing fast. Due to the differences in agricultural mechanization, the gap between birth and mortality is reduced by the provision of contraceptives, reducing the need for natural growth in children's lives.

Phase four: the post transitional phase—Categorized as the postindustrial phase—in the fourth phase of the population transition model (DTM), both births and mortality rates are low and the overall population growth stabilizes. The decline in birth and mortality is mostly found in the fourth-stage countries due to their strong economy, highly educated citizens, adequate health care systems, the migration of rural populations to cities, and the expansion of women's employment opportunities. As a result of these factors, the TFR reduces the level of substitution (4.1 times per female) and the country enters phase 5. It is understood that the DTM Phase 4 country experienced significant economic and social progress that allowed them to reduce the size of the family relative to a few decades ago.

Although both birth and mortality are declining, the fourth stage of the country does have a large population—which is through the progress of stages (one to three). A large population, even if the natural growth rate is small, the average annual growth will remain significant. In addition, the decline in birth rates and mortality rates are not always consistent and does not always decline. Both rates are susceptible to emergencies such as pandemics or environmental disasters.

Having said that, the fourth stage of the DTM is considered to be an ideal country because the overall population growth is gradual. Examples of countries with demographic transition are Argentina, Australia, Canada,

China, Brazil, most parts of Europe, Singapore, Korea, and the United States. Posttransition societies are characterized by low birth rates and low mortality rates. Population growth is negligible and even in a recession.

The high overall population rate is low and the mortality rate is low. Fertility control is widely used, but also a small family's desire. Paramount and fundamental to the understanding of the term "demographic transition" and its situational occurrence around the world is the keen necessity to understand and acknowledge the dynamic history of global population changes. The global population transition reportedly began in the 19th century, and in the economically developed regions (northern regions), the mortality rate declined. At the beginning of the 20th century, the birth rate dropped significantly. Now that the transitions are now more or less done.

5.5 THE COMPONENTS OF DEMOGRAPHIC TRANSITION

Demographic changes ought to be referred to as and should be seen as an important factor in recognizing the population. The demographic transition in most countries is considered to be a major autonomous process and ultimately forms a profound social, economic, and even psychological or conscious basis in various countries. In this chapter, the population will be treated as an independent variable.

Of course, is historical, of course not unilateral; it is undeniable. There are many reasons for this huge change. This study will analyze the population changes as one of them, which is not a trivial change. This chapter aims to promote a more balanced interpretation of social and economic modernization in Europe and other places between 1830 and 1988.

Many arguments assert and make this transition process a pioneer in the transformation of historical democracies. This chapter will also assess the extent of the ongoing process in the world and how these societies can experience their own transition. In order to determine the impact of these processes on national development and social change, demographics must be examined as a factor. Demographic transition is an important part of the process of economic and social modernization and demographic change, and is largely limited by social and economic changes. In most developing countries, especially as a cause, not because of the consequences of its change, but the indispensable force of its existence, the existence and influence of the population transition are relatively small. Ultimately, historians and social scientists tend to conceptualize economic strength, rather than other ways to determine the demographic reality.

5.5.1 *PSYCHOGRAPHIC SEGMENTATION*

Psychology uses demographics to determine the attitudes and tastes of specific groups of people, such as marketing research, consumer satisfaction, and trade expansion.

Psychological diversity expands the diversity equation. In a highly competitive marketplace, this diversity and inclusive thinking model helps to stay ahead of employee recruitment and consumer interest trends.

By definition, psychographic information focuses on people's:

- Tastes and desires
- Opinions and ideas
- Attitudes and mindsets
- Personalities and identity
- Interests and needs

A few examples of psychographic classifications include:

- Political and social views
- Career path and aspirations
- Leisure activities and time

This psychoanalysis is far more than border, place, and marketplace; it strategically allows people to have the opportunity and the ability to read between, rather than assuming that a person's interest or need is likely or may not be based on mass demographic classification.

Psychological segmentation has proven that it is used as a utility market tool for the consumer market. Often, the use of multiple types of variable building segments usually requires demographic data that is mixed by psychology. When mixed with other factors, the social class can be strongly discriminated. Large-scale projects such as ACORN classification can be useful strategies, although they are often expensive and complex. There is not enough evidence to use the personality as the actual segmentation variable. The psychological segmentation of markets and services is influenced by any shortcomings of previous disagreements. The most serious problem is that consumers are constantly changing, so they need to change the breakdown framework to keep up. For example, the above ACORN classification is different from the year of use and the classification before decades. Most organizations and businesses have to do a lot of research to regularly update the model. What happens when there is a major change in the recession?

People's attitudes and circumstances must go through a lot of changes; a fixed frame is difficult to accurately reflect this point. On the other hand, behavior-based programs can capture the results of these changes because they affect the purchase pattern and allow the marketer to respond.

Technically, the division of behavior in each market is not practical, and psychological division and other a priori methods are the only practical methods, because psychology is the study of personality, attitudes, values, interests, habits, hobbies, lifestyles, and views. As with the more objective variables that measure demographic characteristics, even in specific areas, psychology is difficult to measure and change. They are difficult to measure; in general, most companies use proxy variables to determine the interests of people. Psychology can also be used to segment customers, that is, psychological breakdown, rather than the industry's very popular demographics. It is worth noting that the combination of the two helps to better understand the customer, rather than using any one. Psychology helps to develop better product promotion and marketing products and services.

Demography includes information about gender, age, marital status, income, educational level, psychological treatment of life attitude (healthy or irregular lifestyle), appearance (about appearance), food (more like junk food), appreciation of art (the purchase of cultural relics and collectibles), and so on. Demographic information is easier to obtain, and most companies really need to pay a lot of attention and effort to get information about the customer or potential customer's psychological information. Psychological classifications are breakdowns of customers according to personal interest and other factors. This can be a powerful way to market the same products from other people with distinctly different demographics. Each customer has a different psychological composition. The analysis of similar features is the beginning of psychological segmentation. Why use psychological division? When your target market seems to have completely different demographics, it will be a good time to break them down into psychological segments. This is a powerful way to push the same product to a group that is similar to other heterogeneous groups. Based on invisible features such as personality traits, lifestyles, attitudes, expectations and target market activities, the psychological segmentation divides the market into two categories. Psychological segmentation improves the chances of a company's profitable marketing return on investment. By segmenting the market based on psychological characteristics, you can get more niches in ad purchases and maximize the exposure of key departments. As a result, your money costs are more sensible and you should see your advertising costs more efficient or have a higher return on investment.

Companies can also focus more on the motivations of their information, thereby increasing the company's conversion rate. Psychological characteristics also play a key role in developing the company's target market role.

5.5.1.1 HUMAN INTERESTS, VALUES, AND ATTITUDES

Some of the major types of psychographic segmentation include:

Beliefs: The purpose of belief is to accurately present the world with a definition. Therefore, only the formation, retention, and revision of faith are sensitive to the evidence as human beings, so that belief only works. In the definition of faith, "faith is a propositional attitude, characterized by peoples' thoughts on the direction of the world." Most philosophers believe that faith is an inner way of thinking, which can be directly reflective, and the behaviors of believers are different. Thus, faith plays a central role in theoretical reasoning (what is the reason) and therefore in practical reasoning (the reason for doing what). So people need to know what they and others can do and what people can do with what they want. In seeking knowledge of these things, we seek real faith in them. So what we do is what we believe (Bocquet-Appel et al., 2008).

The gap of a generation includes the differences of opinion between two generations. More specifically, compared with the older generation of politics, values, and other matters, you can use the generation gap to describe the differences between the younger generation's behavior, faith, and taste. Although the gap between generations is widespread in all the historical periods, the breadth of the gap has widened in the 20th and 21st century. Breaking the "generation gap" is important as generation gap plays an important role in businesses because most companies always try and attempt to balance the needs and opinions of individuals from different ages.

Businesses must understand the changing demographics of the customer base, as the gender gap may have a huge impact on the businesses and the entire business cycle.

Values: Values are the personal and important elements of people's lives. They are the core beliefs that guide people on how to live in a way that makes sense and satisfies them. Values are the ones that measure people's choice, whether conscious or not. People can use them to attribute their actions to themselves and others. They decide their satisfaction with their choice, even if the decision is not free, but also by other factors. People's values can help them understand the roles they may find in the society which is often considered as a powerful driver for work ethics. A key question can be asked like: Where did they come from? People can share some of their values and that of other people's among one another (such as your friends or family, or your social or national group), and you may have other special values. Values can relate to your personality (e.g., the desire to cooperate with others or management), your needs (e.g., hunger, housing, and security) and your understanding of your own social environment (e.g., environmental protection or political value), and these values can be combined with your background, your experience, and your own evolving self. While some of these values may remain the same throughout one's life, other values will change as one grows older. For example, as personal circumstances change, people will change their own wage attitude, work safety, and working hours of flexibility, such as becoming a parent. The desire to work independently or work with others also becomes less common. Worthy of notice is that values are ideal. But the real world is full of compromise and contingency, we continue to give priority to our values. We can change them through reflections, experiences, or stress to make ourselves consistent with the social environment or the workplace's dominant values.

Attitudes: This is the positive or negative tendency or tendency to have a feeling toward an idea, an object, an individual, or a situation. Attitude affects individual action choices, as well as their responses to challenges, incentives, and rewards (collectively referred to as stimuli). There are four main parts of attitudes, which can be found listed as follows:

(1) Emotions: emotions or feelings
(2) Cognition: conscious beliefs or opinions
(3) Opinion: action tendency
(4) Evaluation: a positive response to the stimulus.

Attitude is the favorable or unfavorable evaluation of a thing or person, manifested by faith, feeling, or intended behavior. This is the fundamental orientation of social orientation something that is beneficial or unfavorable. A belief is an inner feeling that something is true, although this belief may be unconfirmed or unreasonable. For example, when we believe that walking under the ladder will bring bad luck, or when we believe that there is life after death. A value measures the value or importance of a person to things; our values are often reflected in the way we live. For example, we attach importance to freedom of speech, or we attach importance to our families. An attitude is a way in which a person expresses or applies their beliefs and values, and expresses them through words and actions. For example, when we hear the abuse of children and animals, or when I hate school, I am really angry. People have different interests, attitudes, and qualities. For example, some people really care about the environment, while others do not. Some people are very healthy and some are unhealthy, while others are playful. Some people are very serious and task driven. Basically, what is useful for some people might just be useless for others. When traders break the market in terms of these interests and attitudes, there will be psychological segmentations, so they can strategically bring the right product to every market in the market place.

5.5.1.2 GEOGRAPHIC SEGMENTATION

Geographical segmentation is a strategic opportunity for enterprises to apply at the same time in the geographical division of the market. The market can be divided into several aspects in several ways. Marketers can divide markets by region, such as cities, counties, states, regions (such as the West Coast), and national or international regions (e.g., Asia). Marketers can also divide the market into rural, suburban, and urban areas. In addition, they can divide the climate or the population of each region. The advantage of a geographical division of labor is to create effective ways and strategies for larger companies in international or international markets because different regions of different consumers have different needs and cultural characteristics that can and should be targeted specifically for the object. This is also an effective way for small businesses with limited budgets. They can focus; based on their own definition of the field, rather than that the market is not appropriate. The understanding of the geographic area of the business operations helps traders avoid costly marketing funds in a number of favorable ways. It works well in different areas of population density. Consumers in urban environments tend to have different needs and desires than suburban and rural consumers. There are even cultural differences in these three areas. It is easier to divide your market into a geographic area. When the information is collected by the physical location of the customer or other data source, the geographic area is divided into marketing or advertising areas. Marketing products and services companies want to know their product sales are good or bad, according to the planned strategy and tactics to adjust their advertising and sales work. Companies can always refer to population segmentation. Geographical division shows that people with similar demographic and psychological characteristics often live nearby, such as in suburbs, counties, counties, states, regions, and so on. People living in the same geographical boundaries tend to show similar buying patterns. This is partly due to the similarities between the population and the psychological characteristics of the population. This phenomenon is further pursued by local weather and environmental and cultural differences. The division of markets along with geographical boundaries can lead to a more specialized and centralized marketing tools.

5.5.1.3 SOCIOECONOMIC STRATIFICATION

Social stratification refers to the system of social domination of human hierarchy. In the United States, the status, power and wealth of some groups are clearly stronger than those of other groups. These differences lead to social stratification. Social stratification is based on four principles: social stratification is a characteristic of society, not just a reflection of individual differences; social stratification still exists; social stratification is universal (it happens anywhere), but varies (in different societies take different forms); and social stratification involves not only inequality but also faith (inequality is rooted in social philosophy). In the social stratification, the social class is listed in the character category. Social stratification is linked to all aspects of society. By understanding what happens in the social stratification, it helps us to understand social change better. Social stratification is a division of labor based on wealth, social status, and power. Sociologists agree with the concept of class; social stratification is generally inherited from one generation to yet the other generation. Social stratification is at the core of society. Social inequality is a fundamental aspect of all social processes and a position of a person in social stratification, which is a prediction of his or her behavior, attitudes, and opportunities for life. Social stratification is divided into three levels: upper class, middle class, and lower class.

5.6 LABOR AND HUMAN CAPITAL INVESTMENT

Demographic changes and human capital investment have always been a huge of interest to the labor-branded economics. Freeman (1976), in the groundbreaking work of "educated Americans," shows that the eighties and the highly educated baby boomers were depressed by the university wage premium (return to education). Other important studies and research have shown that the sharp increase in the number of young workers in the 1980s also had a negative impact on the wage levels. Recent scholars have formed and created a general framework in which different ages and educational groups are imitated as imperfect alternatives to each other. Human capital investment is known and acknowledged to be the technical process by which an enterprise is used to develop an employee. Enterprises invest in staff training, improve business operations, reduce the waste of unskilled workers, and improve the overall efficiency of staff.

The company calculates the cost of education and training through cost-benefit analysis, and analyzes the total cost of improving employee output and investment benefits.

The two common types of human capital investment include education and training. Education often includes the reimbursing employees for traditional university degrees or paying for seminars or business schools. Education is often used to develop the technical capacity of employees in the fields of accounting, finance, or production. Training is usually conducted by a supervisor who works with the staff to show them how to perform a specific function.

The use of education and training to improve human labor can help companies create competitive advantages just as companies train employees to produce other businesses that can, however, not be used for goods or services. Owners and managers with specific technical skills can perform efficient and intelligent business process replication for employees. Companies need to consider which employees should be trained. While some employees are actively responding to business educational services or training, some employees may not. This response often depends on the motivation of the staff. In the education or training process, employees are more willing and likely to work in the company that is conducive to their growth and desire. Human capital investment, capacity development, and skills are related to unskilled labor theory. Skilled labor usually accomplishes tasks faster and more efficiently than unskilled labor. Skilled employees can also have the expertise to perform certain functions and improve the value of the workplace. The unskilled labor force usually performs the basic or repetitive business task; the value of these business tasks is not very high. Asian countries are in varying degrees in demographic transition. East Asian countries begin early, especially in Japan and China. Changes in population growth and change in size are of course important, but the study is particularly concerned with changes in the age distribution of the population, regardless of the size of the population. The transitional phase of the transition phase begins with a high proportion of children, with a lower proportion of the elderly, and eventually to the opposite: relatively and technically, few children and many older persons. At the early stages of the transition, the proportion of children tends to increase as a result of declining infant and child mortality rates. During the transition period, fertility declines and the proportion of the working population increases by more than half a century, raising the total number of dependents.

This change and the impactful presence of population and aging on the physical capital provide two powerful mechanisms for sustaining or improving the living standards of people.

5.7 DEMOGRAPHIC TRANSITION AND PURCHASING POWER

Rich and developing countries are defined on the basis of demographic transition, the baby boom and the depression after the Second World War, contributed to the low rate of fertility, while life expectancy continued to improve; population age structure underwent tremendous changes.

All of these changes combined formed the short-term fluctuations in age distribution with the long-term trends of the aging population. These changes have created an opportunity to strengthen investment in human capital and physical capital which has promoted economic growth and challenges as well; as the support system for the elderly is under pressure, labor growth slows or turns negative (Ezeh et al., 2012).

Over the past two centuries, the investment in developed countries and wealthy industrial countries in mass formal education has increased significantly. On the one hand, the increase in human capital to replace the reduction in the supply of workers may be closely related to the population that is leading to an aging population (fertility decline and mortality decline). Health is another aspect of human capital and investment which also increases dramatically. Lower fertility and slower population growth have contributed to the human capital investment of each child. As have been observed over time, this has led to an increase in the number of tangible capital per worker. From this perspective of individual behavior, the macroeconomic point of view, there are several ways to view this occurrence from the composition of the aging population.

In all countries asset income grows with age; by suggesting that asset holdings grow stronger with age, reflects the tendency of adults to save and accumulate assets in adulthood. People accumulate assets for many reasons to provide consumption for the elderly, as a remedy for health costs and other unforeseen financial needs, leaving a legacy for future generations, or for any other reason. As the population ages, the proportion of the elderly and the employed population rises, and the age pattern of assets holdings means that the ratio of assets to workers and the general population also increases (Cai, 2010).

In this case, the aging of the population with slower population growth means that the capital of each worker increases. Another assumption is that the capital output ratio remains constant, which is quite correct during the long period of the OECD countries. Under this assumption, the capital labor ratio increases, but it is lower than the constant savings rate assumption. In addition, the savings rate is reduced enough to keep the capital output rate constant and thus has an additional impact on consumption. As the savings rate continues to decline or the capital output rate remains constant, the decline in population growth yields a "second dividend" of higher consumption (Lee, 2003). In the later part of our simulation, it was assumed that the capital output ratio is constant.

5.8 DEMOGRAPHIC TRANSITION AND COMPETITIVE ADVANTAGE

The demographic differences between countries are the source of comparative advantage in the dynamic-process of international trade. As many of the skills are age-dependent, the aging of the population will technically reduce the relative supply, demand, and increase in the relative price of devaluation age skills. As a result, industries that rely on skilled and young workers will be more productive in countries with lower labor. At the same time, countries with older labor will also focus on the use of age-added technologies in industries with comparative advantages. According to the theory of neuroscience and economics, studies have examined the industrial-level intensity measure of age-dependent skills and indicate those countries with higher median age which are particularly engaged in the production of goods with the age-building ability. However, large commodity technology is considered more important. The production and export structure of rapidly aging countries is shifting to industries that use age-increasing skills, with a focus on increasing dependence on age-related devaluation skills. Many countries have observed a significant change in the demographic structure of the population, as aging has led to a shift in the age structure of younger workers to older workers. These changes may have a profound impact on the structure of national economic activity and the relative supply changes relative to age-related skills and the patterns of trade between them.

Recent studies and research on aging have shown and proven that there is a negative correlation between age and certain cognitive abilities, and many studies or research have shown that cognitive decline begins as early as 18 years old.

This means that the aging society has experienced a further decline in certain cognitive skills, which may be the use of these skills while losing competitive advantage. A presentation of a fundamental study of whether the difference in the age structure of the population is the subject of global trade will be valid and relevant. In particular, there is a considerable comparative advantage of novelty and empirical sources: the relative supply of age-dependent skills varies with the age of the population. Human capital is a broad and multifaceted concept that encompasses different types of human investment. However, the key aspects and the prospect of human capital are related to the knowledge and skills that people embody. Human capital has always been a very important determinant of personal and social progress. In the present situation, it is the core tool for the growing global competition in a developing economy that is increasingly competitive and service oriented.

Technical human resources and the capacity assessment on the overall spectrum of development, especially the degree of economic growth, have been very clear and assertive. While there is a known link between economic growth and human capital, it is meaningful to understand the importance of technical labor in the knowledge economy and to understand the stages of economic development. The emerging trend of the global economy shows and illustrates that they are more and more dependent on the development of knowledge-driven industries. Facing the emerging problems and challenges of population aging in developed countries, as well as mobile trends and their possible impact on off-shoring services, focusing on emerging issues and challenges. It has been argued and debated that the constant changes in high-income countries and off-shore outsourcing have provided opportunities for countries such as India and China to increase their pace of economic development.

This has helped to examine, the quality and depth of India's higher education infrastructure and the technical human resources of the system to take advantage of this opportunity. while we can consider or accept the fact that India can use its talent to build or maintain a competitive advantage, and the big question will be whether it can maximize the pace of globalization?

The internal and interspecific competition of food is one of the main factors controlling the evolution of most new community structures and the history of vegetarianism.

The competitive advantage that is considered to be the ability of the entire population or some developmental stages to survive at low food concentrations is related to the individual's "rate of efficiency hypothesis." Individual growth is slow and has proven to have the greatest competitive advantage.

Financially strong economies also welcomed the global economic downturn, experienced a sustained growth. Analysts predict that by 2009, China will surpass the United States as the world's largest economy. In a way of life like a Confucian society like China, consumers are willing to pay the price for the quality of the product or service, especially to improve the way of life. Investors are concerned about this market potential, as China's spending power continues to rise. It has been reported that by 2014, the luxury consumption of Chinese tourists will be more than the world. Chinese people are getting richer, and the middle-class population is expected to rise from 7% to 60% in 10 years. Because the labor costs are relatively low, many companies choose to transfer their business to China. For example, China's notorious Foxconn has more than a dozen factories, for Apple, BlackBerry, Sony, and other global brands to launch products. Despite the pursuit of cheap labor destination, China's literacy rate is also improving. Millions of students graduate from the National University every year and have began to accept English as their second language choice.

This has helped to reduce language barriers and has increased the number of educated workforces. China since the late 1980's began their transport and infrastructure construction. Improvements in transport have led to the vigorous development of national tourism and air cargo. According to Boeing 2014 forecast, air traffic is expected to grow rapidly, doubling in the next two decades. China's leading industries and improved traffic have led to Chinese tourism boost and have encouraged outbound tourism. In addition to air transport, China also has a wide and highly connected road and rail network (more than 80,000 km). According to the "Railway Daily," the Chinese government's goal is to increase the railway network by 2030 to 150,000 km.

5.9 ECONOMICS AND INTEGRATION

The desire to integrate population issues, into the concept of development has long been recognized internationally and domestically. The 1975 United

Nations Population Report states that population issues are the main elements of long-term national planning.

By recognizing and acknowledging that the interrelationship between population, sustained economic growth, and sustainable development is a key factor in integrated strategies. Recognizing that the importance and necessity of integrating population into development initiatives and processes is a key strategy. This key strategy focuses, primarily, on people's needs and expectations as a major consideration in policy and program interventions. It promotes the development of people's own interests and values through economic integration. By scale, the impact of trade regional expansion on the economic development and demographic transition of most countries makes the economy produce and use different levels of technology. A greater integration between regions (broader scale) played an important role in changing the rate of return, leading to the industrial revolution, and even leading to changes in parenting behavior.

While understanding the demographic shift from the mortality rate, their response to income and fertility is also growing. The model uses European historical data for calibration and simulation. Historical evidence to support the rapid rise in economic growth before the integration of ideas. About 300 ago, the living standards of rich countries were only slightly higher than those of poor countries. Today's thing is very different, and their per capita income is 32 times higher.

Understanding the methods applicable to population transformation identification can be analyzed in three ways:

(1) Family fertility decision-making is endogenous, depending on the relationship between life and the overall situation of death.
(2) The size of the body's integration and expansion.
(3) From the Anno Domini 500 simulation to the future, the path to the critical variable depends on the family optimization within the infinite range.

In order and more efforts to achieve the overall development of these areas, an economy must undergo the four structural stages of production and learning techniques. In the first stage, only the technology is still available in the original stage, no learning. In the final stage, balanced growth can only be based on the city, more and more technical returns in use, individuals must be committed to the accumulation of knowledge.

Preindustrial use (both from above) and an industrial stage (using the above three characters). It is important to note that at every stage, people are committed to the support of children, which is a key idea.

This chapter attempts to explore following data. The ever-expanding population growth (in each region) has led to and created a shift from the initial stage to the early stages of industrialization, leading to large-scale growth (economic integration in different regions). The preindustrial shift to the industrialization phase is a turning point, indicating that real growth has begun. Since the end of the Roman Empire, there have been two peaks in population growth in Europe. The population increased from 500 to 1500, slow growth but slow increase. At the beginning of the 14th century, it fluctuated and then descended and was spectacular. A wide range of economic populations, to explore the relationship between population and economy; population affected by economic development, population changes affect the economy. Whether economic resources are sufficient to meet the growing needs of the population, the economic dilemma is always relevant. Other areas of interest are the long-term decline in mortality and the economic impact of the reduction in the number of families associated with the transition (Bloom et al., 2001).

Related contemporary research topics include the causes and consequences of the birth tide as well as the changes in the age structure of the population and its composition of the economic impact. Another important area is the economic cycle and the interaction between marriage, fertility, mortality, and short-term changes in migration. Family and labor market research, immigration patterns, and health and mortality issues are also particularly significant. All these areas are linked to a certain extent and incorporated into a coherent analytical framework, such as studying immigrant family members, their reproductive behavior, entry and exit of the labor market, and their health and death patterns.

Most countries have achieved good results, focusing on specific issues and sectors and analyzing the impact of demographic trends on social service needs at the national and subnational levels; however, the existence, reliability, accuracy, and classification of trainers. Data is the best way to take advantage of population growth in an economy. Increasing efforts are more effective in integrating demographic variables into policy and programs that are valuable factors in the development and implementation of a clear demographic policy process.

The evolution of the assessment transformation will lay a solid foundation for the implementation of technical and sociopolitical legitimacy measures, which may overcome some simple social problems. The model has, however, a high function in resource planning, particularly in response to the challenges and opportunities associated with recent and decentralized public-action strategies. For all these reasons, the government considers it necessary to strengthen the communication and training tools to incorporate demographic factors into its policies and programs.

KEYWORDS

- **population growth**
- **population transfer**
- **fertility**
- **mortality**
- **age structure**

REFERENCES

Bloom, D. E., Canning, D. Sevilla, J. *Economic Growth and the Demographic Transition (No. w8685)*. National Bureau of Economic Research, 2001. Bloom et al 2001. https://www.nber.org/papers/w8685

Bocquet-Appel, J. P.; Bar-Yosef, O.; Eds. *The Neolithic Demographic Transition and Its Consequences*. Springer Science & Business Media, B.V., Springer: Netherlands, 2008.

Cai, F. Demographic Transition, Demographic Dividend, and Lewis Turning Point in China. *China Econ. J.* **2010,** *3* (2), 107–119.

Cho, L. J.; Kobayashi, K.; Eds. *Fertility Transition of the East Asian Populations*; University Press of Hawaii, Honolulu, 1979; pp 9-23. (Monographs of the Center for Southeast Asian Studies, Kyoto University, English-language Series.)

Ezeh, A. C.; Bongaarts, J.; Mberu, B. Global Demographic Trends and Policy Options. *Lancet* **2012,** 142–148.

Freeman, R. B. The Effect of Demographic Factors on Age-Earnings Profiles, . J. Human Res. **1979,** *14* (3), 289–318.

Galor, O.; Weil, D. N. Population, Technology, and Growth: From Malthusian Stagnation to the Demographic Transition and Beyond. *Am. Econ. Rev.* **2000,** *90* (4), 806–828.

Galor, O. The Demographic Transition and the Emergence of Sustained Economic Growth. *J. Eur. Econ. Assoc.* **2005,** *3* (2–3), 494–504.

Handwerker, W. P. The First Demographic Transition: An Analysis of Subsistence Choices and Reproductive Consequences. *Am. Anthropol.* **1983,** *85* (1), 5–27.

Lee, R. The Demographic Transition: Three Centuries of Fundamental Change. *J. Econ. Perspect.* **2003,** *17* (4), 167–190.

United Nations *World Population Prospects: 2006*; United Nations Population Division: New York, 2007.

CHAPTER 6

Effects of Higher Education on the Demographic Change of Third World Countries

BEHİYE ERTAŞ*

Vocational School of Health Services, Bozok University, Yozgat, Turkey

**E-mail: behiyeertas5884@gmail.com*

ABSTRACT

This work has attempted to reveal the changes that higher education has made on third world countries. In this framework, the topics such as activities in higher education and abroad, studies in third world countries and higher education institutions in these countries, and outputs they dealt with are discussed. In this study, in the context of the changes that tertiary education has made on third world countries, in order to be able to be presented in the most versatile way possible; the studies written in this area were handled within the scope of literature review. Evaluations were made on the obtained data and the results were reached.

6.1 INTRODUCTION

This work has attempted to reveal the changes that higher education has made on third world countries. In this framework, the topics such as activities in higher education and abroad, studies in third world countries and higher education institutions in these countries, and outputs they are dealt with are discussed. In this study, the context of the changes that tertiary education has made on third world countries, in order to be able to be presented in the most versatile way possible is discussed. The studies written in this area were handled within the scope of literature review. Evaluations were made on the obtained data and the results were obtained.

6.2 HIGHER EDUCATION CONNECTION AND THIRD WORLD COUNTRIES

Higher education institutions continue their existence as places where individuals can pass through specific education levels and transform their knowledge and experience into their application areas. Higher education is sometimes the educational environment where people are ultimately trained to get career advancement and sometimes profession. Higher education institutions have been developing based on student elections and are shaped according to expectations.

When people choose a higher education institution, not only the location but other factors are also influential. Sometimes there can be more than one higher education institution in a city. There are some factors that attract students and academic staff to a higher education institution. The most important of these is that the students have a wide knowledge and awareness about the institution. In addition, points such as the institution's high learning quality and reputation, education and research status, tuition fees and costs, level of security within the campus/institution, level of internationalization of the institution, living and working environment of the institution, and social and geographical connections are also important (Becker and Kolster, 2012).

In addition to improving people's quality of life, higher education institutions strengthen the mutual exchange of information among countries and increase the intellectual and academic dialogue among cultures. Higher education society also strengthens the interaction and change with research and development and social and technological innovation studies (Çetinsaya, 2014: 141).

The historical process must be known to understand the concept of the third world country. The Western countries had many colonies from 17th to 20th century. Today, most of these colonies declare their independence. After these developments, the map of the world has changed. As a result of these changes, there were places where the majority of the populations in many regions except some countries (North America, Australia, and New Zealand). But in contrast, the majority of Asia, Africa, and South America appear to be dominated by indigenous populations in other regions (Giddens, 2000: 60).

Historically, the separation of certain groups by categorizing the states with some criteria has been expressed in terms of history and sociology. In this grouping, the countries in the first group are industrialized countries included in the United States. Countries in the second category are countries with a much lower level of industrialization. These are often called Third World societies. These societies include China, India, most African countries (such as Nigeria, Ghana, and Algeria), and South American countries (such as Brazil, Peru, and Venezuela). In addition, most of these nations are in the United States and south of Europe, and were named as Southern countries, as opposed to these industrialized and richer Northern countries (Giddens, 2000: 60). In terms of social content, these nations are called "proletarian nations" (Bremond and Geledan, 1984: 28).

The term "Third World" was originally part of the opposition between the three forms of society found in the early 20th century. The "First World" countries are the industrialized countries of Europe, the United States, Australasia (Australia, New Zealand, Tasmania, and Malaysia), and Japan. When we look at first world nations such as Europe, the United States, Australia, New Zealand, Tazmany, Malaysia, Japan, and other industrialized countries, it seems that almost all of them are governed by multiparty, parliamentary government systems. "Second World" countries mean communist societies of Eastern Soviet Union (USSR) and Eastern Europe such as Czechoslovakia, Poland, East Germany, and Hungary. The economies of these societies have a system based on central planning (Giddens, 2000; 60).

When the Third World countries were referred to "third class" that represents the public in the French Revolution. This term was first used to mean a neutral bloc in the late 1940s, in Europe. After the early 1960s, it began to be used in the sense of developing countries outside the (advanced) capitalist countries that make up the "first" world and the communist countries that make up the "second" world (Marshall, 1999: 776).

The "Third World" theme that was introduced by the French population scientist Alfred Sauvy has become a traditional form of talking about underdeveloped societies. According to Adam Smith, these communities, called as primitive, are the communities in which a single production line (especially agriculture) develops against countries where industry, agriculture, and trade develop in a balanced manner. However, the term "underdeveloped country" has actually become a politically used term after the Second World War. This term, which is referred to the underdeveloped societies, then began to be expressed as underdeveloped countries and as still developing countries (Aklin et al. 1988; 77–78).

Another feature that distinguishes these countries from primitive societies is their desire for rapid development and their effort. At the same time, development in the sense of education is the most important problem for them to overcome. The tables comparing the literacy rate among the gender can show the nature of the distribution of education among the gender. Examples from the 1995 United Nations table on literacy rates of some countries are presented: While 21% of men in Tanzania are illiterate, this rate is more than two quarts in women, 43%; 28% and 58% in Malawi; 26% and 50% in Uganda; 14% and 29% in Zambia; 17% and 33% in Congo; 19% and 38% in Lesotho; 20% and 40% in Botswana; 21% and 45% in Tunisia; and 29% and 50% in Saudi Arabia. In Indonesia (10% and 22%) that literacy rate is high, whereas in Bolivia (10% and 24%) and Malaysia (11% and 22%) are half-literally rates. The literacy rate in Morocco is low for both sexes, but even lower for women; in this country, 43% of males and 69% of females are illiterate. Likewise, in Egypt, 36% of men and 61% of women are illiterate; in Nepal 59% of men are 86% of women are illiterate; and in Burkina Faso, 71% of males and 91% of females are illiterate(Özgüç, 1998; 90). The worst situation seems to be in Nigeria 79% of men and 93% of women in this country are deprived of reading and writing (Özgüç, 1998; 90).

Inequality of participation in education, of course, reflects differences in the cultural and economic situation. For example, in Australia, 88% of girls attend secondary school, while in Afghanistan only 4% can continue education after primary school. There is a great disparity in the level of education of women and men in developing countries. While the literacy rate in India varies from region to region, male literacy is higher everywhere. There is a great disparity in North India; rural areas show more inequality than urban areas. Also, many are encouraged by ways such as literacy mobilization but this is not enough (Özgüç, 1998: 91).

6.3 EXTERNAL IMPACT OF HIGHER EDUCATION AND DEMOGRAPHIC CHANGES

All these activities carry the aim of introducing Turkish culture and Turkish language in these countries better. States and some institutions have been various initiatives for this purpose. For this, legal regulations have been made. Education and training programs, textbooks, educational tools, and equipment of the Ministry's foreign schools and institutions have been prepared. These steps are as follows (Web 6):

- Planning social and cultural activities for schools abroad,
- Implementation and support,
- Taking precautions for protecting their national and cultural identities, ensuring that they are in harmony with the society they live in for our citizens and children abroad,

- Opening schools in the countries considered necessary for our citizens who are abroad,
- To open Turkish language teaching centers in order to expand Turkish language and Turkish culture,
- Planning, selecting, and appointing the teaching staff needed by the schools abroad,
- To carry out the necessary education and training services in order to benefit from the education opportunities of the country they are in,
- Conducting studies to ensure compliance with our education system when returning home.

The general directorate responsible for foreign politics within the Ministry also has some duties. These are (Web 5);

- Cooperate with other institutions to take the necessary precautions in terms of determining, developing, and carrying out the strategies and objectives of the higher education policy effectively,
- Cooperate with the relevant units, institutions, and organizations in the determination of the procedures and principles related to the higher education entrance system,
- Submission to the Board of Education.

The Ministry of National Education Abroad Organization (Table 6.1) was established in order to spread and promote Turkish culture in the international scene. We have an abroad organization in 34 countries.

TABLE 6.1 Ministry of National Education Abroad Organization.

Countries		
• Afghanistan	• Germany	• USA
• Albania	• Australia	• Austria
• Azerbaijan	• Belgium	• Bosnia and Herzegovina
• Bulgaria	• Denmark	• France
• Georgia	• Netherlands	• England
• Iran	• Swedish	• Switzerland
• Italy	• Canada	• Kazakhstan
• Kirghizistan	• Kosovo	• TRNC
• Libya	• Macedonia	• Egypt
• Uzbekistan	• Pakistan	• Romania
• Russia	• Saudi Arabia	• Turkmenistan
• Ukraine		

The objectives of the Ministry of National Education's abroad organization in 2016 are to increase the activities aimed at promoting and expanding Turkish language and Turkish cultures in abroad, to update teaching programs and course materials of Turkish language and Turkish culture courses that are taught in abroad in terms of necessity, to increase 17% project production capacities of institution personnel, to increase the level of foreign language competence of individuals by using innovative approaches in education and increase international student–teacher mobility, to increase the level of education and cultural cooperation at the international level compared to the previous year, to increase the knowledge and experience level of employees, to increase cooperation with national and international institutions and organizations that include mobility of employees with the aim of sharing experience, and to expand the domain of Turkish language and Turkish culture in abroad (Web, 5).

In order to promote and expand Turkish language and Turkish cultures abroad, studies have been organized for the children of our citizens and their kin and relatives in abroad for the development of Turkish languages and for strengthening ties with Turkish culture. Turkish language and Turkish culture course and resource book needs are met. Within this scope, in 2016, a total of 125,788 courses and culture books and educational material including 64,213 Turkish language and Turkish culture textbooks prepared by the Ministry in the context of our Far Nears Project, 47,970 textbooks of our compatriots in Western Thrace, 10,305 textbooks and cultural books in various branches, and 3300 educational materials were sent. Updating studies on Turkish language and Turkish culture books used by our children abroad and the textbooks of our compatriots in Western Thrace have been continuing (MEB, 2016: 88).

As Ministry, Turkish language is taught and Turkish culture is introduced in Turkish Education Centers in Turkmenistan, Azerbaijan, Kyrgyzstan, Turkmenistan, and Tajikistan. Preferential Turkish Project was held in cooperation with the Ministry of Education and Ministry of Education and Science of the Republic of Bosnia and Herzegovina Federation in "Bosnia and Herzegovina public schools for students learning as an elective course for students; participated in the "IV Traditional Children's Festival" program held on April 20, 2016 for these students. April 23 National Sovereignty and Children's Day, May 19 Commemoration of Atatürk and Youth and Sports Day, October 29 Republic Day, November 10 Atatürk Commemoration Day, and November 24 Teacher's Day celebrations are held in closed and open areas with the participation of our students and citizens in duty areas of our Education Consultancy and Attachés in abroad (MEB, 2016: 88).

Furthermore, activities were carried out to increase international teacher–student mobility. Six teachers participated to the Pestalozzi In-Service Training of the Council of Europe and participated in trainings in various subjects abroad, sharing knowledge and experience. In total, 100 assistant specialists and general directorate staff were given European Union Seminar on March 8–9 and 10–11, 2016 by Ministry of Education. Modular European Union (EU) Training was held on April 25–27, 2016.

Scholarship work and operations are given priority by our government as well as by foreign governments: interviews and nominations were made for countries such as Switzerland, Israel, Finland, Brunei, Hungary, Slovenia, China, Russia, Belgium, Italy, Romania, Kuwait, Morocco, Greece, Mexico, and Pakistan. Initially, Sri Lanka and Pakistan allocated scholarships to our country to begin in 2016–2017. Russian and Chinese quotas increased. Belgium has included our country in the scholarship again (MEB, 2016: 90).

At the same time, the number of Training Consultants and Education Attachés who have been working in the foreign organization in order to expand the scope of Turkish language and Turkish culture in abroad has been increased to 41 persons. Number of lecturers appointed abroad is 78, number of countries that teachers appointed in abroad is 28, and number of teachers to be appointed abroad is 2117 in total. Higher education activities have been being carried out with different courses, appointments, and academic exchanges in especially most of the African countries such as Nigeria, Ghana, Kenya, Zaire, Tanzania, and Algeria in the third world country category; South American countries such as Brazil, Peru, and Venezuela; and Asian countries such as Burma, Malaysia, and Singapore. A total of 3745 academic staff members were sent to 29 countries within the framework of higher education between 1987 and 2004. Approximately, 50% of them have studied in the United States and 40% have studied in the United Kingdom. Of the submitted candidates, 714 still continue their education. Of these, only 2132 people received doctoral degrees. In total, 352 of the unsuccessful candidates are unable to complete their education and 547 of the unsuccessful candidates are considered to be disrespectful. In the same period, 541 of the 664 candidates sent by Milli Eğitim Bakanlığı (MEB) were trained in the United States. According to a evaluation made by Higher Education Council (YÖK) in 1996, the expenditure of $25,000 per year per research assistant sent abroad shows the source of the state's education abroad (Ünal, 2013).

Yunus Emre Institutes are also an important dimension of foreign education activities. The activities of this foundation in the field of Turkology are important. This foundation is a public foundation that established with the Law No. 5653 on May 05, 2007 to carry out activities for the purpose of promoting the Turkish language, history, culture, and art. These Turkish cultural centers were opened first in Sarajevo on October 17, 2009. Today, 39 Yunus Emre Turkish Cultural Centers (YETKM) have been making activities in 30 countries, including Afghanistan, Germany, Albania, Austria, Azerbaijan, Belgium, Bosnia and Herzegovina, Algeria, Morocco, France, South Africa, Georgia, Holland, England, Iran, Italy, Japan, Montenegro, Turkish Republic of Northern Cyprus (TRNC), Kazakhstan, Kosovo, Lebanon, Hungary, Macedonia, Egypt, Moldova, Poland, Romania, Russian Federation, and Jordan. In these centers, many activities such as concerts, exhibitions, traditional handcraft courses, stage shows, film shows, interviews and signing days, and scientific meetings are held in order to introduce Turkey in the cultural direction. Also in the scope of cultural activities, these centers have been participating in organizations such as fairs and festivals by cooperating with national and international organizations in many countries of the world. Turkish language teaching at the centers is conducted in a total of 12 courses at A1/A2/B1/B2/C1 levels, and each course consists of 72-h training. Approximately 16,000 students were educated in Turkish language with Turkish language courses that were conducted in 28 cultural centers between 2009 and 2013; 300 students from 57 countries attended the Turkish Summer School program in 2013 (Yunus Emre Bülteni, 2013).

In total, 400 students from Azerbaijan, Belarus, Belgium, Bosnia, and Herzegovina, France, South Korea, Georgia, Japan, Kazakhstan, Kyrgyzstan, Macedonia, Yemen, Greece, Poland, and many other countries participated the 2014 Turkish Summer School Program that was held in cooperation with Foreign Turks and Relative Communities Presidency Yurtdışı Türkler ve Akraba Toplulukları Başkanlığ (YTB) and Yunus Emre (YTB) Institute (Demirci, 2014).

One of the most important activities of the Yunus Emre Institute is the Turkology Project. The Turkology Project was initiated by the Turkish Cooperation and Development Administration (TIKA) in 1999 and was first implemented in the 2000–2001 academic year. The establishment of a qualification of Turkish language education, the strengthening of relations between Turkey and other countries, the creation of an existing language capable of speaking Turkish language in the context of education, culture, and social activities to be carried out in these countries, and to promote Turkey and Turkish culture more effectively abroad are aimed within the framework of Turcology Project (Web 6). The protocol was signed between the Yunus

Emre Institute and TİKA on September 20, 2011 and the Turkology Project was transferred to the Yunus Emre Foundation. Following this development, in the scope of Türkoloji Project, Cooperation Protocol was signed with 51 universities in Turkish language and literature/Turkology departments and Turkish courses in 35 different countries (Parlakpınar, 2014).

As of 2012, there is approximately 1700 international academic staff within the Turkish higher education system within the scope of training activities. This number has exceeded 2800 according to the Higher Education Council's April 2014 data. While the number of international academic staff in our country has increased steadily, the proportion of them in total academic staff has increased to only 1.9%. This rate is considerably low compared to developed countries and Turkey is not yet fully compatible with the goals of internationalization of the higher education system. In 1984, a total of 173 international academic staff worked in Turkish universities. The ratio of this number to the total number of academic staff is 0.8%. This number increased to 468 in 1990, 851 in 2000, and 1703 in 2012; in total academic staff ratio was 1.3% in 1990, 1.4% in 1990, 1.3% in 2000, 1.3% in 2012, and 1.9% in 2013 (Çetinsaya, 2014).

6.4 CONCLUSION

Higher education institutions are the institutions where the necessary information is developed for the development of the countries and human resources are raised. Higher education institutions play an important role in the future of countries (Gedikoğlu, 2005). Especially, it has an important role in transferring culture accumulations and values of the country to future generations (Blackburn ve Lawrence, 1995). In addition, higher education institutions will contribute to the effects of changes on students' within the education and training process (Bowen, 1980). Because, when the behaviors, attitudes and value judgments of the students change over time, there will be a process of change in society as well. Thus, as this increases, social change and development will be achieved.

This is not just about a community itself. Now the globalizing world is full of inequalities. In social relations in the world, there is a great disparity between the developed countries and the Third World countries where poverty and misery dominate. In this case, especially in the field of education the disparity shows itself. Cultures open to the world with different activities, projects, and programs in many areas of education and training. Comparing the output of tertiary education in the Third World countries with the output of tertiary education in developed countries, it seems that there is much more to be done as Turkey. In this context, cooperation in education should be not only with close geography and cultures but also with Asian, African, and Latin American countries, developing relations in the field of higher education is an important mission for Turkey and Turkish culture. It is important for Turkey to go to regional cooperation in the field of internationalization and to provide technical assistance and support in the field of higher education. When looking at the activities carried out within the scope of the study, all these activities carry the aim of introducing Turkish culture and Turkish language better in these countries.

KEYWORDS

- **higher education**
- **demographic change**
- **third world countries**
- **Ministry of National Education Abroad Organization**
- **developed countries**

REFERENCES

Becker, R.; Kolster, R. *International Student Recruitment: Policies and Developments in Selected Countries;* Netherlands Organisation for International Cooperation In Higher Education (NUFFIC): Hollanda, 2012.

Blackburn, R. T.; Lawrence J. H. *Faculty at Wor;* The Johns Hopkins University Press: Baltimore and London, 1995.

Bowen, H. R. *Investment in Learning*; Jossey Bass Publishers: San Francisco, 1980.

Bremond, J.; Geledan, A. İktisadi ve Toplumsal Kavramlar Sözlüğü, Çeviren: Ertuğrul Özkök. İstanbul: Remzi Kitabevi, 1984.

Çetinsaya, G. (*Büyüme, kalite, uluslararasılaşma: Türkiye Yükseköğretimi İçin Bir Yol Haritası*, Yükseköğretim Kurulu Yayınları Yayın: Ankara, 2014.

Demirci, C. Türkçeyi, Türkiye'yi Tanıyarak Öğreniyorlar (Haber), 19 Aug 2014. http://www.haberler.com/turkce-yi-turkiye-yi-tanyarak-ogremyorlar- 6393121-haberi/ (accessed Apr 22, 2017).

Gedikoğlu, T. Avrupa Birliği Sürecinde Türk Eğitim Sistemi: Sorunlar ve Çözüm Önerileri. *Mersin Üniversitesi Eğitim Fakültesi Dergisi* **2005,** *1* (1), 66–80.

Giddens, A. *Sosyoloji.* Çev. H. Özel ve C. Güzel. Ankara: Ayraç Yayınev, 2000.

Marshall, G. Sosyoloji Sözlüğü. Çev. O. Akınhay ve D. Kömürcü Ankara: Bilim ve Sanat Yayınları, 1999.

MEB. *2016 Yılı İdare Faaliyet Raporu.* Ankara: MEB, 2016

Mohrmana, K.; Mab, W.; Bakerc, D. The Research University in Transition: The Emerging Global Model. *Higher Educ. Policy* **2008,** 21, 5–27.
OECD. *Education at a glance, 2011*, OECD: Paris, 2011.
Özbek, N. *Türkiye'deki yabancıların öğrenim ve öğretim özgürlüğü*, Mülkiyeliler Birliği Vakfı Yayınları: Ankara, 2000.
Özgüç, N. *Kadınların Coğrafyası;* Çantay Kitabevi: İstanbul, 1998.
Parlakpınar, A. G. M. *Yunus Emre Enstitüsü'nün Türkoloji Alanındaki Faaliyetleri, Proceedings Book*, 3rd International Conference on Language and Literature "Turkish in Europa" Tirana/Albania, Proceedings Book, v. 2, 274, 2014.
Teichler, U. The Changing Debate on Internationalisation of Higher Education. *Higher Educ.* **2004,** 48, 5–26.
TİKA http://www.tika.gov.tr/tr/sayfa/hakkimizda-14649 E.T (accessed June 18, 2017).
Ünal, M. (2013). Gelişmekte Olan Ülkelerde Beyin Göçü (Bg) http://Www.Ayk.Gov.Tr/Wp-Content/Uploads/2015/01/%C3%9CNAL-Mesud-GEL%C4%B0%C5%9EMEKTE-OLAN-%C3%9CLKELERDE-BEY%C4%B0N-G%C3%96%C3%87%C3%9C-BG.Pdf.
WEB 1, http://www.studyinturkey.gov.tr/.
Web 2, https://erasmusplus.ibb.gov.tr/proje-nedir-2/.
WEB3,http://www.yok.gov.tr/documents/757816/1684767/mevlana_kitapcik.pdf/410e03de-63e8-42e8-9684-ad0abb35fcb3.
Web 6, Yunus Emre Enstitüsü Türkoloji Koordinatörlüğü Tanıtım Kitabı. Merdiven Reklam Tasarım, Ankara, Aralık 2013.
Yükseköğretim Kurulu 2016-2020 Stratejik Planı, Ankara.
Yunus Emre Bülteni (2013) Ekim, s.1.

CHAPTER 7

The Impact of Demographic Change in Turkey on the Higher Education System

GÖNÜL ŞAYİR*

Department of Educational Sciences, Faculty of Education, Division of Educational Administration, Hacettepe University, Ankara, Turkey

**E-mail: gonulsayir@gmail.com*

ABSTRACT

In this study, higher education system and the historical development of higher education in Turkey were examined. One of the organizations which have the most important tasks in the development of countries is higher education organizations. When higher education organizations are mentioned, universities come to mind. At first, concept of the university emerged with the establishment of Bologna Universities. Previously, universities addressed elite people; in 19th century universities showed general public service an understanding and has been continuing developments in this direction. Historical process of the higher education system in Turkey can be examined in three periods as madrasahs, Ottoman, and Republican periods, as well as can be considered as the models that applied in universities. From this point of view, the Turkish higher education system is examined in three periods as the classical period which cannot be described as a complete model until 1773, as "Continental European Model" between 1773–1981 and "Anglo-Saxon Model" after 1981 and over. As a result reforms and changes have been made in the higher education system from time to time. Referring to the reforms and changes to show parallels with the political events of the changes carried out in higher education in Turkey and the Turkish shaping demographic characteristics are noteworthy.

7.1 INTRODUCTION

One of the organizations which have the most important tasks in the development of countries is higher education organizations. In Turkey, higher education is defined as organizations that enable people who have successfully completed secondary education to continue their education (Korkut, 1984). In order for individuals to enter higher education organizations, it is necessary to enter the examination in accordance with the agreed principles of higher education board. According to the results of the exam, the students are placed in a higher education organization among their own preferences, taking into consideration their secondary education achievements.

The beginning of higher education can be seen from Plato and Aristotle periods. It can be said that the "Academy" which was founded by Plato in 387 BC is an example to the university. Afterwards, Lyceum which was founded by Aristotle who was educated in the Academy is also among the first examples of the university (Demirer et al., 2000). But some researchers have also pointed out that the first university was founded in medieval times. For example, Haskins argues that the origins of modern-day universities are based on the universities of Paris and Bologna, and not Athens or Rome (Timur, 2000).

Higher education organizations have important duties such as raising highly qualified human power, producing solutions for the problems of the country, and ensuring the progress of science and technology (Korkut, 1984; Akın, 2003). However, preparing the necessary conditions for their countries to compete with the World among the purposes of the higher education organizations is noteworthy. The general purpose of higher education in Law is stated in the following (no. 2547 sayılı kanun, 1981)

i. To make education, scientific researches, publishing, and consultancy based on the needs of society, principles, and objectives of development plans and at various levels based on secondary education in an order based on contemporary civilization and education and training principles,
ii. To cultivate human power in the areas and numbers that the country needs by using its own specialized power resources in a rational, productive, and economical way in line with principles and objectives of development plans and national education policies with plans and programs which were made by Higher Education Council,

iii. To disseminate the scientific data that raise the welfare level of the Turkish society and enlighten the public through words, writing, and other means,
iv. To enable the society to be educated in the fields of industrialization and modernization in agriculture, through formal, informal, lifelong, and open education,
v. To make scientific research on the problems related to the progress and development of the country in scientific, cultural, social, and economic directions by collaborating with other institutions and by making proposals to public institutions, to present the results to the public, to announce the opinions and suggestions by concluding the investigations and researches requested by the public institutions,
vi. To take preventions to contribute to organizations that undertake formal, informal, lifelong, and open education services for education–training mobilization,
vii. To contribute and to the raise of the professional staff which are convenient to the development of the agriculture and industry and their needs in the region, to provide modernization with industry, agriculture and health services and other services, to carry out activities, and programs to increase production,
viii. To cooperate with related institutions and to make proposals to solve environmental problems,
ix. To produce, develop, use, and disseminate educational technology, to develop the education and training principles for the implementation of higher education,
x. To establish the revolving fund enterprises, to operate efficiently and to make necessary arrangements for the development of these activities

It is seen that higher education has very important 10 goals in terms of the development of the society. When higher education organizations are mentioned, universities come in mind. In this sense, it is thought that it is important to mention the concept of university. The origin of the word of the university is based on the word "üniversalis" in Latin. "Üniversalis" is defined as "a group of individuals with independent legal entities and common interests" (Gürüz et al., 1994; Şimşek, 2006). The Turkish Language Institute defines the concept of university as "educational institution which has scientific autonomy and public legal personality and which is composed of organizations and units such as faculties, institutes, and colleges that provide education, training, scientific research, and publications." From these definitions, universities can be regarded as institutions that research for the development of the society in the direction of traditions and characteristics of their own society and as a result increase their level of development in cultural, social and economic dimensions. At first, concept of the university emerged with the establishment of Bologna Universities. Previously, universities addressed elite people, in 19th century universities showed general public service an understanding and has been continuing developments in this direction (Timur, 2000; Şimşek, 2006). When the literature is examined, it seems that importance of the universities around the world has increased for the last 200 years and universities are among the indicators of development (Karataş, 2006).

In the light of these explanations, there is a need for a good university education and training that will enable students to acquire these qualifications in order for them to gain the desired high qualifications. For this purpose, reforms and changes have been made in the higher education system from time to time. When the historical development of higher education is examined, this situation will become clearer. For this reason, the history of higher education in this study has been examined in three parts.

7.2 MADRASAH'S PERIOD

When the history of higher education in Turkey is examined, it is seen that there are researchers who argue that there is a process from madrasa toward universities.

In this process, it can be said that the first official madrasa was opened in Nişabur in the time of Great Seljuk, and then under the leadership of the Vizier Nizamülmülk, regular madrasahs which were applied specific programs opened in the name of Nizamiye Medreseleri in the centers such as Baghdad. Later, madrasahs continued to develop in the 14th and 15th centuries in Egypt and North Africa (Gürüz, 2003). Previously, religious officers were educated in madrasahs, in 11th century madrasahs began to give education in areas such as Arabic, philosophy, history, engineering, chemistry, mathematics, medical information, geography, and poetry. For example, El Harezmi who is the foremost of mathematics and medical and metaphysical commentator was educated in Farabi Turkistan Madrasa (Gürüz, 2003).

7.3 HIGHER EDUCATION SYSTEM IN OTTOMAN PERIOD

The subject of the beginning of higher education in the Ottoman period is a controversial. While some of the researchers accepted the madrasahs as university and took the history of the higher education in Ottoman to the Fatih period, some of them accepted Darülfünun as the first university. However, when looking at the literature, it is seen that the majority of the researchers and the Council of Higher Education (YÖK) stated the beginning of university history as Darülfünun. In the light of this explanation, madrasahs are mentioned as important in the higher education in the Ottoman period. In the early periods when the Ottoman Empire was founded, they developed the education system through the madrasa system that they took from the Seljuks. In addition to the madrasahs established by the Seljuks such as Kayseri, Sivas, and Konya, they have opened new madrasahs in the conquered areas. In the Ottoman Empire, the madrasahs were first established in 1331 in İznik, followed by other madrasahs established in Bursa and Edirne. Although these madrasahs, when the origin of higher education is mentioned in the Ottomans, Fatih Madrasa, which was founded by Fatih Sultan Mehmet after the conquest of Istanbul, is generally expressed by some researchers.

Another madrasa that attracted the attention after the Fatih Madrasa was Süleymaniye Madrasa which was built between 1550 and 1559 by Mimar Sinan during Kanuni Sultan Süleyman period. Both the Fatih and Suleymaniye Madrasahs conducted various studies to advance the development of education and science. In addition to this, the Fatih Kanunnamesi were put into action in the time of Fatih Sultan Mehmet who determined the rules and principles regarding the academic structure, administration, education program, and academic personnel rights in the madrasahs. Later, Kanuni Sultan Süleyman made some changes and additions to the Fatih Kanunnamesi. As a result, the first source of the legislation of higher education was accepted as these fermans published by these two sultans. The understanding of the madrasa continued to the same structure from Kanuni Sultan Süleyman until III Mustafa. However, after the defeat of the Ottoman navy by the Russian navy, Mühendishane-i Bahri-i Hümayun was founded by Sultan III. Mustafa in 1773, and this school brought changes in the structure of the madrasah. The first higher education institution was formed in Western style with the Mühendishane-i Bahri-i Hümayun. Then in 1795, Sultan III Selim established Mühendishane-i Berri-i Hümayun. These two schools are the foundation of today's Istanbul Technical University (Erdem, 2004).

After the laws of Fatih Sultan Mehmet and Kanuni Sultan Süleyman about the higher education, the second development is accepted as published law of III Selim in 1795 which included the working procedures and academic progress conditions of academic staff in detail.

In the Ottoman Empire, the first higher education institution in modern style was Darülfünun which started preparatory works in 1845 and was opened in 1863. A different higher education system has been introduced from the madrasa with Darülfünun. The lessons were given by senior officials in the form of conferences at Nuri Efendi Mansion. Darülfünun closed in 1865 with the burning of the mansion (Akyüz, 2014). The subject of Darülfünun came into agenda in 1869 again in the Maarif-i Umumiye Nizamnamesi and in 1870 "Darülfünun-u Osmani" was established in Istanbul. Darülfünun-u Osmani had three faculties: philosophy–literature, law, and science–mathematics. But it was closed again in 1872 (Koçer, 1979). Moreover, civilian higher education institutions were noteworthy in this period. The first one of these organizations was Mekteb-i Mülkiye which was established in 1859 in order to train bureaucrats such as district governorships and directorates, and the other one was Mekteb-i Tıbbiye-i Mülkiye which was established in 1867 to educate doctors (Karataş, 2006). Later, in 1900, Darülfünun-u Şahane was opened. The most important feature of Darülfünun-u Şahane is that it has not been closed and it has come up until today. Darülfünun-u Şahane was named "Istanbul Darülfünunu" in 1912 with a very important transformation. The general manager in today's words "rector" was appointed by the recommendation of the Minister of National Education and the Council of Ministers and the approval of the Sultan. By the year 1915, "İnas Darülfünunu," which was for girls, was opened for 3 years. İnas Darülfünun was closed in 1920 and transferred to Darülfünun (Akyüz, 2014).

As a result, many practices related to higher education have been included in the Ottoman period but higher education was occasionally hitched due to the wars. But there is a fact that during the Ottoman period, higher education was given importance and initiatives were made in order to develop it.

7.4 HIGHER EDUCATION SYSTEM IN REPUBLIC PERIOD

There have been radical changes in the education system as well as every other area with the establishment of the Republic in 1923. Significant changes have also been made in the dimension of higher education. The first of these was the abolition of all education institutions to the Ministry of Education by the law of Tevhid-i Tedrisat in 1924 and the closure of the madrasahs (Doğramacı, 1996). Unlike the Ottoman Empire, it was started in Ankara to promote higher education in the republican period. We can give examples such as the opening of the Military School in 1923, the opening of the Law School and the Musical Teachers' College in 1925, and the opening of the Gazi Secondary Teachers' School and the Education Institute in 1927 (Fidan and Erden, 1993).

The problem of Darülfünun came to the agenda again in the 1930s. It has been argued that Darülfünun is a school or a faculty, and it has been emphasized by everyone that Darülfünun needs to be developed. Works to develop the institution was started and Prof. Albert Malche was invited in 1932 (Akyüz, 2014). Despite these developments, the desired result could not be obtained, and the Istanbul Darülfünun, which was described as the first reform of higher education, was abolished by the Turkish Grand National Assembly in 1933 by Law no. 2252. The reason for the removal of Darülfünun is that the numbers of the scholars are low and that they cannot deal adequately with the institution due to their work outside the university and that scientific cooperation cannot be achieved. In the same year, a new university under the name of Istanbul University was established. (Doğramacı, 1996).

Another important point about the Law no. 2252 of 1933 is that the concepts such as university, faculty, dean, and rector were used for the first time (Korkut, 2003). In addition, administrative autonomy is restricted with this law. Before 1933, the directors were elected and from this date, they started to come to work with the appointment (Akın, 2003). There was no radical change until 1946. But the Second World War had upset all the balances in the world and has caused changes in the country's policies. Universities have taken their share from this exchange. Some arrangements have been made as a result of the Second World War politics in Turkey, and multiparty life has been started on behalf of democratization. The Law no. 4936 on universities was accepted with the transition to multiparty life, in 1946. The most important factor in the preparation of this law was the failure of universities to meet demand due to rapid population increase (Timur, 2000). The main purpose was to bring "autonomy" to the universities with the Law no. 4936 in 1946, and autonomy was ensured by Article 120 of the 1961 Constitution. The innovations introduced by this law can be expressed as follows (Akın, 2003):

- Autonomy was given to universities. In addition, the decision of the deans to come work by election was taken. In addition, it is stated that it is necessary to inform the decisions of the universities to the

Ministry of National Education and that the head of the universities is the Minister of National Education.

- It has been suggested that a "University Council" should be established in order to provide cooperation between universities.
- Autonomy and legal personality are recognized in faculties.
- While the purposes of the universities are stated, research has also been given importance.
- Faculty member was first recognized as a profession and subjected to a separate legal regulation (Hatipoğlu, 1998).
- In the management of the universities, the boards have been in the forefront and it has been revealed that the "Professors Board" should be established in the faculties.

After the law issued in 1946, it was suggested that universities should be founded nationwide; many universities were established between 1946 and 1973. For example, the "Black Sea Technical University" in 1955 with the Law no. 6594 and the "Ege University" in the same year with the Law no. 6595, "Middle East Technical University" with the law number 7307 with the support of the United Nations in 1956, and "Atatürk University" with the Law no. 6990 in 1957 were established.

After the military movement of 1960, some legal regulations were made in 1946. For example, for the first time in the 1961 Constitution, an article related to universities was added. According to 120th Article of the Constitution, these decisions are as follows:

- Universities are established only by the State and by law. Universities are public entities with scientific, administrative, and public autonomy.
- Universities are managed and supervised by bodies established by their respective authorized faculty members; provisions on State Universities established in accordance with special law are reserved.
- University organs, faculty members, and assistants cannot be removed from their duties at any time by authorities outside the university.
- University faculty members and their assistants can freely make research and publish it.
- The establishment and applications of universities, their organs and their elections, the duties and authorities, the supervision of faculty members and their assistants by the university organs, are regulated by law according to these principles.

- The prohibition of membership in political parties does not apply to university faculty members and their assistants. However, they cannot take management duties except for the general centers of the parties.

When it came to 1973, the Law of Universities no. 1750 was accepted. The Higher Education Council was established in order to provide coordination among the universities and to make various evaluations with this law, but it was canceled by the Constitutional Court in 1975. Most of the members of the Higher Education Council are nominated by the government and fewer than universities are listed as the reasons for cancelation. In summary, the decisions of the Law no. 1750 can be expressed as (Hatipoğlu, 1998; Akın, 2003):

- YÖK was established for the first time with this law.
- The tasks of the Inter-University Council have been clarified.
- "Faculty Board" was established instead of "Professor Board" in faculties and the rate of representation of all faculty members was increased.
- It is stated that the instructors can be employed with contract, and the age limit for retirement was decided as 70.
- Faculty members have been banned from membership in political parties.
- Articles that give the universities legal personality were protected.

By the 1980s, the number of universities had increased to 19, but it was observed that these were not developed because of infrastructure problems of the regions that they opened (Akın, 2003). Again at the beginning of the 1980s, demand for higher education increased while anarchy and terror continued to increase. So severe acts took place in universities and chaos developed in the country (Hatipoğlu, 1998). As a result, the September 12th Military Attack was held. After the attack, the Higher Education Law no. 2547 in 1981, which was related to universities, was regulated before the 1982 Constitution entered into force. The changes in universities brought about by Law no. 2547 can be summarized as follows:

- The Higher Education Council has been reestablished. YÖK is defined as an organization with autonomy and public legal personality, and broad authority is given to it. These powers include supervising the use of resources allocated to universities, examining university activity reports, conducting disciplinary proceedings of rectors, and terminating academic connections of faculty members to the university on the recommendation of the rector.

- Administrative autonomy of universities has been abolished (Korkut, 2003; Tekeli, 2003)
- The removal of managerial autonomy has also transformed scientific autonomy into formalism.
- A way to establish a higher education institution has been opened to nonprofit organizations. Foundation universities started to open in Turkey with this decision.
- Weekly course load of faculty members increased from at least 6 h to 10 h.
- All academies, colleges, conservatories, and educational institutes are bound to this law and the difference of university and college disappeared (Tekeli, 2003).
- Assistantship status has been abolished, whereas research assistantship and assistant professorship statutes have been introduced.
- The rostrums of faculties were removed, the number of departments was reduced, and departments were established.
- It is stated that research funds will be set up for the researches to be made to the university budgets.
- Atatürk Principles and Revolution History, Turkish Language, and Foreign Language are compulsory courses in all higher education institutions (Fidan and Erden, 1993).

Before the university reform in 1981, the Turkish higher education system consisted of five types of institutions. These institutions are:

1. Universities,
2. Academies affiliated to the Ministry of National Education,
3. Two-year vocational colleges and conservatories which have affiliated some part of them to other ministries, most of them to the Ministry of National Education,
4. Three-year educational institutions affiliated to the Ministry of National Education,
5. Common Higher Education Institution (YAYKUR) that teaches with letters.

These five types of institutions were assembled in 1981 under the name of YÖK with Higher Education Law no. 2547. Law no. 2547 is secured by Articles 130 and 131 of the 1982 Constitution. Article 130 of the Constitution states that YÖK is authorized to remove academic staff from duty and appoint deans, where as the President appoints the rectors. In Article 131, the functions and structure of YÖK are mentioned (1982 Constitution). Bilkent University which is the first foundation university was opened in Ankara in 1984 after the Law no. 2547. The year that the actual university numbers began to increase dramatically has been 1992. Thanks to the newly opened universities; the spread of the universities to the Anatolian has been ensured and important steps have been taken to eliminate the inequality in the regional distribution of the universities.

Since 1981, regulations have also continued to make occasional adjustments and are still in progress. For example, the Higher Education Law have been regulated especially during the last years and the opening of the new universities in various areas of Anatolia has been continuing, some faculties were formed with a system called "secondary education" whose fees were higher than normal education with the Law no. 3843 in 1992 (Akyüz, 2014).

Historical process of the higher education system in Turkey can be examined in three periods as madrasahs, Ottoman, and Republican periods, as well as can be considered as the models that applied in universities. From this point of view, the Turkish higher education system is examined in three periods as the classical period which cannot be described as a complete model until 1773, as "Continental European Model" between 1773 and 1981 and "Anglo-Saxon Model" after 1981.

The characteristics of the higher education system in the Continental European Model can be stated as follows: universities are directly affiliated to the Ministry of Education; the rector is seen as a symbolic leader and rector's duration of duty is short (YÖK, 1996); the rector is responsible for the administration of the university but administrative and financial authorities are at government representatives (Mayntz, 2002).

The "Anglo-Saxon Model" was started to apply by Law no. 2547 in 1981 in the Turkish higher education system (Erdem, 2004). In the Anglo-Saxon model, there are intermediaries that are responsible for supervision, coordination, and central planning. The rector is appointed by these intermediaries and is considered as an executive body; the duration of duty of rector is longer than the Continental European Model. In addition, Ministries of Education have influence on universities (Ertekin, 2006).

7.5 THE IMPACT OF DEMOGRAPHIC CHANGE IN TURKEY ON THE HIGHER EDUCATION SYSTEM

When we look at the history of higher education in Turkey, it is seen that the point is noteworthy. This point is that the changes in higher education are parallel to the political events in Turkey which are shaped according to Turkish demographic characteristics. If this topic is to be examined a little deeper, it can be summarized as follows:

Darülfünun, which assumed university duty in the Ottoman state, was criticized in 1930s for not showing expected development and not reflecting the ideological state. As a result, it was removed with the reason that it faced the regime (Katoğlu, 2000: 424; Öncü, 2002: 521–525).

When we look at another event, the first National Turkish Student Association was established in 1916, and this association was represented at international meetings to Turkey after 1926. But after the March 12, 1971 Military Coup, this association was closed. Because, it was thought that the students who are in this foundation are actively involved in this corp. When we examined the reforms, higher education which defined the first university reform of the Republican era with the Law no. 2252, issued in 1933, has given more importance to universal concepts. For example, this law focuses on the concepts of research, dissemination of research, and teaching. Considering this result, it can be considered that changes in Europe and America are influential among the reasons for Turkish higher education changes.

Another important point is that the changes made in the universities after especially 1946 are on the same dates with military coups. When we look at the legislations that affect universities, it is not a coincidence that the amendments to the laws were in the years after the military coup and the military coup were in power (İnan, 1989: 24–27). For example, the aim of education in the Law no. 1750, which emerged after March 12, 1971, emphasized more on national goals (Timur, 2000: 291–293). After the coup of September 12, 1980, Higher Education Law no. 2547 was issued in 1981 and the universities were attached to the Higher Education Council (Gök, 1998: 79). When we consider these reforms, it has been observed that YÖK has been changed from time to time and it is thought that changes in the structure of YÖK should be considered when we discuss in detail the changes in economic, social, and political structure. Because the student structure, the structure of society, and the expectations of the country from the universities have changed a great deal.

KEYWORDS

- **higher education**
- **education system**
- **history of higher education**
- **Continental European Model**
- **Bologna Universities**

REFERENCES

Akın, F. 1933'ten bu yana Türkiye'de üniversite. *Mülkiye Dergisi* **2003,** *27* (240), 207.

Akyüz, Y. *Türk eğitim tarihi: M.Ö.1000 - M.S. 2014. (26. baskı).* Pegem Akademi Yayıncılık: Ankara, 2014.

Demirer, Y., Duran, M. ve Orhangazi, Ö. *Ateş altındaki üniversite.* Özgür Üniversite Defterleri, Türkiye ve Ortadoğu Forumu Vakfı: Ankara, 5, 2000; p 13.

Doğramacı, İ. Yükseköğretimin gelişimi-üniversite yönetiminde özerklik ve demokrasi. *Yeni Türkiye* **1996,** (7), 340–341.

Erdem, A. R. Türkiye'de yükseköğretim ve değişimi. *Bilim, Eğitim ve Düşünce Dergisi* **2004,** *4* (2), 299–314.

Ertekin, A. Üniversite yönetimi ve organizasyonu ve Trakya Üniversitesi'nde bir uygulama (Yüksek lisans tezi). Trakya Üniversitesi Sosyal Bilimler Enstitüsü: Edirne, 2006.

Fidan, N.; ve Erden, M. *Eğitime giriş.* Meteksan: Ankara, 1993.

Gök, F. Cumhuriyet Dönemi'nde eğitim sistemi ve üniversiteler. İktisat Dergisi **1998,** *383*, 75–80.

Gürüz, K. Dünyada ve Türkiye'de Yükseköğretim tarihçe ve bugünkü sevk ve idare Sistemleri, ÖSYM Yayınları: Ankara, 2003.

Gürüz, K., Şuhubi, E., Şengör, C., Türker, K. ve Yurtsever, E. (1994). Türkiye' de ve Dünya' da Yükseköğretim, Bilim ve Teknoloji, TÜSİAD Yayınları No: T/94, İstanbul.

Hatipoğlu, T. *Türkiye üniversite tarihi;* Selvi Yayınevi: Ankara, 1998.

İnan, N. In *Türkiye'de Üniversitelerin ve Yükseköğretim Kavramının Nitelikleri.* Yükseköğretimde Değişmeler, Türk Eğitim Derneği 12. Eğitim Toplantısı, Türk Eğitim Derneği Yayını, 1989; pp 21–41.

Karataş, S. *Türkiye'de yükseköğretim finansman politikaları ve kurum başarısına etkisi.* Afyon Kocatepe Üniversitesi Sosyal Bilimler Enstitüsü (Yayımlanmamış Doktora Tezi), Afyonkarahisar, 2006.

Katoğlu, M. *Cumhuriyet Türkiye'sinde Eğitim, Kültür,* Sanat. Akşin, S.; Ed., Türkiye Tarihi 4-Çağdaş Türkiye 1908-1980, 6. Baskı; Cem Yayınevi: Istanbul), 2000.

Koçer, H. A. *Türk üniversitelerinde örgütsel gelişme, üniversite yönetiminin uluslararası sorunları.* Ankara Üniversitesi Eğitim Fakültesi Yayınları, Ankara Üniversitesi Eğitim Fakültesi Yayınları, Ankara, 1979.

Korkut, H. Üniversiteler – Cumhuriyet döneminde eğitim; Milli Eğitim Basımevi: Istanbul, 1984.

Korkut, H. Türkiye'de Cumhuriyet Döneminde Üniversite reformları. *Milli Eğitim Dergisi* **2003,** *160*, 184.

Mayntz, R. University Councils: An Institutional Innovation in German Universities. *Eur. J. Educ.* **2002,** *37* (1), 21–28.
Öncü, A. *Akademisyenler: Üniversite reformu söyleminde Batı.* Kocabaşoğlu, U.;Ed.; Modernleşme ve Batıcılık, İletişim Yayınları: Istanbul, 2002.
Şimşek, Ş. *Kırmızı çizgi.* YÖK, Ankara, 2006.
Tekeli, İ. Dünya'da ve Türkiye'de üniversite üzerinde konuşmanın değişik yolları. *Toplum ve Bilim Dergisi* **2003,** *97,* 123–143.
Timur, T. *Toplumsal değişme ve üniversiteler.* İmge Yayınları: Ankara, 2000.
Türkiye Cumhuriyeti Anayasası. 1961. https://www.tbmm.gov.tr/anayasa/anayasa_2016.pdf (accessed Oct 24, 2016).
Türkiye Cumhuriyeti Anayasası.1982. https://www.tbmm.gov.tr/anayasa/anayasa_2016.pdf (accessed Oct 24, 2016).
Yüksek Öğretim Kurulu. Üniversite yönetim sistemleri ve özerklik kavramı. YÖK: Ankara, 1996.
2547 Sayılı Yükseköğretim Kanunu. *Resmi Gazete*; 1981; Vol. 21, p 17506 (Tertip: 5, p 3). http://www.yok.gov.tr/content/view/435/183/lang,tr/ (accessed Mar 11, 2016).

CHAPTER 8

Are Demographic Transition and Economic Growth Triggering Each Other?

ŞUAY NİLHAN AÇIKALIN[1] and HALİME GÜNGÖR[2]

[1] *Department of International Relations, Middle East Technical University, Ankara, Turkey*

[2] *Department of Educational Sciences, Faculty of Education, Division of Educational Administration, Hacettepe University, Ankara, Turkey*

[*] *Corresponding author. E-mail: suaynilhan@gmail.com*

ABSTRACT

The concept of demographic transition, which is as old as the human history, tells us that throughout the process of world evolution, human beings are relocating according to the circumstances that they need to day rather than live in a fixed place. As we look at the fertility control and demographic transition in developing and developed countries we can say that in developed countries fertility control strategies have been started before in developing countries, so fertility rates in developing countries are higher than developed countries. All in all, developing countries have problems on demographic transition and of course economic growth which is effected by demographic transition. Fertility rate is high on low-income countries and parents have not enough money to cover needing of their children, so labor force is high in these countries. Unemployment rates are high and because of highness of labor force can be employed in low wages. This situation has negative impact on labor market and economic performance, so educated and young people have to work in bad conditions with low wages. Mortality rates are high in these countries because of insufficiency of public health services that are provided by the state. Nonlinear dynamics in life is also available for countries as living systems, so any unexpected input like mass migration or mass deaths due to war or terrorist attacks affect the order of the system. Quality of life and high life conditions are more humane in developed countries, so all countries should complete the demographic transition period to live in more humane conditions. At this point, we wonder that dealing with chaos in that country is related to economic growth of that country or what? Are demographic transition and economic growth triggering each other?

8.1 INTRODUCTION

Demographic transition and economic growth are interrelated concepts that indicate countries development level. Fertility, mortality, marriages, and social mobility as fundamental concepts of demographic transition affect citizens' quality of life of any countries. Nonlinear dynamics in life is also available for countries as living systems, so any unexpected input like mass migration or mass deaths due to war or terrorist attacks affect the order of the system. As Erçetin et al. (2013) said that there are three main reasons why the future of dynamic systems cannot be predicted:

1. The state of a system is completely unknown, in other words there is no analytical solution,
2. The initial conditions cannot be determined precisely,
3. Minor changes in the initial condition can lead to significant changes that we consider to be sensitive to initial conditions—butterfly effect.

So these country or region transformed chaos into dynamics and confronted with a rapid change process. As a result, there are two outputs at the end of this process: chaos transform the order or chaos created more chaos. At this point, we wonder that dealing with chaos in that country is related to economic growth of that country or what? Are demographic transition and economic growth triggering each other?

8.2 DEMOGRAPHIC TRANSITION

Concept of demographic transition, that is as old as the history of human history, tells us that throughout the process of world evolution, human beings are relocating according to the circumstances that they need to day rather than live in a fixed place. Demography can be defined as the statistical analysis of the changes in the population over time, size, composition, and spatial distribution of population, and also functioning of five processes: birth rate, death rate, marriage, migration, and social mobility (Kabeer, 1996).

Demographic transition is related with change in age structure through reduced birth and death rates (Basu et al., 2013). If we look at the World in general terms of fertility and mortality, we can say that past trends of fertility and mortality are increasing at a very young population in high-fertility countries in South and led to a more elderly population in North (Bongaarts, 2009).

As we look at the fertility control and demographic transition in developing and developed countries we can say that in developed countries, fertility-control strategies have been started before the developing countries, so fertility rates in developing countries are higher than developed countries. Thus, demographic transition is a process involving the transition from young-aged population structure to old-age population structure (Joe et al., 2015). The world is in the midst of a major demographic transition in terms of slowing of population growth and changing age structure of population with share of young falling and that of elderly rising (Callen et al., 2004).

Demographic transition can occur in asynchronously in terms of death and birth rates, such as firstly, death rates decline and then birth rates decline; population growth and demographic transition is a transformation period. There is a period of rapid demographic change in the world and in countries at an unprecedented rate (Bongaarts, 2009). Speed of the evolution period of the World is rapid in the last half century with scientific and technological developments; therefore, these rapid evolution process reverberate our lives and societies' and countries' lives and structure. Furthermore, the dominant view is that demographic change is still going on at large, leading to a major "demographic bonus" in the pace of development over the next decades (Attanasio et al., 2006).

Demography is not destiny, therefore, human society can do in response to changes and adaptations of individuals, to transform the challenges of aging population changes and growth opportunities are equally effective and is the key (Bloom and Luca, 2016). How can we define and explain *demographic transition theory*?

8.2.1 DEMOGRAPHIC TRANSITION THEORY

Demographer Warren Thompson developed the demographic transition model (DTM) based on the observation and has created four categories originally to describe the population changes and was considered to be universal and applicable to all countries, stage for which all should perform the same steps of assuming depends on the level of their own development (Gerőházi et al., 2011). So, model consists of at least four different stages and countries are progressing effectively from high fertility and low lifespan to low fertility and demographic transition as far as their lifespan extends (World Bank Group, 2016).

DTM describes the transition of populations from high to low birth and death rates (World Bank Group, 2016) and this transition generally parallels countries' economic development (Szreter, 1993). We can say that demographic transition period of the countries is related with development level of the countries. So, the formation and duration of the demographic transition is different in various regions of the world and it has been 50–150 years since this transition was completed (Iqbal et al., 2015). According to Theory, demographic transition has three distinct phases (Joe et al., 2015; Williamson, 1997; Kirk, 1996; Schultz, 2001; Alleyne and Cohen, 2002) as follows:

1. In the first phase: *high children dependency ratio*
2. In the second phase: *high working-age population ratio*
3. In the third phase: *high old-age dependency ratio* (Fig. 8.1).

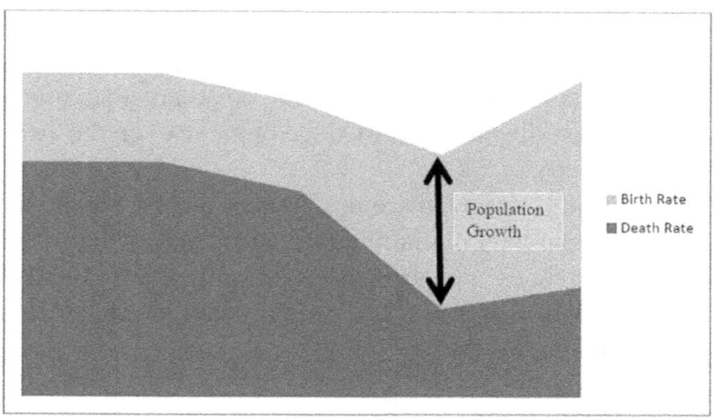

FIGURE 8.1 The demographic transition.
Source: Alleyne and Cohen (2002).

Fertility rates and mortality rates are high when population tends to be younger and the population growth is stable but low in the first stage; as mortality rates decline, population growth accelerates in the case of high fertility rates, an increasing number of young and rising young dependency in the second stage; fertility rates also begin to fall and population growth slows down in the third stage; fertility rates and mortality rates reach low levels in the fourth stage (World Bank Group, 2016) (Fig. 8.2).

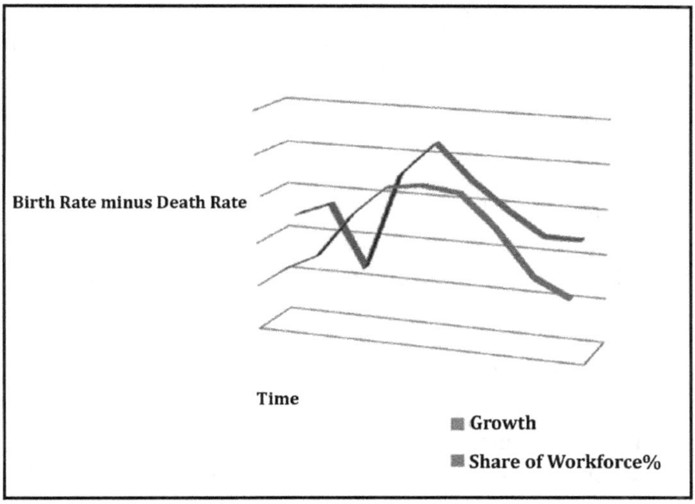

FIGURE 8.2 Demographic transition and population growth.
Source: Bloom and Williamson (1998).

Demographic transition process in developed countries realized smooth and spontaneous nature when experiencing industrialization, urbanization (Fang and Dewen, 2006), and modernization.

Despite, many developed countries seem to be in final stages of transition, rapid economic growth, and human development, the implementation of effective policies in majority of African countries seems to be at second stage of transition (Basu et al., 2013). Furthermore, Turkey is also at second stage of transition as a developing country.

Beyond fertility and mortality rates, demographic transition has increased life expectancy, the years of healthy living, proportion of retirees, population aging, and migration rates (Lee and Mason, 2007; Suntoo, 2012), whereas it has caused difficulties in designing and implementing pension systems in developing countries, including low financial literacy, and public budget constraints (Bloom and McKinnon, 2014). The process of demographic transition of developed countries has completed. As we focus on demographic transition process of the countries, we conclude:

-In Western and Eastern Europe, at the end of 19th century and beginning of 20th century, population growth declined (Oded, 2005). Latin America began its transition in late 1960s (Iqbal et al., 2015). Countries in Europe are generally in the last stage of demographic transition and most of countries in Latin America and Asia are in second stage of demographic transition.

Demographic transition has triggered effect on economic growth. Rapid economic growth and human development policies in demographic transition can be implemented with four mechanisms that can be defined as the phenomenon of demographic dividend (Basu et al., 2013):

1. The increased labor supply.
2. The increase in savings.
3. The human capital.
4. The increase in domestic demand resulting from increased per capita gross domestic product (GDP) and a decreasing dependency ratio.

8.3 DEMOGRAPHIC TRANSITION AND ECONOMIC GROWTH

There are different countries and regions at different stages of this demographic transition period, but the important question is how the ongoing demographic change will affect economic performance in coming years (Callen et al., 2004); *the impact of demographic transition in developing economies (the South) is dependent on demographic trends in the developed world (the North)* (Attanasio et al., 2006).

Direct econometric estimates of life expectancy, economic growth, income distribution, and poverty can allow one to simulate consequences of recent and predictable improvements in life expectancy over poverty (Alleyne and Cohen, 2002). In the last decade, two important developments have been recorded concerning the impact of demographics on national economic performance (Bloom, 2011):

1 the influence of a changing population's age structure and

2 population health.

Bloom and Canning (2001) are few in exploring the impact of demographic transition on economic growth. Large percentage of elderly population and/or children increase burden to take more responsibility for dependents to community and reduce productive skills of economy, thus affecting economic growth negatively (Fang and Dewen, 2006).

Demographic opportunity window that some of poorest countries of our world are currently entering offers a unique opportunity for economic and for a restricted number of years, size of working age population is at its maximum compared to size of dependent population and hence a high productive capacity goes together with low caring costs for the young and the old (van der Ven and Smits, 2001). According to Samans et al. (2017) the inclusive development index 2017 rankings of countries (according to 2017 data of World Economic Forum), there are four types of countries in terms of economy: countries in advanced economy, upper–middle income countries, lower–middle income countries, and low-income countries. There are 30 countries in advanced economy and these countries have more than 17,000 GDP per capita; there are 26 countries in upper–middle income countries which have 6000 and 16,999 GDP per capita; there are 37 countries in lower–middle income countries which have 1320 and 5999 GDP per capita; and there are 16 countries in low-income countries which have less than 1320 GDP per capita.

According rankings of World Economic Forum 2017, "Turkey's score of 4.30 places it 20th on the IDI. It has the highest labor productivity among this group, high GDP per capita and living standards, and low poverty. In terms of Framework indicators these good outcomes are driven by strengths such as relatively high competition among companies, which ensures that large individual firms do not dominate the economy and stifle activity. Turkey also benefits from a fairly sophisticated financial sector, which adds to this business dynamism by providing investment. On the other hand, the unemployment rate is somewhat high, particularly among the young. These points to the continuing need to strengthen the education system, especially to make outcomes more equitable for students from all income groups. Expanding female participation in the labor force is also a priority, alongside reducing the wide gender gap in pay" (Samans et al., 2017: 55).

8.4 CONCLUSION

On the whole, developing countries have problems on demographic transition and of course economic growth which is effected by demographic transition.

- Fertility rate is high in low-income countries, and parents does have not enough money to satisfy the needs of their children, so labor force is high in these countries.

- Unemployment rates are high, and because of the availability of high labor force, they can be employed in low wages.
- This situation has negative impact on labor market and economic performance, so educated and young people have to work in bad conditions with low wages.
- Mortality rates are high in these countries because of insufficiency of public health services that are provided by the state.

Quality of life is high and living conditions are more humane in developed countries, so all countries should complete the demographic transition period to live in more humane conditions.

KEYWORDS

- **demographic transition**
- **economic growth**
- **developing countries**
- **developed countries**
- **low-income countries**

REFERENCES

Afzal, M. Population Growth and Economic Development in Pakistan. *Open Demogr. J.* **2009,** *2*, 1–7.

Alleyne, G. A. O.; Cohen, D. *Health, Economic Growth, and Poverty Reduction.* The Report of Working Group 1 of the Commission on Macroeconomics and Health, World Health Organization, Geneva, 2002.

Attanasio, O.; Kitao, S.; Violante, G. Quantifying the Effects of the Demographic Transition in Developing Economies. *Adv. Macroecon.* **2006,** *6* (1), 1–44. DOI:10.2202/1534-6013.1298 (accessed Apr 23, 2017).

Basu, T.; Barik, D.; Arokiasamy, P. Demographic Determinants of Economic Growth in BRICS and Selected Developed Countries. XXVII IUSSP International Population Conference (IUSSP 2013). August 26–31, 2013 at BEXCO, Busan, Republic of Korea.

Bloom, D. E.. *Population Dynamics in India and Implications for Economic Growth.* Program on the Global Demography of Aging (PGDA) Working Paper No. 65, 2011.

Bloom, D.; Canning, D. Cumulative Causality, Economic Growth, and the Demographic Transition. In *Population Matters: Demographic Change, Economic Growth, and Poverty in the Developing World*; Birdsall, A. C., Sinding, S., Eds.; Oxford University Press: Oxford, 2001.

Bloom, D. E.; Finlay, J. E. *Demographic Change and Economic Growth in Asia*. Program on the Global Demography of Aging, Working Paper Series, 41, 2008.

Bloom, D. E. Luca, D. L. *The Global Demography of Aging: Facts, Explanations, Future*. Program on the Global Demography of Aging at Harvard University Working Paper No. 130, 2016.

Bloom, D. E.; McKinnon, R. The Design and Implementation of Public Pension Systems in Developing Countries: Issues and Options. In *International Handbook on Ageing and Public Policy*, Harper, S., Hamblin, K., Eds.; Edward Elgar, 2014, pp 108–130.

Bloom, D. E.; Williamson, J. G. Demographic Transitions and Economic Miracles in Emerging Asia. *World Bank Econ. Rev.* **1998,** *12* (3), 419–455.

Bongaarts, J. Human Population Growth and the Demographic Transition. *Philos. Trans. R. Soc. Lond. B Biol. Sci.* **2009,** *364* (1532), 2985–2990. DOI: 10.1098/rstb.2009.0137.

Callen, T.; Batini, N.; Spatafora, N. How will Demographic Change Affect the Global Economy? (Chapter 3) In *World Economic and Financial Surveys, IMF World Economies Outlook (WEO), The Global Demographic Transitions*, September 2004, International Monetary Fund, Washington DC, 2004.

Erçetin, Ş. Ş.; Açıkalın, Ş. N.; Bülbül, M. Ş.. A Multi-Dimensional Approach to Leadership in Chaotic Environments. In *Chaos and Complexity Theory for Management: Nonlinear Dynamics,* Banerjee, S., Ed., IGI Global: USA, 2013; pp 89–104. DOI:10.4018/978-1-4666-2509-9.ch005.

Fang, C.; Dewen, W. In *Demographic Transition and Economic Growth in China*. Presentation at the International Conference on the Dragon and the Elephant: China and India's Economic Reform, July 1–2, Shangai, China, 2006.

Feng, W. The Future of a Demographic Overachiever: Long-term Implications of the Demographic Transition in China. *Popul. Develop. Rev.* **2011,** *37* (Suppl), 173–190.

Gerőházi, É.; Hegedüs, J.; Szemző, H Tosics, I.. *The Impact of European Demographic Trends on Regional and Urban Development*. Synthesis Report. Issued Within the Framework of the Hungarian Presidency of the Council of the European Union, Metropolitan Research Institute: Budapest, 2011.

Gomez, R.; de Cos, P. H. Demographic Maturity and Economic Performance: The Effect of Demographic Transition on Per Capita GDP Growth. The Banco de Espana. 2003. www.bde.es. (accessed May 18, 2017)

Iqbal, K.; Yasmin, N.; and Yaseen, M. R. Impact of Demographic Transition on Economic Growth of Pakistan. *J. Finance Econ.* **2015,** *3* (2), 44–50. http://pubs.sciepub.com/jfe/3/2/3 Science and Education Publishing. DOI: 10.12691/jfe-3-2-3.

Joe, W.; Dash, A. K.; Agrawal, P. *Demographic Transition, Savings, and Economic Growth in China and India*. IEG Working Paper No. 351, 2015.

Kabeer, N. Gender, Demographic Transition and the Economics of Family Size: Population Policy for a Human-centred Development. Occasional Paper 7, June 1996. United Nations Research Institute for Social Development, United Nations Development Programme, 1996.

Kirk, D.. Demographic Transition Theory. *Popul. Stud.* **1996.** *50* (1996), 361–-387

Lee, Ronald, The Demographic Transition: Three Centuries of Fundamental Change. *J. Econ. Perspect.* **1996,** *17* (4), 167–190.

Lee, R.; Mason, A. *Population Aging, Wealth, and Economic Growth: Demographic Dividends and Public Policy*. UN World Economic and Social Survey Background Paper, 2007.

Mason, A. Will Population Change Sustain the Asian Economic Miracle? *Asia Pacific Issues* **1997,** *33*, 1–8.

Mason, A. *Demographic Transition and Demographic Dividends in Developed and Developing Countries*. United Nations Expert Group Meeting on Social and Economic Implications of Changing Population Age Structures (Mexico City), 2005.

Oded, G. The Demographic Transition and the Emergence of Sustained Economic Growth. *J. Eur. Econ. Assoc.* **2005,** *3*, 494–504.

Reher, D. S. Economic and Social Implications of the Demographic Transition. In *Demographic Transition and Its Consequences*; Lee, R. D., Reher, D. S., Eds. (*Population and Development Review*, A Supplement to Vol. 37); Population Council, New York, 2011; pp 11–33.

Samans, R.; Blanke, J.; Drzeniek Hanouz, M.; Corrigan, G. *The Inclusive Growth and Development Report 2017*, World Economic Forum, 2017.

Schultz, T. P. *The Fertility Transition: Economic Explanations*. Center Discussion Paper No. 833 Yale University Economic Growth Center: New Haven, CT, 2001.

Suntoo, R. Population Ageing and the Theory of Demographic Transition: The Case of Mauritius. *Univ. Mauritius Res. J.* **2012,** *18*, 1–18.

Szreter, S. The Idea of Demographic Transition and the Study of Fertility Change: A Critical Intellectual History. *Popul. Develop. Rev.* ***1993,*** *19* (4), 659–701.

van der Ven, R. Smits, J. *The Demographic Window of Opportunity: Age Structure and Sub-National Economic Growth in Developing Countries*. NiCE Working Paper 11–102. Nijmegen Center for Economics (NiCE), Institute for Management Research, Radboud University Nijmegen, 2001.

Williamson, J. *Growth, Distribution and Demography: Some Lessons from History*, NBER Working Paper No.6244. National Bureau Economic Research: Cambridge, MA, 1997.

World Bank Group *Global Monitoring Report 2015/2016: Development Goals in an Era of Demographic Change*. Washington, DC: World Bank, 2016. DOI: 10.1596/978-1-4648-0669-8. License: Creative Commons Attribution CC BY 3.0 IGO.

CHAPTER 9

An Analysis of the Demographic Transition in Norway and Niger in Terms of Educational Indicators

SELDA ÇELİK*

Department of Educational Sciences, Faculty of Education, Division of Educational Administration, Hacettepe University, Ankara, Turkey

E-mail: seldaceliks@gmail.com

ABSTRACT

One of the most important factors determining the economic and social development levels of countries is education. Today, societies link up with the level of education and the economic development of their countries. Education is the basis of economic development. The purpose of the education is to meet all kinds of demands and needs of both the country and the public. Education, on the one hand, meets these demands and needs while at the same time influence the demographic transition in the country in parallel with the development of the country. The level of development of the country indirectly affects the proportion of the population in that country, the birth and death rates of the baby, the average life expectancy. Individuals change their own lives and living standards in proportion to the education they receive, as well as they contribute the change in the society where they live at the macro level. From this point of view, it is aimed to show how education affects the demographic transition in comparison with the education indicators in Norway which is in the highest human development index, and in Niger which is in the underdeveloped human development index in this section. Starting from this aim, firstly the concepts of education and development, and demographical transition and later the properties of developed and underdeveloped countries are explained. In the study, Norway and Niger are compared and evaluated in terms of sex education, schooling rates, adult literacy, education expenditure, fertility, and death and population growth rates.

9.1 INTRODUCTION

As a result of the advancements in science and technology, the circulation of scientific knowledge, and the cultivation of individuals with highly creative potential countries have undergone significant changes in the economic and political fields. This tendency to constantly innovate and develop among countries, particularly in the field of education, has accelerated unlimited competition for the purpose of more information and newer technology. Parallel to the rapidity of time, new entrants and technological breakthroughs have also affected the systems of education. The purpose of the education is to raise a healthy society with a high level of knowledge, both individual and universal. All innovation and development initiatives in education are of interest to all segments of society. Because education is responsible for any change in the society, it must adapt the change before other systems. Prosperity and happiness of a country are up to whether people have a qualitative and continuous education and whether people contribute to the economic growth with this knowledge. For this, the most important driving force of socioeconomic development and the most important element of the productivity increase is the education level of the society.

The world is in the middle of a demographic transition process that affects the size of the population and the age structure (Batini, 2006: 4; Lee and Reher, 2011). Demographic transition (Hacettepe University Institute of Population Studies, HUNEE, 2010), which is seen as the result of "Modernization Theory," is related to the socio-economic change and modernization process of the world population (Reher and Sanz-Gimeno, 2007). By 20th century, the growth rate of the world population was slow, the age structure was fairly constant, and the percentage of people living above 65 years in the population was lower. This has shifted to another stage in the demographic transition process in the second half of the century, with the average expectation of life and the rate of population growth increasing in the first half of the 20th century: There has been a marked decline in fertility and mortality rates worldwide.

In this context, the aim of this chapter is to emphasize the importance of education on demographic transition. Societies link up the increase on the level of education with efficiency, and believe that the individual contributes to the society that he/she lives in the amount of education he or she has received. The level of qualification of educational activities also shows that the rate of population growth and the mortality rates of the country are changing due to the influence of the individual on the quality of the economic, social, political, and cultural development of the society in which the individual lives. Research reveals that there is linear relationship between educational level and demographic transition. There are few studies on the relationship between education and demographic transition in the field. For this reason, it is considered that the examination of the relationship between education and demographic transition at the present time is likely to contribute to the field. In this study, it is aimed to examine the relationship between education and demographic transformation and to show how education changes demographic transition by taking into account the data of the Norway and Niger countries in the Human Development Index 2016 Report, each of which is in a different human development category.

9.2 BASIC CONCEPTS

9.2.1 EDUCATION AND DEVELOPMENT

Education is a difficult concept to define because it is an ongoing phenomenon from birth to death of the individual and includes political, social, cultural, and individual dimensions at the same moment. Education includes all social processes that are effective in helping individuals acquire the standards, beliefs, and ways of living of society. In other words, it includes attitudes, abilities, and other forms of behavior developed within the society in which the person lives. It is a social process that provides optimal social competence and individual development within a selected and controlled environment (especially school). Education is a planned sequence of events that allows people to evolve their behavior according to predetermined principles. Education is the process of intentionally changing the behavior of an individual through his own experience (Peker and Altınışık, 2012).

Education gives people the ability to produce, understand, and interpret knowledge while giving information on the one hand. The education given to individuals is an investment in human capital; because through education, people become more productive throughout life. Education is different from other investments because it is an investment made on people. Today, human capital has become the most important production factor of the rapidly changing and globalizing world. Almost all the societies that have been increasingly developing and are prosperous by help of technological advances in the world are owed that statement to the investments they make for the growth of human capital. Today, the most important measure of being a strong and respectable country in the international arena is a healthy and stable economy. This way of having a healthy and steady growth economy also passes through education.

In classical economic theory, the concept of capital was explained by physical capital consisting of machinery, hardwares, and other equipments. However, the growing importance of the effects of the personal and social traits on production led to the acceptance of positive values and hence the emergence of the concept of human capital. The positive values mentioned here are the knowledge, skills, and experiences in the workforce. In internal growth models, the importance of human capital is emphasized seriously. The concept of human capital, which constitutes the main source of economic growth, is used to express concepts such as knowledge, skills, abilities, health status, place in social relations, and level of education that a person or society has. Education has developed the capacity and productivity of the workforce and it also contributed to the increase in national income with this way and even the opinion that the majority of the rapid growth rate in the United States that had lived in the past stemmed from the investments made in education was also supported by classical economists (Taş and Yenilmez, 2007).

Microeffects of human capital investments are seen onindividuals, families, and companies. These microeffects are concentrated in specific areas, affecting macroeconomic areas of the country. The increase in personal income as a result of educational investments may cause positive effects on family and firm as well as affect income distribution, technological development, labor market, agricultural and industrial sectors, productivity, and regional development differences at various rates (Yumuşak, 2000). While human capital contributes to the social dimension of the development process by creating a change of behavior, it also contributes to economic development as an input of the production process (Özgür, 2005). Thus, the formation, development, and continuity of the human capital will exactly accelerate the socioeconomic development and make the societies and the countries in which they live reach prosperity.

In order for a country to be industrialized and developed, that society must have certain minimum conditions. We can list these conditions in the following way (Eroğlu, 2001):

- constant and stable political environment,
- high transportation facilities,
- good communication and finance environment, and
- an adequate level of education.

If these four conditions are fulfilled, it can be said that the most important steps in ensuring social development also occur.

Scientific research reveals that there are linear relations between the educational level and "the economic growth, political, and social development" which are elements of development. The contribution of human resource, especially to social healing and thus to economic development, is great. Social benefits of education within the context of social cohesion are less crime, participation in democratization and governance, protection of individual health, and greater income for the individual. With this feature, education not only benefits the individual but also benefits the society and reduces public welfare costs.

Education constitutes the main means of planning toward economic life, general development, and employment (Aydın, 2000). The cooperation between the education system and the economy is seen as an important condition in increasing the efficiency and preventing the wasting of human resources. In recent years, it has become important that the relations between economic growth and human development proposed by many economists are reciprocal, that human development accelerates economic growth by increasing labor productivity and that economic growth carries high levels of human development by increasing health, education, and social expenditures through income increase.

Development is also an attempt of political power to increase the welfare levels of the individual and to change the structure ofthe society by pursuing certain economic policies (Adem, 1993). Nowadays, the level of development of the countries is measured with—besides the amount of national income—education, social, cultural, and political conditions. Economic development can be expressed by per capita goods and services units, and also per capita education and health expenditures are important measures of development. Parallel to these, values such as literacy and schooling rate and average life span are also indicators of the level of development of a country. All of these statements place the human in the center of development. The economic and cultural environment created by human thought, talents, and education level contribute to the economy as an input of the production process by developing innovation and creativity.

It can be clearly seen that there is a very strong relationship between education and economic development in the light of allthis information. As the relationship between human capital and development becomes clearer, the importance of investments in education has increased considerably throughout the world.

9.2.2 DEMOGRAPHIC TRANSFORMATION

The Demographic Transformation Theory foresees a transition from high fertility and mortality rates to a new situation in which mortality rates have fallen and births are deliberately controlled. This theory, which was developed by Warren Thompson (1929) and Adolphe Landry (1934), was further systemized by studies of Kingsley Davis (1945 and 1963), Dudley Kirk (1945), and Frank Notestein (1945 and 1953) between the years 1940 and 1960 (Hirschman, 2001). After analyzing the demographic data between the years 1908 and 1927, Thompson (1929) noted that countries split into three groups: those experiencing rapid declines in birth and mortality rates (North and Western European countries and the United States), those with less or mild severe signs (Italy, Spain, and the United States; between 1908 and 1927 Central Europe), and those who have experienced no decline in these indicators over the years covered by the study. This work by Thompson (1929) is important in that it determined that the demographic transition even in the early 20th century followed various geographical paths at various stages. In the first years of Demographic Transformation Theory, it was thought that high fertility was developed in response to high mortality, but in the following years experts have shown that changes in demographic behavior are due to economic development and modernization. Countries with low birth and death rates have been found to be in good economic development and less developed countries have high birth and death rates (Caldwell, 2001). Demographic Transformation Theory can be expressed with the analysis of the changes in birth and death based on data of many years, especially in western European countries (Rowland, 2006; HUNEE, 2010).

The general characteristics of the four different stages of demographic transition in relation to socioeconomic development are as follows: In general, in economies such in which there is no health and education services, there are low-level technologies, there is a prevalent rural life, and there are high birth and mortality rates that fluctuate over years, death rates can increase birth rates (starting phase/ phase 1). In this case, population growth is generally less than 1%, or in other words, low. When the economy changes, when the health services and facilities increase and technological conditions improve, the death rates do start falling (transition phase 1/phase 2), but birth rates remain high. This causes the rate of population growth to be generally between 2% and 4%, in other words reaching high levels. If development is provided in a certain way, this leads to an increase in educational opportunities, an increase in living standards, and acceleration of urbanization; and if the height of population growth is beginning to be seen as a problem by individuals/couples from this stage, birth control occurs, and in this case, decrease in birth rate can be seen (second transition phase/phase 3). The last stage/phase 4: the fact that birth rates are slightly higher than the death rates leads to a stable population growth in which the rate of population growth is around 1% or less. This last stage is exemplified by the industrialized countries of today (Weinstein and Pillai, 2001; Rowland, 2006; Yaukey et al., 2007; Başar, 2010).

The theory was formulated taking into account some European countries where data could be obtained in the years when the theory was first established. As the standard of living increased, the mortality rate decreased and, after a few decades, the birth rates were low enough to be called low, though rarely as low as the death rates. The reason why the decline in birth rates lagged behind the declines in mortality rates can be explained by the fact that this adaptation of the population took time to actually believe that births should be less or lower. Decreasing mortality was not difficult, as a response to the importance of the vast majority of people in maintaining the survival of society, but on the contrary, social norms based on compensating for high mortality rates continued to encourage high fertility and it was not easy to transform the norms into low fertility, even in poverty. As a process brought by industrial and urban life, family life has lost its significance and form, large family pressure has weakened, and finally birth rates have decreased. The large/traditional family format is actually a social institution where parents bring their children to the world and raise them, and when they are old they perceive some kind of old-age insurance to look after them. The transformation from a labor force to an urban-industrial area where compulsory education is made can be associated with an economic development model in which low mortality exists. At this point, what people understand and realize from infant mortality is the belief in the need to have fewer children or to have a certain number of children (Weeks, 1992).

With the increase in women's literacy over time, countries can be said to become richer; and if more investments can be made on human capital, this could be a part of the development process of the country. Even if there are no other variables, the fact that the education of women alone has a considerable influence on the number of children they can have suggests that the relationship between education and fertility also has significance. Of course it can be emphasized here that education is part of the development process as one of the catalysts for demographic transformation (Basu, 2002).

9.3 CHARACTERISTICS OF DEVELOPED COUNTRIES

The concept of developed country is the term used for countries that have shown a high level of development according to certain criteria. From these criteria "economic criteria" play a more dominant role in the evaluations made. The "national income per capita," one of the economic criteria, is the most used criterion for determining the level of development. While making international comparisons, the amount of Gross National Product (GNP) per capita, which is a universal economic criterion, is examined and found in various perspectives. Another economic criterion is the level of "industrialization" of the countries. The countries where the industries dominate are seen as developed countries. Moreover, the "income inequality" rate is seen as another economic indicator.

In developed countries, the 20% of the lowest income population receives 6–8% of the income, while the highest income segment receives 40% of the income. Income distribution values, in terms of percentages expressed by various developed countries can be given in this regard: Germany, respectively, 7.0 and 40.3; Italy 6.8 and 41.0; Finland 6.3 and 37.6, France 5.6 and 41.9; Japan is 8.7 and 37.5, and finally the Netherlands is 8.2 and 36.9. In addition, the tendency to save in developed countries is high due to the high income. In such countries, the amount of informal economy is also very low. The informal economy, which includes activities such as showing incomplete taxable income, not being affiliated with any social security institution, and activities strictly prohibited by law (such as drug and gun smuggling, check-bill collection, bribery, usury, etc.), creates income transfer considerably rather than creating added value (Murat, 2013).

It is incomplete to express the level of development of countries only by linking them with the economic criteria, because it excludes social structure. In order to be able to determine the social and political characteristics of the developed countries in a healthy way, the social structure of the country should be observed very well and the social criteria should be determined correctly. The specified social criteria can be set out by observing and evaluating in detail. It is a little difficult to arrive at the final judgment of the social criteria. Because social structures, on the one hand, reflect the processes of their original formation, on the other hand they are constantly changing in a dynamic social formation.

Social characteristics of developed countries include;

- balanced and healthy nutrition,
- adequate health conditions,
- optimal population growth,
- low dependency ratio,
- regular urbanization,
- the place of women in social life,
- the prevailing middle class,
- higher education level, higher literacy rate,
- political stability, and
- the completion of democratic institutions and regulations. (Han and Kaya, 2004).

An important criterion for describing the development of countries, including those mentioned above, is "Human Development Index." It is stated in United Nations Development Program's (UNDP) human development reports published every year since 1990 that the real richness of a society can be provided by citizens who can benefit from a healthy and free living environment that allow for a community that people will enjoy (UNDP, 1995). Human development is defined as the process of increasing people's choices, expressing a long and healthy life, providing information, and the conditions for a good standard of living. It is also difficult to say if a country has a high GNP, even if it is in a low level of education, that country is a developed country.

Human development, defined as the process of increasing the options in front of people, focuses on raising individuals to an acceptable quality of life by putting people at the center of development. In the 1990 Human Development Report, UNDP identified human development as: "Human development is the process of multiplying people's choices. Basically these options can be unlimited and change over time. However, three of them are very important in all phases of development; to live a long and healthy life, to learn and to reach the resources necessary for a good standard of living. In the absence of these options, many other options are not possible."

9.3.1 NORWAY

Norway is a European country located in northern Europe and the western part of the Scandinavian Peninsula. Norway has Sweden in the south east, Finland and Russia in the east. The North Sea is located in northern Norway, the Norwegian Sea in the west, and the Baltic Sea in the south. The three sides of Norway are surrounded by seas. In the north and west of Norway there are the islands that are under its government.

Norway is a country governed by the monarchy system having an advanced economy. The Norwegian king is the head of the Norwegian army and churches. The Norwegian king had the right to veto legislation prepared by the government in parliament. But today it has only symbolic significance and the law recommended in parliament means the approval of the king. The Norwegian king, by the way, has the right to choose the prime minister, ministers, and other senior executives elected by the created government. Having a parliamentary system, Norway has very successfully carried out its own external, social, financial, and economic policies. The government intervenes in the main sectors of the country's economy in the production and distribution of fisheries, oil, natural gas and other energy.

Today, the Norwegian population reaches 4.6 million and the annual increase is around 0.4%. 500,000 people live in the capital Oslo. In addition to Oslo, there are major cities such as Bergen, Trondheim, Tromso, Stavanger, and Kristiansand. Norwegian is used in the Norway. There are two different forms, Bokmal and Nynorsk.

While Norway is a country with scarce resources, it has been rapidly enriched by oil and natural gas revenues since the 1970s. Norway is one of the world's leading exporters of oil and gas, and today Norway is among the richest countries in per capita nationalincome. Norway has been in the first place for 5 years in the Human Development Index published by the UNDP.

9.4 CHARACTERISTICS OF UNDERDEVELOPED COUNTRIES

When the phenomenon of underdevelopment comes into question, the question of why these countries are underdeveloped has to be discussed first. The criteria set out for developed countries are also used to explain underdeveloped countries. The low GNP per capita, not being sufficiently industrialized, and inequality in income distribution are some of these criteria. In addition, relative poverty is used to explain underdevelopment concepts such as the share of agriculture in GNP, technology, basic social indicators, population growth, income distribution, and foreign trade dependency ratio. Also, insufficient democracy, transparency, and accountability in underdeveloped countries can lead to unregistered economic activities. Other reasons causing the formation of the informal economy can be listed as follows (Özsoylu, 1995):

- unequal distribution of national income,
- high inflation,
- the crisis in the agricultural sector,
- flexible production systems and globalization,
- tax injustice and resistance to taxation, and
- social structure and moral norms.

Another feature of underdeveloped countries is that the population grows faster than developed countries.

Birth rates in developed countries are around 10–15 per thousand, while in developing countries this rate is around 40 (World Bank WDI Online Database, 2017). Population is triggered by high birth rates. The high fertility rate leads the inactive and consumer population aged 0–15 years to go up, which causes an increase in the dependency rate. Dependency rate is the ratio of inactive population to active population. In developed countries the dependency rate is low and the economic burden of the active population is rather low compared to underdeveloped countries.

According to the UNDP Human Development Index, countries are categorized according to average life span, level of knowledge, and living standards. According to human development index, countries are classified as very high human development rate, high human development rate, and medium and low human development rate (UNDP, 2016). Niger is among the countries with low human development rate according to the classification.

9.4.1 NIGER

It is one of the countries of the western part of Africa with a population of about 17 million and a large territory. It was occupied by the French colonists in the 19th century and became a separate state after the borders of the country were established as a colony in 1922. In December 1958, the Republic of Niger was established and on August 3, 1960 it became an independent country. Immediately after winning his independence, it became a member of the United Nations on September 20, 1960 (Sebou, 2014).

The Niger economy has begun to develop after the discovery of uranium. The average life expectancy is 58.4 years. The crude death rate was 1.3 per thousand, the child mortality rate was 63.2 per thousand, the maternal mortality rate was 590 per 100 thousands, the incidence of acute malnutrition is 16.7%, and the rate of infant feeding only breast milk is 26.9%. 50.1% of the country's population is female and 49.9% is male. Niger is made up of 8 regions and 21% of the people live in urban areas and 79% live in rural areas. The total population consists of 55.4% Hausa, 21% Djerma, 8.5% Fulani, and 9.3% Tuareg tribes. Almost all of the people are Muslims. The people of Niger speak more often their own language. Hausa and Djerma are more common than local languages. Literacy rate in Niger is low. The biggest reason why it is 58.6% is economical. Niger is very rich in terms of underground resources. It has uranium, petroleum, phosphate, coal, and zinc deposits. The uranium mine, which is refreshing the country's economy, is quite copious. But itsoperation, technique, and industry are dependent on Western countries, France, China, USA, and Japan. The majority of exports and imports are done with France. The basis of exports is uranium, which is extracted in the Arlit, Akouta, and Agadez regions in the desert region. Niger also generates a large income from meat production. The food, textile, and cement industries are available to meet the needs of the country. Fishing is another important source of income. The fact that the country with a closed land has not a suitable transport system is one of the most important disadvantages for the economy (NBS, 2012).

9.5 COMPARISON OF NORWEGIAN AND NIGER EDUCATIONAL INDICATORS

In this section, the proportions of Norwegian and Niger countries in the human development index, education expenditures, average and expected schooling rates, adult literacy rates, and school enrollment (primary, secondary, and higher education) will be compared. In Figure 9.1, Norway and Niger Human Development Index (2016) ranking is showed.

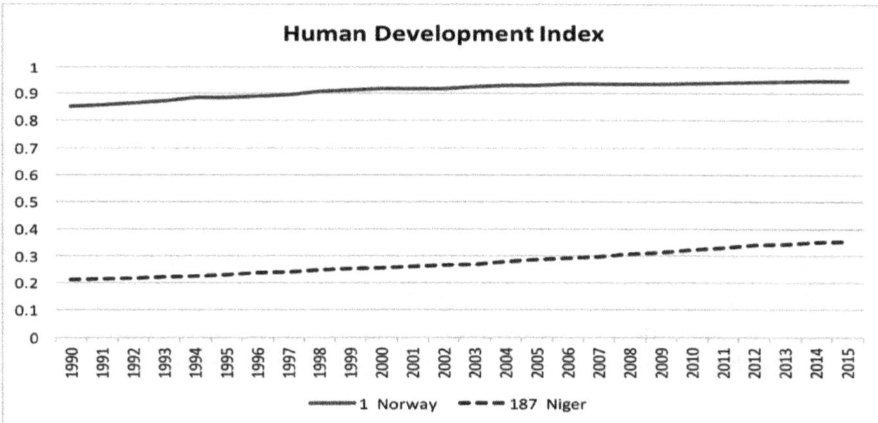

FIGURE 9.1 Norway and Niger Human Development Index rankings (2016).
Source: http://hdr.undp.org/en/data.

The Human Development Index gives a value between 0 and 1 according to the development of the countries in the ranking. As can be seen in Figure 9.1, Norway is in the highest category of human development (rank 1), showing an upward (close to 1) improvement since 1990. When we look at Niger, it is in the low human development category in the order of Human Development Index (187th place), though it shows an upward trend between 1990–2015. This is because of Niger's three basic dimensions of human development; a long and healthy life, access to knowledge, and a lack of a decent living standard. Figure 9.2 compares expenditure on education in both countries in 2016.

Government expenditure on education (% of GDP)		
	Norway	*Niger*
1990	6.3	2.9
1995	7.7	2.9
2000	6.5	3.2
2005	6.9	-
2010	6.8	3.7
2011	6.5	4.2
2012	7.4	4.3
2013	-	4.9
2014	-	6.8

FIGURE 9.2 Educated expenditure (% of GDP, 2016).
Source: http://hdr.undp.org/en/data.

Looking at Figure 9.2, it can be seen that the two countries have increased in education rates over the years. The most striking point, however, is that the rate of education in Niger's economy has increased from 2.9% to 6.8% over the past years. Increases in both Norwegian and Niger's educational expenditures are expected to positively impact educational indicators.

One of the educational indicators of the countries continues to be the school-attending period and the average school year. In Figure 9.3, Norway and Niger, one of the educational indicators, were compared with the expected years of schooling and the mean years of schooling.

Looking at this figure, there is an increasing trend in expected years of schooling and mean years of schooling data between the mentioned years. When we compare two countries, the big difference between the countries is seen. In the last year, the expected years of schooling are around 6 years, which corresponds to the level of primary education. In Norway, this expectation is over 18 years, which means that the level of education of the people in the country is at least the level of higher education. And also another striking factor is the difference between the sexes in the charts. In Niger, male individuals have higher expected years of schooling and mean years of schooling rates, while in Norway these rates are favored by women. Another educational indicator is the proportion of the population who has received at least secondary education in the country. This ratio is examined in Figure 9.4.

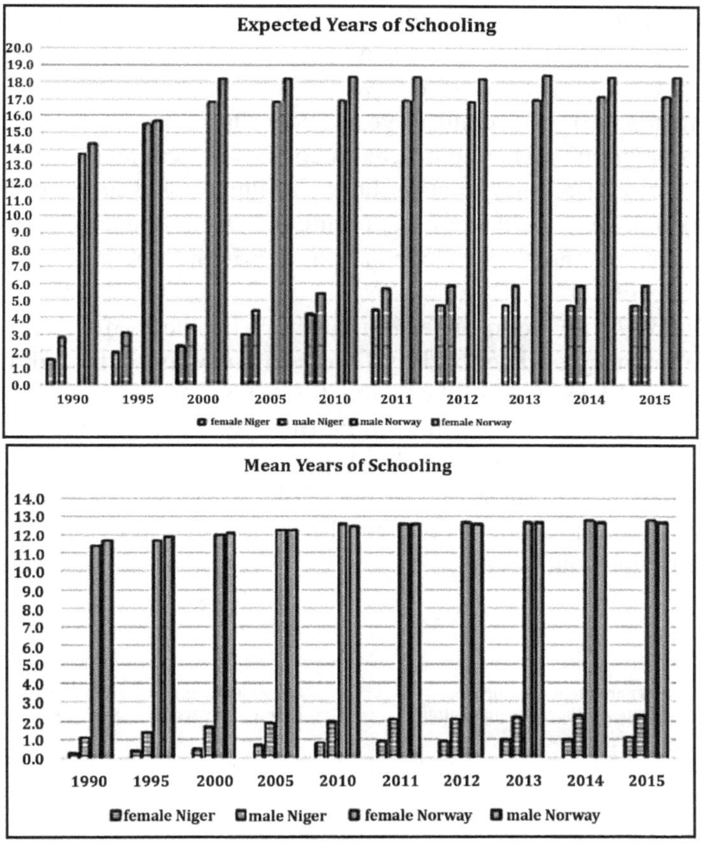

FIGURE 9.3 (See color insert.) Expected years of schooling and mean years of schooling.

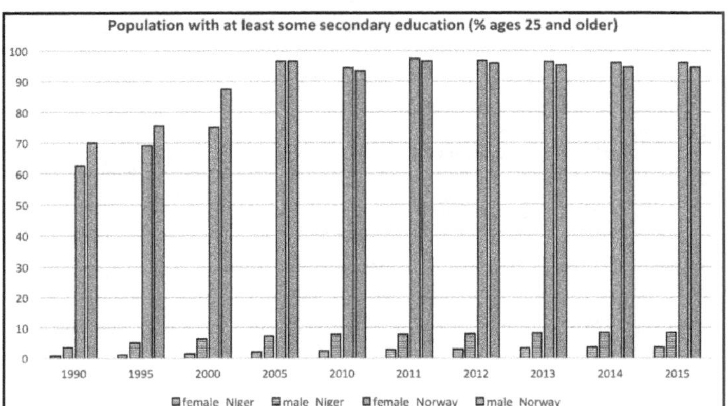

FIGURE 9.4 (See color insert.) Proportion of population who completed at least secondary level in the country (% ages 25 and older).

A large majority of the population in Niger cannot reach the secondary level, as can be predicted from the average attendance rates of the previous charts and from Figure 9.4. In Norway, it is not surprising that the average school attendance rate is about 13 years according to the 2015 year data, and that the population that has completed secondary education is close to 100%. In the face of educational levels, school enrollment rates are also one of the indicators that we should examine. Figure 9.5 shows the school enrollment rates of the Niger and Norway populations in primary, secondary, and higher education levels.

In Figure 9.5, we can observe that there is a schooling rate of over 100% in primary and secondary levels in Norway. Although this ratio has fallen to the level of higher education, it has been observed that this ratio has increased to 80% since 2005. In Niger, these rates are quite low. In Niger, the rate of primary schooling has risen from 30% to 70%. At the secondary level, this rate is around 30%. The reason for this increase in Niger school enrollment rates can be interpreted in the last decade by the incremental budget.

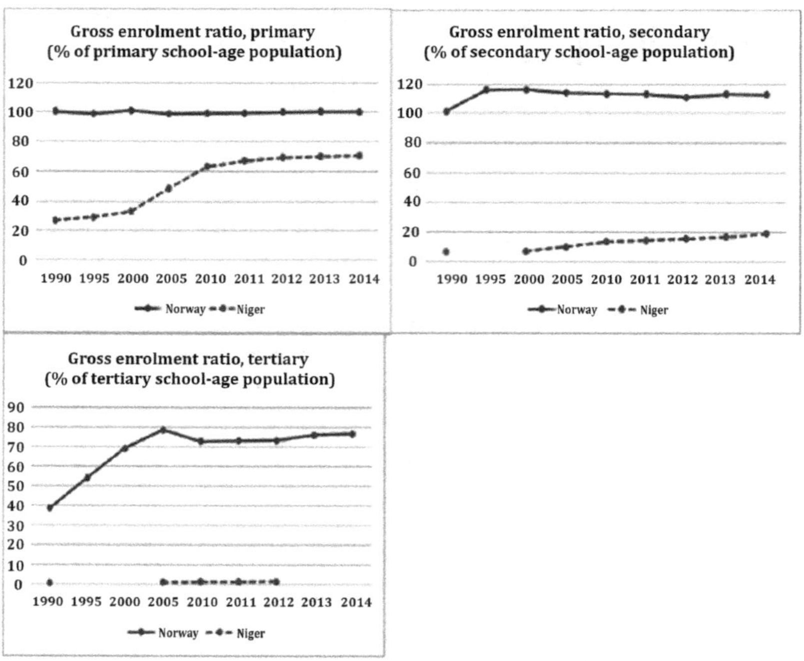

FIGURE 9.5 School enrollment rates (primary, secondary, and higher education).

9.6 COMPARISON BETWEEN NORWAY AND NIGER IN TERMS OF DEMOGRAPHIC TRANSITION

The demographic transition also emerges as another concept that affects the level of human development of the countries. Variables such as adult mortality rate, infant mortality rate, and average life expectancy will be examined and interpreted in the following figures (Figs. 9.6–9.8).

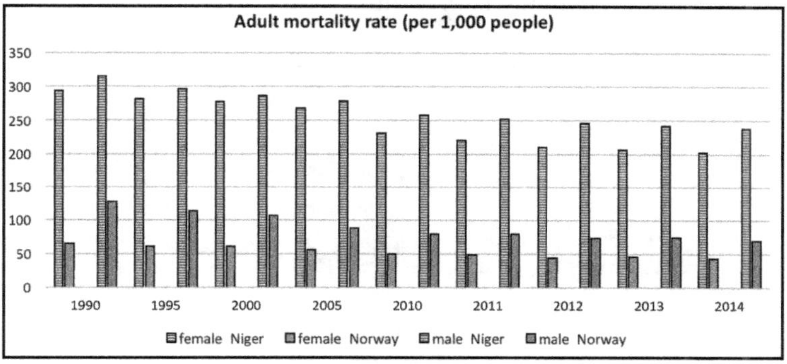

FIGURE 9.6 (See color insert.) Adult mortality rate.

Figure 9.6 shows the adult mortality rates. The adult mortality rate in Niger is about 200 of women per 1000 people and about 250 of men per 1000 people in the last couple of years, while the same rate is, respectively, around 40 and 60 in Norway. These ratios have important implications for the level of human development and development of the countries. Another important indicator, infant mortality rates, was examined in Figure 9.7.

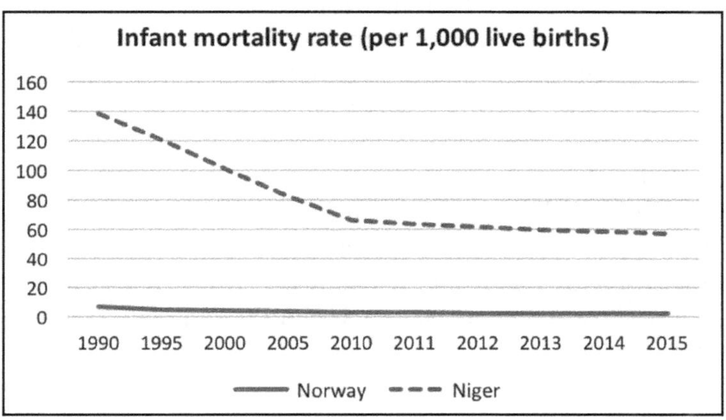

FIGURE 9.7 Infant mortality rates.

Although infant mortality rates in Niger show a rapid decline in the past few years, nearly 60 of every thousand newborn babies die each year. In Norway, this number is too small to be tried. Thehigh level of these numbers in Niger can also be related to the level of development of the country and the inadequate health conditions.

The average life expectancy is also an indicator that can relate to human development, development, and investments in health. In Figure 9.8, countries' average life expectancy is presented.

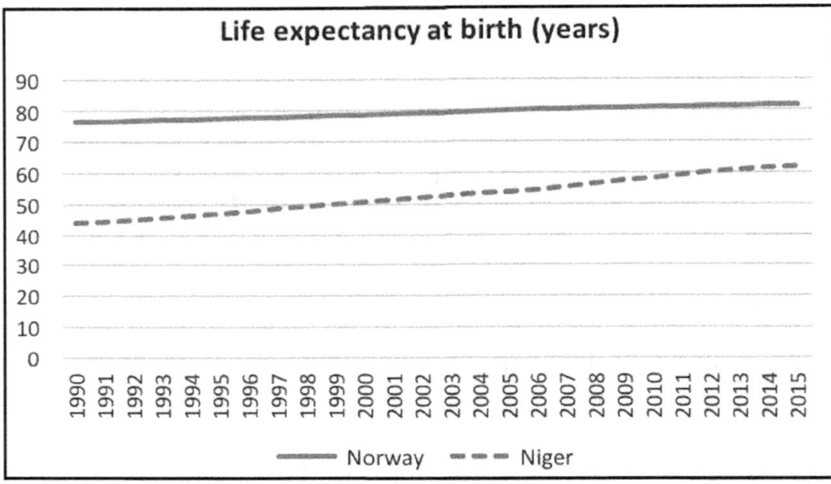

FIGURE 9.8 Life expectancy.

As it can be guessed from the level of human development of nations, the expectation of life in Norway is quite higher than Niger (63); almost 80 years. Finally, when we look at the total population growth, it is seen that the total population has not increased much between 1990 and 2015 in Norway, but it has increased about three-fold in the given year in Niger.

Figure 9.9 and other indicators, it can be said that the countries with the highest level of human development have a positive dominance in the indicators of education and demographic transition, the population growth rates are low; and that in countries with a low level of human development, these rates are negative and the population increases are more rapid.

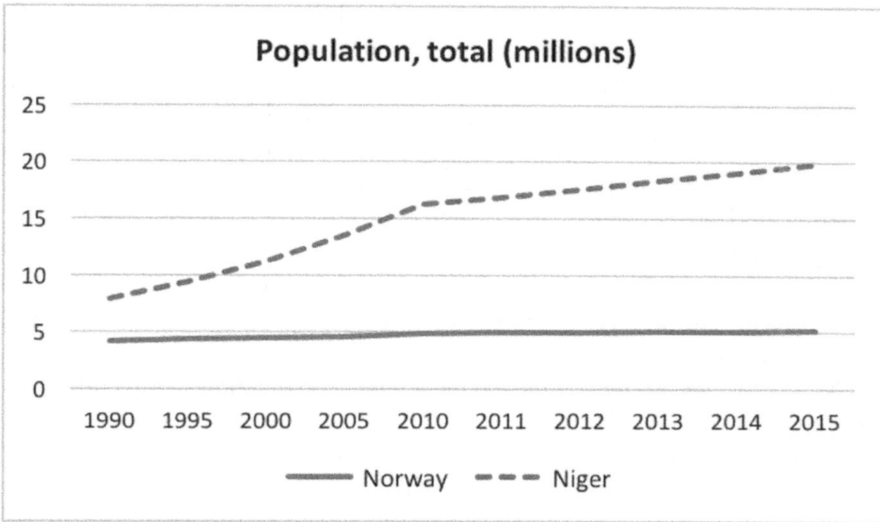

FIGURE 9.9 Population (million).

9.7 CONCLUSION

The development of a country depends upon its people who have good and continuous education, and also, it relies on the contribution they can make to economic growth by the knowledge and skills that they gain with the education. The individual who is the brains of development must be equipped with awareness, search, study, learning, and thinking. Education is obliged to give new behaviors to the individuals so that they can adapt to the changes in their environment. The ability of the education system to fulfill this obligation can only be achieved by rapidly adapting to changing information and technology. One of the tasks of the education is both to let people adapt the changes in the environment and to let them reach the capacity to create a change. The demographic structure of the country is also influenced by education. As individuals become more conscious with the level of education they receive, they will make their communities more livable for themselves and their surroundings. As the average life expectancy increases in societies composed of healthier and more aware individuals, there will be further reductions in infants birth and mortality rates. When we look at the two countries covered in the study, Norway has the highest human development, ranking first in the 2016 Human Development

Index. Norway has an advanced economy; but this alone is not enough to explain the development of the country. When it comes to human development, Norway meets criteria such as long and healthy life (average life expectancy); information access: higher average years of schooling of the adult population; that individuals (aged 25 years and older) are able to receive education for as long as they can get for a lifetime; and the length of time they are expected to attend the school. In Niger, the opposite is true: literacy rate and mean years of schooling rates are very low. This situation is an obstacle for individuals to become conscious and to live healthier. In Niger, both population growth rate and birth and death rates are high.

As a result, individuals increase their level of education while increasing their productivity, contributing to the economic, social, political, and cultural development of the community they live in. In doing so, it also affects the demographic transformation process of the country, causing changes in the population growth rate and birth and death rates of the country. We can say that a well-educated society is a more qualified life.

KEYWORDS

- **education**
- **education and development**
- **demographic transition**
- **developed countries**
- **less developed countries**

REFERENCES

Adem, M. *Our National Education Policy and Financing*; Ankara University Educational Sciences Faculty Publications: Ankara, 1993.

Altınışık, İ.; Peker, H. S. The Effect of Education on Economic Development, the Influence of Education on Economic Development. *J. Socio-tech. Soc. Tech. Res.* **2012,** *1* (4), 1–13.

Basu, A. M. Why Does not Education Lead to Lower Fertility? A Critical Review of Some of the Possibilities. *World Develop.***2002,** *30*, 1779–1790.

Başar, E. *Demographic Entry*; Özbaran Offset Printing: Ankara, 2010.

Batini, N.; Callen, T.; McKibbin; W. The Global Impact of Demographic Transition. International Monetary Fund (IMF). Working paper, No: WP/06/9, 2006.

Behar, C. Demographic Developments and Complementarities: Aging, Labor and Migration. *Turkish Stud.* **2006,** *1* (7), 17–31 (Boğaziçi University, İstanbul).

Caldwell; J. C. The Globalization of Fertility Behavior. *Popul. Develop. Rev.* **2001,** *27*, 93–115.

Eroğlu, A. Educational Resources in Turkey and Education in Economic Development Role. Unpublished Master's Thesis, Marmara University: İstanbul, 2001.

Galor, O. *The Demographic Transition: Causes Consequences*; Brown University, 2010.

Han, E.; Kaya, A. A. *Economic Development and Growth*, 1st ed.; Anadolu University Publishing: Eskişehir, 2004.

Hirschman, C. Comment: Globalization and Theories of Fertility Decline, Global Fertility Transition. In *Population and Development Review*; Bulatao, R. A., Casterline, J. B., Eds.; Supplement to Vol. 27, Population Council: New York, 2001; pp 116–125.

HÜNEE (Hacettepe University Institute of Population Studies). *Turkey's Demographic Transformation*; HÜNEE: Ankara, 2010.

https://tr.wikipedia.org/wiki/az_geli%c5%9fmi%c5%9f_%c3%bclk. (accessed May 9, 2017)

https://tr.wikipedia.org/wiki/geli%c5%9fmi%c5%9f_%c3%bclk.

https://tr.wikipedia.org/wiki/nij.

http://users.hist.umn.edu/~ruggles/hist3797/demographic%20transit ion.pdf.

http://www.nigerianstat.gov.ng/.

Kirk, D. Demographic Transition Theory. *Popul. Stud.* **1996,** *50* (3), 361–387, (Great Britain).

Lee, R. D. Reher, D. S. Introduction: The Landscape of the Demographic Transition its Aftermath. *Popul. Develop. Rev.* **2011,** *37* (Supplement), 1–7.

Özgür, E. Educational Policies and Education for Development in Turkey Return on Investment. Unpublished Graduate Thesis, Eskişehir Osmangazi University, 2005.

Özsoylu, A. F. *Informal Economy in Turkey*; Bağlam Publications: İstanbul, 1995.

Reher, D. S.; Sanz-Gimeno, A. Rethinking Historical Reproductive Change: Insights from Longitudinal Data for a Spanish Town. *Popul. Develop. Rev.* **2007,** *33*, 703–727.

Rowland, D. T. *Demographic Methods and Concepts*, Oxford University Press: New York, 2006.

Rowland, D. T. *Demographic Methods and Concepts*; Oxford University Press: Oxford, 1946.

Seybou, A. A. The Situation of Family Planning Services in Niger's Capital Niamey Factors that Affect Services. Unpublished Graduate Thesis, Trakya University, Institute of Health Sciences, 2013.

Sezer, N. On the Employment of Privatization Practices in Developing Countries: Examples from Bangladesh, Nigeria, Argentina and Turkey. Unpublished PhD Thesis, Istanbul University, Institute of Social Sciences, 2013.

Taş, U. Yenilmez, F. The Role of Education in Turkey on Development and Education Return on Investment. *Eskisehir Osmangazi Univ. Soc. Sci.* **2008,** *9* (1), 155–185.

The World Bank WDI Online Database, Dated March 21, 2017. It Was Taken from http://data.worldbank.org/indicator.

UNDP. *Human Development Report*; Oxford University: New York, 1994, 1995, 2000.

Weeks, J. R. *Population: An Introduction to Concepts and Issues*, 5th ed.; Wadsworth Publishing Company: Belmont, 1992.

United Nations Development Program (UNDP), March 21, 2017, on http://hdr.undp.org/en/statistics.

Weinstein, J.; Pillai, V. K. *Demography: The Science of Population*; Allyn and Bacon Press: Boston, 2001.

Yaukey, D.; Anderton, D. L.; Lundquist, J. H. *Demography: The Study of Human Population*, Waveland Press: Illinois, 2007.

Yumuşak, İ. G. The Theory of Human Capital and the Economic Development of the Human, Capital Role. Unpublished Doctoral Thesis, Istanbul University, 2000.

Yüksel, Y. Demographic Transition Process and Social Policies in Turkey. Expertise Thesis, State Planning Organization, 2007.

CHAPTER 10

Analyzing the Academic, Social, and Cultural Aptitudes of Syrian Students in a Chaotic Context

YENER AKMAN* and GÜLŞAH İMAMOĞLU AKMAN

Ministry of National Education, Ankara, Turkey

**Corresponding author. E-mail: yenerakman26@gmail.com*

ABSTRACT

There have been intricacies in North Africa due to various reasons in 2010. This situation has also affected Syria. As a result, increasing violence has caused that the Syrian people to emigrate to different countries. In this process, Turkey has become a country that is the most Syrian immigrants live. As of 2017, the number of people who emigrated from Syria to Turkey can be noted that reached nearly 3.5 million. Syrian immigrants in Turkey is seen as important to society to adapt in the quickest way. Effective provision of this harmony is socialized through education. The prolongation or ineffectiveness of this process can lead to cultural incompatibility, resulting in inter-communal problems and the development of a chaotic environment. In this research, the academic, social and cultural adaptations of the Syrian immigrants have been discussed through the views of a working group in the district of Altındağ. A phenomenological approach has been adopted in the research. According to the results of the research, it is determined that the academic background of Syrian immigrants is weak. It has been determined that there are problems especially at the point of learning and speaking Turkish. It has also been observed that this situation negatively affects the process of social and cultural adaptation. As a result, despite the socialization policies about the Syrian immigrants in Turkey, it can not at the desired point about the academic, social and cultural adaptation.

10.1 INTRODUCTION

Humans have been migrating to different regions of the world for various reasons since the earliest days of their existence. The place of migration can be a different country as well as another city or region within a country (Aksoy, 2012). The countries of migration are usually the neighboring countries. The causes of migration can be divided into two categories as "voluntary" and "obligatory." Mostly economic reasons underlie voluntary migration. The desire for better education, desire to have better medical and work conditions, for instance, can trigger voluntary migration. In the case of obligatory migration, on the other hand, there are various factors obliging people to migrate. The factors include chaos, natural disasters, and wars (Duruel, 2017).

A review of literature has demonstrated that migrating individuals are considered in the framework of different concepts. The individuals migrating in the context of law are referred to as *immigrants*, while those who wish to return to their country again, who do not think that they have life safety and think that they are treated unfairly are called *refugees*, and the individuals who go to another country to gain the status of refugees but whose status as refugee is not certain according to legal authorities are called *asylum seekers* (Deniz, 2014). According to Ünal (2014), societies face significant changes in consequence of migrations. Migration can be considered as a global phenomenon having changes, uncertainties, and chaos in it. The effective management of changes in a society caused by migration will help to overcome the uncertainty/unpredictability for the future with minimum harm. Uncertainty in relation to what an impact on a system will change or where it will go is considered as "Chaos Theory" in social sciences (Gleick, 2000, p 6). Unpredictability is on the basis of Chaos Theory. It is stressed in the theory that a minor action or an action which is ignored can lead to bigger and more important actions. Thus, social, economic, and cultural differentiation in societies receiving migration can have the potential to create a chaotic atmosphere. Benefits to be obtained from the differentiation by the citizens of the country of migration and citizens of the country from which there is migration are dependent on the implementation process of migration policies. In this process, the policies to support the development of social and cultural adaptation should be implemented rapidly and effectively.

Movements of migration occur in the world for various reasons. Those movements occur as asylum seeking, refuging in a country followed mostly by economic concerns, and then as moving to another place in the same country. According to "International Migration Report 2015" announced by the United Nations (UN), the number of international immigrants has reached 244 million with a great increase. Accordingly, the majority of immigrants are in European, Asian, and North American countries. Turkey, on the other hand, can be considered as one of the countries through which the greatest number of immigrants transit due to its geopolitical position (Kıratlı, 2011). Mass protests labeled as "Arab Spring" in 2010 spread into the Middle East and North Africa in a short time. The activities reaching Syria in 2011 were longer and more destructive there than in the other countries. Conflicts and civil war in Syria prevented people from meeting their essential needs and their right to live was threatened. The fact that Turkey had the longest borders with Syria and that Turkey applied "open door" policy from the very beginning of migrations caused a great number of immigrants to settle in Turkey. The process beginning with the arrival of a group of 200–300 Syrians in Turkey in April 2011 continued with the settlement of millions of Syrians in Turkey (Akşit et al., 2015; Yazgan et al., 2015). According to the General Directorate of Immigration Authority (2017), the number of Syrians under temporary protection in Turkey reached 3,303,113 in November 2017. While 227,332 of them stay in 21 temporary refuge centers established in 10 provinces, the remaining 3,075,781 Syrians has scattered in various cities of Turkey. Moreover, approximately 50% of them are children of 0–18 years old. Intercultural adaptation can be attained and developed through education. Socializing over 1,500,000 Syrian children could bring benefits to everyone in the short term as well as long term. Turkey is aware of her responsibilities for the education of Syrian children and has been developing policies for this. The educational rights of Syrian children were considered on legal basis with the regulation of "educational–instructional services for foreigners" declared in September 2014 by the Ministry of National Education (MNE). Today, educational service is offered to Syrians students inside camps, outside camps (temporary centers for education, public schools), and through private schools. However, it is observed that the great majority of those children are still out of the educational process (Disaster and Emergency Management Centre, 2013; Political, Economic and Social Research Foundation, 2016). According to Disaster and Emergency Management Centre (2013), while 83% of the students inside the camps attend education, only 14% of Syrian children all over Turkey receive continuous education. Considering the chaos in Syria continuing for over 6 years and the time necessary for repairing the destruction in the country when the chaos ends, it may be predicted that the Syrians immigrated into Turkey will stay in

Turkey for years. Therefore, it is vital to start the process of adaptation into Turkey via education for Syrian children who drifted away from their country, lost their relatives, and came to Turkey suffering from traumas. Chaotic situations can be hindered in this way, and future which is more desirable for Syrians and for Turkish society can be built. In this process, teachers' consciousness of Syrian students' academic, social, and cultural aptitudes and their differences can be regarded as an element facilitating the process of education.

Review of literature makes it clear that various studies concerning Syrian students are available (Akpınar, 2017; Gözübüyük Tamer, 2017; Kağnıcı, 2017; Özer et al., 2016; Uzun and Bütün, 2016). The studies in general stress determining the problems Syrian students encounter in the process of education and developing policies for the solution of these problems. According to the data coming from the General Directorate of Immigration Authority (GİGM, 2016, p 78), almost half of the 70,000 Syrian immigrants living in Ankara (Press Announcement Association, 2017) live in Altındağ district. Therefore, the research was conducted in Altındağ. This study aims to determine the views on Syrian students' academic, social, and cultural aptitudes held by teachers working in schools located in Altındağ district. Additionally, based on the findings, this study also investigates how the education of Syrian students could affect Altındağ district.

10.2 METHOD

10.2.1 RESEARCH MODEL

This study analyzes the views of teachers working in public schools of MNE concerning Syrian students' academic, social, and cultural properties in a descriptive approach. The study employs phenomenology design-one of qualitative research methods. Phenomenology design can be referred to as a research design "focusing on phenomena of which we are aware but about which we have no in-depth and detailed knowledge" (Yıldırım and Şimşek, 2005).

10.2.2 STUDY GROUP

The study group was composed of 13 teachers teaching in public schools located in Altındağ district of Ankara in 2017–2018 academic year. The data analyzed were collected through interviews. According to Stewart and Cash (1985), the technique of interview is "the process of mutual and interactive communication which is performed for a pre-determined purpose and which is based on asking and answering questions" (Yıldırım and Şimşek, 2005). Care was taken with balanced distribution of teachers according to different stages of schooling (such as primary school, secondary school, and high school) in establishing the study group. The research data were collected in November of 2017–2018 academic year. Demographic properties about the study group are shown in Table 10.1.

TABLE 10.1 Descriptive Properties of the Study Group.

Codes	Teaching experience	Branches	Levels of education	Types of schools
Ö1	5 years	Guidance and psychological counseling	Postgraduate	Secondary school
Ö2	10 years	Information technologies	Postgraduate	Secondary school
Ö3	12 years	Primary school mathematics	Postgraduate	Secondary school
Ö4	12 years	Physical education	Graduate	Secondary school
Ö5	7 years	Science and technologies	Graduate	Secondary school
Ö6	30 years	Elementary school teaching	Postgraduate	Elementary school
Ö7	30 years	Elementary school teaching	Graduate	Elementary school
Ö8	23 years	Elementary school teaching	Graduate	Elementary school
Ö9	21 years	Elementary school teaching	Graduate	Elementary school
Ö10	13 years	Secondary school mathematics	Postgraduate	High school
Ö11	21 years	Guidance and psychological counseling	Graduate	High school
Ö12	17 years	Physical education	Graduate	High school
Ö13	12 years	Turkish literature	Graduate	High school

According to Table 10.1, the participating teachers' teaching experience is between 5 and 30 years. In addition to that four (30.7%) of the participants teach in elementary schools, five (38.4%) of them teach in secondary schools, and four (30.7%) teach in high schools. Five teachers (38.4%) hold a postgraduate degree while eight (61.6%) hold a graduate degree. As to their branches, four of them are elementary school teachers, two are guidance and psychological counseling teachers, two are physical education teachers, one is a primary school mathematics teacher, one is a science and technologies teacher, one is a secondary school mathematics teacher, and one is a Turkish literature teacher.

10.2.3 PROCEDURES

Studies in relevant literature analyzing the process of Syrian students' educational and social adaptation were taken into consideration in preparing the interview form in this study. The questions in the form were prepared so as to find Syrian students' academic, social, and academic levels and their differences in this respect. The semistructured form prepared as a draft was first analyzed and discussed with two teachers in terms of how much it could serve to the purpose of the study. After that, the draft form was given its final shape on the basis of views stated by an academician who was an expert of the field. Prior to the interviews, teachers to be included in the research were contacted, and the interviews were done on a day convenient for the teachers. The data collected through the interviews were recorded in writing with the teachers' consent.

The research data were put to descriptive analysis. The data collected in the interviews were primarily stored in a computer. The themes were distinguished within the framework of the questions in the interview form, and the participants' responses having similar associations were given in relevant themes. Special care was taken with confidentiality of identity and the teachers were labeled in codes such as Ö1, Ö2, etc. Besides, care was also taken with detailed reporting and quoting directly from the teachers' views, which had significant effects on the validity of the research. And finally, the teachers' views were reclassified into relevant themes by a different expert so as to determine the degree of reliability of the research. The reliability was found to be 0.90 according to the formula [Reliability = Agreement/(Agreement + Disagreement) × 100] suggested by Miles and Huberman (1994). According to Yıldırım and Şimşek (2005), the value indicated that the research was reliable enough.

10.3 FINDINGS

Teachers' views on Syrian students were divided into academic, social, and cultural themes according to the findings obtained, and they are shown in Table 10.2. Table 10.2 shows teachers' views on Syrian students in the theme of academic.

TABLE 10.2 Teachers' Views on Syrian Students in the Theme of Academic.

	Themes	*f*	%
1	They have very low academic proficiency	13	26
2	Comprehensible but reading and writing are too weak	10	20
3	Majority of students attend classes	9	18
4	Families do not support the students academically	7	14
5	The majority did not receive any education in the past	6	12
6	They write down what they see rather than what they listen to	5	10

As is clear from Table 10.2, all of the participants stress that Syrian students are academically inadequate and that they could be comprehended but they have weak reading and writing skills. Accordingly, the participant coded as Ö1 stated his/her view as "... they have very low proficiency, most of them can understand when they hear something in Turkish, they can say something in Turkish; but no writing. Some of them change the place of letters when they write...." Another participant coded as Ö2 said "... most of them came here when they were little children. Therefore, they can speak Turkish but they have problems in writing...." In support of this view, participant Ö3 said "... they have problems in perception. They can understand but there are problems in reading and writing...." The participant Ö10 emphasized problems in writing and said "... the skill in which they are the best is understanding, which is followed by speaking; and the skill in which they are the weakest is writing because of the alphabet we use...."An elementary school teacher, Ö6, on the other hand, said "they were registered to the fourth grade by considering their age but they do not recognize the letters of the alphabet..." and thus called our attention to those students' academic inadequacy. The fact that the majority of those students were not exposed to education before can be one of the causes of their academic incompetence. In this context, the participant Ö1 said "... the students in our school are generally the students who came from the underdeveloped regions of Syria. Almost all of the 5th graders had never gone to school before" while the participant Ö5 said "... due to their age, most of the students start secondary school without going to elementary school..." and Ö9 said "...their readiness is weak...." In addition to that, Ö5 also said

"... they are weak in writing by understanding. They can write by looking..." Ö8 said "... they write what they see, just like drawing a picture but they do not understand...." In terms of family support, Ö2 said "... they do not have regular life and I don't think that they receive family support..." and Ö11 said "... their families' level of education is low, they have to study hard...." Differing views were found in terms of students' attendance in classes. Ö1 said "... 25 out of 31 students in our school regularly attend classes but the others do not have regular attendance...," Ö7 said "... our students are very young and therefore their parents bring them to school...," whereas Ö5 said "...they sometimes come to school, they have attendance problems. Some of the students did not attend classes more than seven days...." A high school teacher, Ö10 said "... students with attendance problems are dismissed from school after 20 day-absenteeism; because they are paid money for education...." And Ö12 said "they receive financial help from the government as they attend school. There is no attendance problem...."

Table 10.3 shows teachers' views on students in the theme of social.

TABLE 10.3 Teachers' Views on Students in the Theme of Social.

	Themes	*f*	%
1	Mostly Syrian students are together	13	23.6
2	Syrian students speak in Arabic with each other	12	21.8
3	Very few of Syrian students socialize with Turkish students	7	12.7
4	Teachers overlook Syrian students and do not care about them enough	7	12.7
5	Most of Syrian students are introvert and not sociable	6	10.9
6	There are very few fights Syrian students are involved in	5	9
7	Syrian students do not get along with each other and with Turkish students	5	9

According to Table 10.3, it may be said that Syrian students are usually together with Turkish students but they do not have much communication with Turkish students. Ö1 said "... some of them are lonely, some talk only to Syrian students, very few of them have socialised with Turks..." in this respect. Ö2, in support of this, said "...Syrian children usually socialise with each other...." Another participant, Ö4 said "... they have formed their own groups. They play together in a certain part of the school garden. They do not go out of their own group"; Ö11, on the other hand said "...what can they do? They are in search of support from children who are just like them. They have seen a war, they have suffered from problems, and this is worrying...." The teachers also thought that Syrian students were introvert, quiet, they kept to themselves, and that they were not involved in fights. These findings were found to be consistent with the statement made by Ö5 "... there may be language problems, Syrians are introvert...," with the statement made by Ö3 as "...there are no fights with Turkish students, nobody interferes with each other...," and with the statement made by Ö1 as "... incidents of fights between Turkish and Syrian students are very rare ...." However, the statement made by Ö12 as "Syrian students occasionally have fights both with each other and with Turkish students in groups..." reflects a different view. Besides, students communicate with each other mostly in Arabic; as is clear from the statements made by Ö6 as "... they have difficulty in communicating in Turkish and they keep away from others and therefore this causes a vicious circle" and by Ö7 as "... Turkish is not spoken in their homes and this situation makes it difficult for them to learn Turkish...." On analyzing teachers' approaches toward Syrian students, it was found that Syrian students were usually ignored. This was a finding supported by the statements made by Ö1 as "... most of the teachers just ignore...," by Ö2 as "... most of the teachers do sufficiently not take care of those students...," by Ö3 as "... they are ignored. They escape attention during classes...," and by Ö9 as "it seems as if Syrian students experience what inclusive students experience...."

Table 10.4 shows teachers' thoughts on Syrian students in the theme of cultural.

TABLE 10.4 Teachers' Views on Syrian Students in the Theme of Cultural.

	Themes	*f*	%
1	There are no cultural differences in clothing	13	34.2
2	They try to speak in Turkish if they are in a difficult situation	10	26.3
3	Their patterns of behavior have not been established yet	8	21
4	They try to conform to the traditions and customs of society	7	18.4

As is evident from Table 10.4, the majority of teachers pointed out that those students did not have the effects of Arabic culture in terms of clothing and that they wore clothes similar to those of Turkish students. Besides, the teachers also observed that those students made efforts to adapt into the traditions and customs of Turkish society. It was found that the findings were supported by the views stated by Ö1 as "...there are no problems in terms of clothing...," by Ö2 as "... there are no distinctive differences, they try to adjust to the society...," and by Ö12 as "the clothes they are given as charity and they themselves buy are the same as ours, maybe the only difference is in cleanliness...."Another participant, Ö1 explained his/her view as "... there are no problems in clothing...." and Ö10 said "their sense of decency is different from ours, they are not familiar with the concept of 'private sphere'...." Ö2 said "Syrian students communicate in Turkish when they are in a difficult situation. They speak in Arabic between them but I do not think they understand each other...." In the same way, Ö1 said "they tend not to speak in Turkish...." Another participant, Ö6 said "...they move to the side of the window and search for the source of the noise when there is a noise out of the school. All the students obey classroom rules but they do not...," and Ö8 said "... lawlessness is dominant in their behaviours but I don't believe it is intentional, maybe it is instinctive..." and thus complained that those students' behaviors were not established yet.

10.4 DISCUSSION

This study analyzed the views of teachers working in schools located in Altındağ district of Ankara which Syrian students attend concerning those students' academic, social, and cultural competence and their adaptation. What social, economic, and cultural differences would be caused by demographic changes to arise in the future in Altındağ district was analyzed in the light of those views. The views held by the teachers were classified into three themes according to the findings.

The first theme distinguished was academic theme. The thought held in general in this theme was that Syrian students had low levels of academic proficiency. The other findings obtained in this study signaled the factors shedding light on academic incompetence. It was stated that the Syrian students migrating to Altındağ had not received adequate formal education not only in Altındağ but also in their country. In this respect, a participant stressed that those students could not communicate well with each other and that one of the reasons for this was the fact that they could not learn their native language, Arabic, accurately. That is to say, this situation makes us think that they have problems in the grammar and vocabulary of their native language. The fact that they did not have good command of their native language can make it difficult for them to learn another language and can signal a chaotic process. The problems encountered included considerable lack in especially reading and writing skills due to being registered to a higher grade level because of their age without receiving adequate education in elementary and secondary schools in their country or in Turkey. While this situation makes it difficult for those students to follow lessons, it also affects their motivation to learn in a negative way and will reduce their academic achievement. We can refer to education as an important factor raising the social status of disadvantaged individuals. Besides, the educational proficiency of immigrants will influence possibilities for employment in the future and thus it will also affect their economic position in positive or negative ways (Papademetriou, 2005). Stark and Bloom (1985) also point out that immigrants improve their economic status through education and that economic development also facilitates the process of adaptation into society. Financial difficulties which the families of those students have—the primary problem for many of them in Altındağ—makes it difficult to monitor those students' academic achievement or to provide them with educational support in their home. In addition to that, the low level of those parents' educational status plays negative roles in helping their children with their education. Apart from that, language problems those families encounter can also cause them to keep away from school environment (Gözübüyük Tamer, 2017). In support of this, according to Lee (2011), the educational levels of immigrant families directly influence the importance attached to education. Considering those drawbacks as a whole, it is not thought that it will be easy for Syrian students to attain academically adequate levels in the short term.

The second theme in this study was labeled as social theme. On analyzing the teachers' views, it was found that Syrian students mostly acted jointly with their groups of friends. This situation can be explained as in the following: the process of adaptation into a different country by overcoming the difficulties of a civil war can pose problems. Being together with other Syrian students having a similar cultural background can enable those students to reach social support in this process in a country whose language they are not familiar with. The fact that those students spoke in Arabic when they were together—which was stated by the participants—was supportive of this. However, it may also be said that this situation will have negative effects on the process of coalescence with Turkish students. This negativity may cause harm to communication and thus it may also cause harm to the development of social adaptation (Bosch-Supan, 1987). It may also lead to experiencing undesirable chaotic events. News items appeared in printed media in relation to the tension and fights occurring in this respect (Hürriyet, 2014; Milliyet, 2017). Attaining social adaptation will also enable the development of tolerance between diversities within society (linguistic, religious, racial, and gender diversities). It may be thought that chaos can arise in a society where there is no tolerance for differences.

Effective communication can be mentioned as one of the most important components of controlling and managing chaos. Yet, educational process plays an important role in the development of communication (De Jong, 2000). On the other hand, it was observed that Syrian students—who had additional traumas—refrained from communicating with other Syrian students as well as with Turkish students and tended to be quiet, isolated, and introvert. Teachers can be considered as the most important elements to remove this negativity in a short time. However, an analysis of teachers' views demonstrated that teachers mostly ignored Syrian children. This situation can result in negative interaction between teachers and Syrian students. According to Giani (2006), negative teacher–student relations are influential in immigrant students' dropping out or in their absenteeism from school. Besides, the reasons underlying teachers' inadequate care for those students include teachers' belief that it would be a waste of time to involve students having undeveloped linguistic abilities in classes. In addition to that, teachers' lack of knowledge about multicultural education and about educational approaches toward immigrants can also trigger ignoring such students. These are the findings which are also supported by the ones obtained by Ereş (2014). The author highlights the fact that students coming from immigrant families are academically inadequate, that there are no consistencies between their knowledge and grade levels, and that they have adaptation problems. Er and Bayındır (2015) and Sakız (2016) stress the fact that the majority of teachers are not knowledgeable about how to educate immigrant students. The views included in the social theme point to the fact that the process of Syrian students' adaptation into Turkey is still continuing in difficulty.

The final theme in this study was labeled as cultural. In this theme, it was emphasized that Syrian students adopted Turkish students' style of clothing rather than the one specific to their culture. It was stated that Syrian students made efforts to display behaviors compatible with Turkish traditions and customs since they were in the process of adapting into a different society. Yet, it was pointed out by some of the participants that Syrian students' patterns of behavior had not been established fully yet and that the approaches of their own society were reflected into their behaviors. Emphasis was laid to the fact that Syrian students usually violated classroom rules and that they responded naturally to the events occurring outside the classroom (such as standing up, going to the window, and looking out of the window when they hear the sound of a horn). In this process, adapting socially without problems and in a short time is due to mutual interaction between the two communities. The interaction can be developed effectively through channels of communication. Forming groups of only Syrian students and having weak interactions with Turkish students are the factors making the process of adaptation difficult for Syrian students. They were observed to communicate in Turkish only when they were in a difficult situation or when they were the only Syrian student in the classroom. This was a finding compatible with the one obtained by Gözübüyük Tamer (2017). The study mentioned suggests that it was difficult for Syrian students to adjust to Turkish students and to benefit from classes without solving the language problem. Some other studies supported the finding that language problem was one of the most fundamental problems (Deniz et al., 2016; Karasu, 2016). It was found that communication in Turkish was at the point of termination when a second Syrian student came to the classroom and that they switched to Arabic. Negativities mentioned make us think that Syrian students have difficulty in the process of education in Turkey. According to the European Union (EU) reports (2012), it was found that immigrant students in Europe also had those difficulties. The process of learning another language and of social and cultural adaptation determines the effectiveness of education in Europe.

10.5 CONCLUSION AND RECOMMENDATIONS

It may be supposed that the interaction and harmony between Syrian immigrants and Turkish citizens in Altındağ district of Ankara—where the biggest number of Syrian immigrants live in Ankara—would play roles in the economic, social, and cultural differentiation of the region. Especially the fact that a large portion of this mass is below 18 years old indicates the importance of education. It can turn into an advantage for Turkey to implement the right migrant policies in this process. Wrong policies, on the other hand, can cause a hornet's nest and can bring Turkey to a chaos. It should be emphasized that the only way to overcome the chaos with minimum harm is by means of education. While education, as an element of human capital, makes it easier for individuals to get a job and assures that they work on better economic conditions, it is also an important political instrument in reducing unemployment and poverty (Çalışkan et al., 2013). It may be stated that education of higher quality and higher level would bring more earning and more comfortable life. It is also reported in various news sites that the amount of money spent for immigrants in such areas as education, accommodation, health, security, and public order since 2011 has climbed up to 100 billion Turkish Liras (Milliyet, 2017; Sözcü, 2017). The fact that the above mentioned amount of money spent creates added value in human capital will cause a reduction in expenditures in the future or transferring the expenditures into other areas. However, the fact that the Syrian families and their children living in Turkey have low levels of education, that the majority of them work in jobs requiring no qualification and having no social security to survive, and that almost half of those children at school age do not go to school (Gür, 2016) does not make us think positively for the future. This situation still causes Syrian families to face continued financial problems and it will also influence Turkey's need for qualified employees to support economic growth. Besides, birth rates are also remarkable in analyzing the effects of population change on economic growth. Accordingly, the fact that Syrian families have high rates of birth can be considered as another element along with imbalance of income level which will make economic growth more difficult (Erdoğan, 2017).

The findings indicate that Syrian students were largely incompetent in academic, social, and cultural aspects. Language problem stands in front of us as a problem for both families and students to overcome primarily. While lack of command in Turkish language makes it difficult for Syrian students to follow lessons in schools, it also poses difficulties in sharing with other students. This situation can result in getting away from schools and failure to benefit from education—through which socializing is the easiest. However, education can be considered as the only element to assure the development of socialization and cultural adaptation of Syrian students—who have low income levels—and to open a new door to the future. Improvement can be made through education and thus chaos, both in their personal realm and in their social adaptation, can be overcome with minimum harm and an equalitarian balance of life full of wealth can be achieved.

Based on the results obtained, the following recommendations could be made:

1. Syrian students' proficiency in Turkish can be determined and their levels of Turkish proficiency can be raised in private classrooms or in schools opened specially for this purpose.
2. It should be made sure that Syrian students are educated in grade levels appropriate to their academic aptitudes.
3. Their school attendance should be checked.
4. In-school and out-of-the-school activities should be organized so that Syrian students can socialize.
5. Policies for the employment of Syrians should be implemented.
6. Prospective teachers, practicing teachers, and school directors should be offered training in multicultural education and in managing the differences.

KEYWORDS

- **Syrian students**
- **aptitude**
- **teacher**
- **chaos**
- **education**

REFERENCES

Afet ve Acil Durum Yönetimi Başkanlığı (AFAD). *Türkiye'deki Suriyeli Sığınmacılar, 2013 Saha Araştırması Sonuçları.* https://www.afad.gov.tr/upload/Node/3925/xfiles/syrian-refugees-in-turkey-2013_baski_30_12_2013_tr.pdf (accessed Nov 10, 2017).

Akpınar, T. Türkiye'deki Suriyeli Mülteci Cocukların ve Kadınların Sosyal Politika. Bağlamında Yaşadıkları Sorunlar. *Balkan ve Yakın Doğu Sosyal Bilimler Dergisi* **2017,** *3* (3), 16–29.

Akşit, G.; Bozok, M.; Bozok, N. Zorunlu göç, Sorunlu Karşılaşmalar: Hisar Köyü, Nevşehir'deki Suriyeli Göçmenler Örneği. *Maltepe Üniversitesi Fen-Edebiyat Fakültesi Dergisi* **2015,** *1* (2), 92–116.

Aksoy, Z. Uluslararası göç ve Kültürlerarası Iletişim. *J. Int. Soc. Res.* **2012,** *5* (20), 292–313.

Avrupa Birliği (AB). *Migration and Education*; Conference Proceeding Book, Larnaca, October 15–17, 2012. http://www.eunec.eu/sites/www.eunec.eu/files/attachment/files/report.pdf (accessed Nov 15, 2017).

Basın İlan Kurumu. *130 Bin Misafirimiz var*; 2017. http://www.bik.gov.tr/130-bin-misafirimiz-var/ (accessed Dec 1, 2017).

Birleşmiş Milletler (BM). *International Migration Report 2015: Highlights Key Facts*; 2016. http://www.un.org/en/development/desa/population/migration/publications/migrationreport/docs/MigrationReport2015_Highlights.pdf (accessed Nov 12, 2017).

Bosch-Supan, A. *The Role of Education: Mobility Increasing or Mobility Impeding?* NBER Working Paper No: 2329, 1987.

Çalışkan, Ş.; Karabacak, M.; ve Meçik, O. Türkiye'de Eğitim-ekonomik Büyüme Ilişkisi: 1923–2011 (Kantitatif bir Yaklaşım). *Yönetim Bilimleri Dergisi* **2013,** *11* (21), 29–48.

De Jong, G. F. Expectations, Gender, and Norms in Migration Decision-making. *Popul. Stud.* **2000,** *54*, 307–319.

Deniz, A. Ç.; Hülür, A. B.; Ekinci, Y. Göç, Strateji ve Taktik: Suriyeli Sığınmacıların Gündelik Hayat Deneyimleri. *Uluslararası Sosyal Araştırmalar Dergisi* **2016,** *9* (42), 1077–1087.

Deniz, T. Uluslararası Göç Sorunu Perspektifinde Türkiye. *TSA Dergisi* **2014,** *18* (1), 175–204.

Duruel, M. Suriyeli Sığınmacıların Türk Emek Piyasasına Etkileri Fırsatlar ve Tehditler. *Uluslararası Ekonomik Araştırmalar Dergisi* **2017,** *3* (2), 207–222.

Emin, M. N. *Türkiye'deki Suriyeli Çocukların Eğitimi: Temel Eğitim Politikaları*. SETA, 2016; pp 153. http://file.setav.org/Files/Pdf/20160309195808_turkiyedeki-suriyeli-cocuklarin-egitimi-pdf.pdf (accessed Nov 10, 2017).

Er, A. R.; Bayındır, N. İlkokula Giden Mülteci Çocuklara Yönelik Sınıf Öğretmenlerinin Pedagojik Yaklaşımları. *Uluslararası Sosyal ve Eğitim Bilimleri Dergisi* **2015,** *2* (4), 175–185.

Erdoğan, M. *Türkiye'deki Suriyeliler*; 2017. https://mmuraterdogan.com/2016/06/06/raporlar/ (accessed Dec 3, 2017).

Ereş, F. *Göçmen Öğrencilere Eğitim Veren Öğretmenlerin Sorunları: Tokat Örneği*. Ulusal Eğitim Bilimleri Kurultayı, Kocaeli Üniversitesi, 2014.

Giani, L. *Migration and Education: Child Migrants in Bangladesh*; *2006*. https://www.sussex.ac.uk/webteam/gateway/file.php?name=mwp33.pdf&site=252 (accessed Nov 11, 2017).

Gleick, J. *Kaos: Yeni bir Bilim Teorisi* (Çev. F. Üçcan); Tübitak Popüler Bilim Kitapları: İstanbul, 2000.

Göç İdaresi Genel Müdürlüğü. *2016 Türkiye Göç Raporu*; 2016. http://www.goc.gov.tr/files/files/2016_goc_raporu_.pdf (accessed Nov 10, 2017).

Göç İdaresi Genel Müdürlüğü. *Göç Istatistikleri, Geçici Koruma*; 2017. http://www.goc.gov.tr/icerik6/gecici-koruma_363_378_4713_icerik (accessed Nov 11, 2017).

Gözübüyük Tamer, M. Geçici Koruma Kapsamındaki Suriyeli Çocuklarin Trabzon Devlet Okullarındaki Durumu. *Göç Dergisi* **2017,** *4* (1), 119–152.

Gür, A. *Suriyelilerin Entegrasyonu*; 2016. http://www.milliyet.com.tr/suriyelilerin-entegrasyonu-siyaset-ydetay-2363971 (accessed Dec 2, 2017).

Hürriyet. *Ankara'da Suriyeliler ve Mahalleli Arasında Gerginlik*; 2014. http://www.hurriyet.com.tr/ankarada-suriyeliler-ile-mahalleli-arasinda-gerginlik-26375256 (accessed Nov 17, 2017).

Kağnıcı, D. Y. Suriyeli Mülteci Çocukların Kültürel Uyum Sürecinde Okul Psikolojik Danışmanlarına Düşen rol ve Sorumluluklar. İlköğretim Online **2017,** *16* (4), 1768–1776.

Karasu, M. A. Şanlıurfa'da Yaşayan Suriyeli Sığınmacıların Kentle Uyum Sorunu. *Süleyman Demirel Üniversitesi İktisadi ve İdari Bilimler Fakültesi Dergisi* **2016,** *21*, 995–1014.

Kıratlı, T. *Mülteci ve Sığınmacıların Toplumsal Görünümü (Türkiye Örneği)*. Yayımlanmamış Yüksek Lisans Tezi. Süleyman Demirel Üniversitesi Sosyal Bilimler Enstitüsü, Isparta, 2011.

Lee, G. Understanding the Process of Educational Assimilation for Refugee and Non-refugee Immigrant Students: A Pilot Study of a Community College. Master Thesis, Graduate School of Vanderbilt University, 2011.

Miles, M. B.; Huberman, A. M. *Qualitative Data Analysis*, 2nd ed.; Sage Publications: Thousand Oaks, CA, 1994.

Milli Eğitim Bakanlığı (MEB). *Yabancılara Yönelik Eğitim-öğretim Hizmetleri Genelgesi*; 2014. http://www.ilkergedik.net/uploads/dokuman/2014-21%20say__1__%20genelge.pdf (accessed Nov 12, 2017).

Milliyet. *Akdağ'dan Kılıçdaroğlu'na "Suriyelilere 30 milyar dolarlık yardım" Yanıtı*; 2017. http://www.milliyet.com.tr/akdag-dan-kilicdaroglu-na-siyaset-2563525/ (accessed Dec 3, 2017).

Milliyet. *Başkent'te Suriyeli Kavgası*, 2017. http://www.milliyet.com.tr/baskent-te-suriyeli-kavgasi-ankara-yerelhaber-2163555/ (accessed Nov 17, 2017).

Özer, Y. Y.; Komşuoğlu, A.; Ateşok, Z. Ö. Türkiye'deki Suriyeli Çocukların Eğitimi: Sorunlar ve Çözüm Önerileri. *Akademik Sosyal Araştırmalar Dergisi* **2016,** *4* (37), 76–110.

Papademetriou, D. In *Education and Immigrant Integration in the United States and Canada*, Proceedings of Conference. The Migration Policy Institute, 2005.

Sakız, H. Göçmen Çocuklar ve okul Kültürleri: Bir Bütünleştirme Önerisi. *Göç Dergisi* **2016,** *3* (1), 65–81.

Sözcü. *Suriyelilere Giden Para 30 Milyar Dolara Çıktı*; 2017. http://www.sozcu.com.tr/2017/gundem/suriyelilere-giden-para-34-milyar-dolara-cikti-2121423/ (accessed Dec 3, 2017).

Stark, O., Bloom, D. The New Economics of Labor Migration. *Am. Econ. Rev.* **1985,** *75*, 173–178.

Ünal, S. Türkiye'nin Beklenmedik Konukları: «Öteki» Bağlamında Yabancı Göçmen ve Mülteci Deneyimi. *Z. Welt Türk. (J. World Turks)* **2014,** *6* (3), 65–89.

Uzun, E. M.; Bütün, E. Okul Öncesi Eğitim Kurumlarındaki Suriyeli Sığınmacı Çocukların Karşılaştıkları Sorunlar Hakkında Öğretmen Görüşleri. *Uluslararası Erken Çocukluk Eğitimi Çalışmaları Dergisi* **2016,** *1* (1), 72–83.

Yazgan, P.; Eroğlu Utku, D.; Sirkeci, I. Syrian Crisis and Migration. *Migr. Lett.* **2015,** *12* (3), 181–192.

Yıldırım, A.; Şimşek, H. *Sosyal Bilimlerde nitel Araştırma Yöntemleri*; Seçkin Yayıncılık, 2005.

CHAPTER 11

Women in Turkey: Labor Force and Social Life

ŞEFİKA ŞULE ERÇETİN[1] and LEYLA YILMAZ FINDIK[2]

[1]*Department of Educational Sciences, Faculty of Education, Division of Educational Administration, Hacettepe University, Ankara, Turkey*

[2]*Student Affairs Department, Hacettepe University Rectorate, Ankara, Turkey*

[*]*Corresponding author. E-mail: sefikasule@gmail.com*

ABSTRACT

The demographic transition in women has a positive role in enhancing women's position in the society. The consequences of the demographic transition for women resulted in more women in the labor force and more educated women in society. However, Turkey still has a long way to go in approaching the level of most developed countries. All the statistics reveal that Turkey still seems as a male-dominant society. Analysis related to women in labor force and social life is essential indicators for the development of integrated strategies to promote nationwide fundamental principles. Turkey needs to set policies and strategy and then monitor the progress toward goals and to adjust policies where needed. In this chapter, we examine labor force and social life of women in Turkey.

11.1 INTRODUCTION

The demographic transition model has emerged as a classification of populations differentiated by different fertility and mortality combinations (Kırk, 1996). Secular shift in fertility and mortality from high to low refers to demographic transition. The spread of demographic government, the industrial revolution, the increase in urbanization, and the increase in educational levels of human populations are evaluated as most important changes which affects human society in the last century (Lee and Reher, 2011). All countries experience a change in population because of the demographic transition. The demographic transition varies between regions and countries. Because of the divergent trends worldwide in population, growth, and aging, demographic transition are intertwined with science, technology, economy, cultural change, social and political processes (Willekens, 2014). Poverty, economic growth, health, fragility, and human development have all affected by demographic change (Canning, Raja, and Yazbeck, 2015).

Contrary to demographic transition referring long-lasting and accompanying by weak growth rates in most developed countries, the transition from high mortality and high fertility rates too low mortality and low fertility levels in developing countries has just started in the second half of the 20th century. Demographic transition affects the demography of societies "*from many children and few elderly to few children and many elderly; from short life to long; from life-long demands on women to raise young children to the concentration of these demands in a small part of a adulthood; from horizontally rich kin networks to vertically rich ones*" (Lee and Reher, 2011, pp.1). The entire process of transition has generated four important changes in the society such as reduction in fertility leads to changes in population age structure, increasing life expectancy, a dramatic liberation in the time spent by mothers in bearing and rearing children, return to investments in human capital and expands older age years (Lee and Reher, 2011).

In this demographic transition process, education received a lot of attention as education is improved to increase the likelihood and pace of the transition because of its impact on fertility and mortality rates. Education and women participation in education is an important driver as it plays a particular role in demographic transition progress in developing countries (Goujon, 2003).

11.2 WOMEN IN DEMOGRAPHIC TRANSITION

11.2.1 POPULATION BY AGE GROUP AND SEX

Age and gender structure of the population gives many clues about the demographic characteristics of a country. Age structural dynamics are assessed for the development planning. Age structure is used for demographic analyses and these analyses provide long-term detail information to shape the needs of present and future population (Riyaza, 2000). Men outnumber women worldwide however women outnumber men in older age groups (United Nations, 2015).

Female population constituted 49.8%, whereas male population constituted 50.2% of the total population in Turkey in 2016. The graph indicates that females live longer than males in Turkey. This expansive age pyramids for Turkey indicate that Turkey has a young and growing population. Turkey has a larger percentage of the population in the younger and working age cohorts. This type of populations is typically representative of the developing nation (Fig. 11.1).

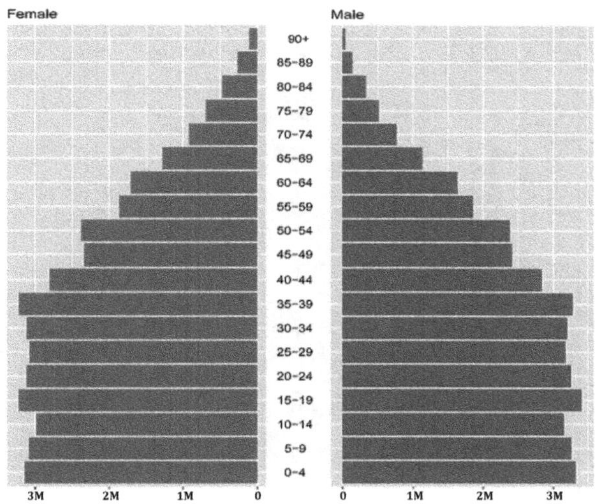

FIGURE 11.1 (See color insert.) Population age pyramid for Turkey, 2016.
Source: TurkStat, Address Based Population Registration System (ABPRS), 2015, 2016.

11.2.2 HIGHEST LEVEL OF EDUCATION COMPLETED IN TURKEY

Most nations have witnessed remarkable progress in participation in education since 20 years. The gender gap has narrowed in most countries; however, some developing countries could not reach gender parity to the detriment of women and gender disparities are even broader at the tertiary level. Female participation in higher education overall has increased globally and at present surpasses male participation in almost all developed countries and in half of developing countries. Statistics indicates that women are

TABLE 11.1 Highest Level of Education Completed in Turkey since 1975.

	Illiterate		Literate without diploma		Primary school		Primary education		Junior high school and equivalent		High school and equivalent		Higher education		Master/ doctorate		Unknown[a]	
Years	**M**	**F**	**M**	**F**	**M**	**F**	**M**	**F**	**M**	**F**	**M**	**F**	**M**	**F**	**M**	**F**	**M**	**F**
1975	29.1	65.6	12.6	5.1	45.2	23.9	–	–	4.5	2.1	5.2	2.5	3.0	0.7	–	–	0.3	0.1
1980	24.8	62.4	10.7	5.6	46.7	24.7	–	–	5.4	2.2	6.6	3.5	5.7	1.6	–	–	0.1	0.0
1985	16.3	45.9	9.2	8.4	53.0	36.1	–	–	6.4	2.8	9.0	4.9	6.0	1.8	–	–	0.1	0.1
1990	13.2	40.2	5.7	5.6	54.9	41.6	–	–	7.9	3.6	10.8	6.1	7.3	2.8	–	–	0.1	0.1
2000	7.0	27.4	5.3	7.5	50.3	45.2	–	–	11.2	5.3	16.0	9.1	10.2	5.4	–	–	0.0	0.1
2010	3.1	13.9	3.9	7.1	33.1	40.1	10.6	6.1	9.3	5.5	21.3	13.8	11.9	7.7	1.4	0.9	5.4	4.9
2015	1.8	9.0	3.5	8.8	27.2	35.9	13.6	9.2	9.5	5.9	23.5	15.6	17.9	13.1	2.0	1.4	1.1	1.1

[a]Unknown includes unknown for literacy and education level.

Source: TurkStat, National Education Statistics Database, 2008–2015.

still underrepresented in the more advanced degree programs in worldwide (United Nations, 2015).

In 1975, illiteracy rate for male and female were 29.1% and 65.6%, respectively. The female illiteracy rate is the highest in all categories and in all periods. This illiteracy rate had a tendency of decreasing through the passing years and decreased 1.8% for male and 9.0% for female in 2015 (Table 11.1).

The proportion of male graduated from higher education was 3.0% and the proportion of higher educated female was 0.7% in 1975. This proportion of the higher educated population has increased for 30 years and in 2015, the proportion of male graduated from higher education is 17.9% and the proportion of female having higher education is 13.1%.

Turkey has no male or female population who graduated from master or doctorate programs till the 2000s. After 2000s, Turkey has a population graduated from master or doctorate programs. About 1.4% male population and 1.1% female population have master or doctorate education.

All level of education completed has gradually increased by the passing years; however, these proportion of education completed has not been enough when compared with developing countries (ILO, 2016b).

11.3 WOMEN IN LABOR FORCE IN TURKEY

The labor force participation rate is defined as a measure of the promotion of a country's working-age population that engages actively in the labor market by working or looking for work. The proportion of labor force by sex and age group gives a profile of the distribution of the labor force within a country (ILO, 2016b). Women in the labor force is an important driver for growth and development and female participation rates present the potential of a country to grow more rapidly (Verik, 2014). Female participation in the labor force has increased in most developing countries over the past 20 years (World Bank, 2012). There is a tendency to increase female participation in the labor force with economic development however this tendency is not straightforward in every country. Economic and social factors including economic growth, education, and social norms affect this variation in the level of country (Verik, 2014).

The governments participating in the Fourth World Conference declared to advance the goals of equality, development, and peace for all women everywhere. In the conference action plans on gender equality and gender's empowerment especially equality for women and men in the labor force has been announced and women's equal access to employment has been emphasized (United Nations, 1995). Since that time governments have been dealing with achieving these missions. Most of these sentiments declared in the conference have been reiterated in the more recent worldwide. There has been improvement related to gender equality with the passing years: gender gap in education has narrowed and in some countries women outnumber men in tertiary education; half of the world's women join the labor force, this is situation converse that of 20 years ago; women have an equal voice to men in public and private spheres (United Nations, 2015).

ILO report indicates that the rate of female labor force participation has increased since the 1980s and in most countries the women participation in the labor force has been much higher than the median in 1980. This means a general increase in female participation in countries' economic. This increase in women participation in the labor force is also evaluated as a shrinking of the distance between countries with low and high levels of participations. On the other hand, male labor force participation rates have shown a tendency to decrease since 1980. This is interpreted as gender differentials in labor force participation rate have been decreased over time (ILO, 2010).

11.3.1 LABOR FORCE STATUS BY EDUCATIONAL LEVEL IN TURKEY

The share of labor force with a tertiary education has increased in the last 15 years and the key indicators presents that people with tertiary education are less likely to be unemployed than people having lower levels of education (ILO, 2016b).

TABLE 11.2 Labor Force Participation Rate for 15+ Age in Turkey, 2015.

	Total	Illiterate	Less than high school	High school	Vocational high school	Higher education
Male	71.6	30.9	68.8	71.3	81.1	86.2
Female	31.5	16.1	26.6	32.7	40.8	71.6
Total	51.3	18.6	48.1	54.1	65.4	79.8

Source: TurkStat, Labour Force Statistics, 2014–2015.

The total labor force participation rate for aged 15 and over is 51.3 in 2015 in Turkey. The male participation rate in labor force is 71.6%, whereas female participation rate in labor force is 31.5%. The female participation rate in the labor force is less than half of male participation rate. The female participation rate in the labor force in Turkey is the approximately the lowest rate in European Union Countries as average female participation rate is 66% in European Union Countries (Worldbank, 2014).

Table 11.2 shows that the labor force participation of females with higher education is 71.6% in 2015 in Turkey and this is a large female proportion of labor force participation when compared with other education status. Labor force participation of females having vocational high school degrees is 40.8%; this proportion is 32.7% for females with high school degrees; 26.6% for females having less than high school degrees and 16.1% for illiterate females.

11.3.2 EMPLOYMENT RATE IN TURKEY

Employment is the central objective of macroeconomic and social policies. It is well known that education is at the core of the effort to increase a country's productivity and to improve people's accessing employment. People's educational attainment and labor force trends are related and these indicators shape individual and national well-being (ILO, 2016b).

Higher levels of education are associated with greater labor market success. An increase in tertiary level education in the labor force could facilitate an expansion in production of higher value-added goods and services and a speeding up of productivity growth, thereby these result in supporting economic growth and development. Higher educational attainment could enhance the opportunities for individuals to enter a better position in the labor force and protect people from unemployment. In addition, a higher level of education could affect the quality of working conditions (ILO, 2016b).

TABLE 11.3 Employment Rate for 15+ Age in Turkey, 2015.

	Total	Illiterate	Less than high school	High school	Vocational high school	Higher education
Male	65.0	27.4	61.9	64.5	74.9	79.6
Female	27.5	15.6	23.9	26.0	33.4	59.9
Total	46.0	17.6	43.3	47.3	58.7	71.0

Source: TurkStat, Labour Force Statistics, 2014, 2015.

The total employment rate of the population aged 15 and over is 46%: male employment rate is 65% and the female unemployment rate is 27.5 in 2015. This shows that employment rate for male is two times more than female employment rate (Table 11.3).

Higher educated people in Turkey have the highest employment rate when compared with the other education status. The employment rate for higher educated people is 71%; the proportion of higher educated female is 59.9, whereas the proportion of male population having higher education degrees is 79.6. The employment rate of population for both male and female is decreasing in lower education status. The employment rate for illiterate female is 15.6 and this rate is 27.4% for the illiterate male population.

The current rise of female participation in the labor force denotes positive consequences of the demographic transition in women status. Education seems to be an effective tool for protecting people from unemployment. Especially in most high-income countries, higher educational attainment tends to protect works from unemployment (ILO, 2016b).

The unemployment rate in a country is an important indicator for a better understanding of composition of the jobless population and so to target unemployment policies. Assessing the educational profile of unemployed people is also another key indicator as it provides essential insights related to the extent of the mismatch between supply and demand in the labor market (ILO, 2016b).

The overall trend in many countries indicates that in the labor force, people with a tertiary education have the lowest likelihood of being unemployed (ILO, 2016). However, women are more likely than men to be employed or to be contributing family workers; this usually means that women have no access to monetary income (United Nations, 2015).

TABLE 11.4 Unemployment Rate for 15+ Age in Turkey, 2015.

	Total	Illiterate	Less than high school	High school	Vocational high school	Higher education
Male	9.2	11.3	10.0	9.5	7.7	7.6
Female	12.6	2.9	10.1	20.3	18.1	16.3
Total	10.3	5.3	10.0	12.4	10.2	11.0

Source: TurkStat, Labour Force Statistics, 2014–2015.

The total unemployment rate of population aged 15 and over in Turkey in 2015 is 10.3%. Female unemployment rate aged 15 and over is 12.6% and male unemployment rate for the same group is 9.2%. The unemployment rate for higher educated female is 16.3% and 20.3% for female having high school degrees and these are the highest unemployment rate in all education status when compared with male unemployment rate (Table 11.4).

11.3.3 AVERAGE ANNUAL CORE BUSINESS EARNINGS BY EDUCATION LEVEL AND GENDER IN TURKEY

Higher educated people are associated with higher salaries. In general, it has been seen that higher educated people earn more than the other people (ILO, 2016b).

TABLE 11.5 The Average Annual Core Business Earnings by Education Level and Sex in Turkey, 2015.

	Male	Female	Total
Illiterate	3141	2152	2645
Without diploma	3951	2682	3605
Under high school	5509	3100	5083
High school	7397	5070	6877
Higher education	11980	9194	10944
Total	7130	5647	6765

Euro is calculated as 3.18 according to exchange rate on December 31, 2015.

Source: TurkStat, Income and Living Conditions Survey, 2015.

The average annual core business earnings is 6765€; 5647€ for female and 7130€ for male workers. There is a big difference between male and female annual earnings. Higher educated male earns 11980 € and females graduated from higher education earn 9194 €. There is a big income difference between male and female and the annual core business earnings of male is much more than female earnings. Both female and male graduated from higher education earn more income among other education status (Table 11.5).

Women participation in the labor force still seems open to improvement as women in most countries earn less than men (Verik, 2014).

11.3.4 ATTITUDES TOWARD WOMEN IN THE WORKFORCE THROUGHOUT BY EDUCATIONAL STATUS IN TURKEY

Table 11.6 shows the results of the proportion of individuals who approved females' working or not in Turkey according to 2016 survey results. The proportion of higher educated people who find females' working appropriately is 93.2% and this is the highest appropriate rate in all types of the education status. The proportion rate of individuals who found females' working appropriately is decreasing according to the education status. 78.2% of the individuals having no school completed do not find females' working appropriately.

TABLE 11.6 Attitudes Toward Women in the Workforce Throughout by Educational Status in Turkey, 2016.

	Appropriate	Not Appropriate
No school completed	78.2	21.8
Primary school	83.0	17.0
Primary education/junior high school or vocational junior high school	82.8	17.2
High school or vocational high school	88.5	11.5
Higher education school, faculty, master/doctorate	93.2	6.8

Source: TurkStat, Family Structure Survey, 2016.

11.4 WOMEN IN SOCIAL LIFE IN TURKEY

Demographic transition results in radical changes in women's economic and social roles. These fundamental alterations in women's lives lead to a host of changes influencing women and their role in society as well as affecting children and men. Both increasing in educational levels and changes in women's role in society have roots in demographic transition. This transformation results in important social changes such as life expectancy, marriage age, and women status in the labor force (Lee and Reher, 2011).

11.4.1 LIFE EXPECTANCY BY SEX AND AGE IN TURKEY

Average life expectancy at birth is 78 years. Males live 75.3 years and females live 80.7 years. This means that females live longer than males and the difference in life expectancy at birth is 5.4 years (TurkStat, Women in Statistics, 2016).

11.4.2 AGE AT FIRST MARRIAGE BY SEX IN TURKEY

Worldwide, marriage pattern has changed over the past two decades. Both men and women get married later. This has been evaluated as a reflection of increases in education levels, later entry into the labor force, greater economic independence of women, and a rise in informal unions. Women prefer getting married a few years earlier than men, at age 25 on average, compared to 29 for men. This later marriage age is considered as opportunities for education and empowerment for women, on the other hand, this later marriage age exposes women to the risk of domestic violence and social isolation (United Nations, 2015).

Child marriage, which is identified as a formal marriage before the age of 18, seems still a worldwide problem even though marriage before the age of 18 is not allowed by law. Girls are more likely to marry at a young age than boys. Globally in 2010, 26% of women aged 20–24 were married before the age of 18 (United Nations, 2015).

TABLE 11.7 Age at First Marriage by Sex in Turkey, 2016.

	Under 18	18–19	20–24	25–29	30–34	35+
Turkey	17.9	16.9	37.5	21.2	5.1	1.4
Male	5.6	9.9	41.3	32.7	8.3	2.1
Female	28.2	22.8	34.3	11.5	2.4	0.7

Source: TurkStat, Family Structure Survey, 2016.

Table 11.7 indicates that approximately 35% of the first marriages were realized under ages of 20–24 in Turkey according to family structure survey, 2016. In total, 37.5% of the first marriages were realized between the ages of 20% and 24. 51% of the first marriages in female were realized under the ages of 20–24 in Turkey. 34.3% of the first marriages in female were realized between the ages of 20 and 24 in Turkey.

11.4.3 GENERAL LEVEL OF HAPPINESS IN TURKEY

About 61.3% of people living in Turkey declared themselves as happy in 2016 according to the life satisfaction survey. The proportion of female feeling happy has increased since 2006 and 64.5% of female declared that they were happy; 58.1% of male felt themselves as happy and this proportion of male declaring themselves as happy has increased since 2006 too (Table 11.8).

TABLE 11.8 General Level of Happiness, 2006–2016.

	2006			2016		
	Male	**Female**	**Total**	**Male**	**Female**	**Total**
Happy	55.2	60.5	57.8	58.1	64.5	61.3
Neither happy, nor unhappy	32.1	28.5	30.3	30.2	26.4	28.3
Unhappy	12.8	11.0	11.9	11.7	9.1	10.4

Source: TurkStat, Life Satisfaction Survey, 2006–2016.

11.4.4 ENVIRONMENTAL SECURITY IN TURKEY

Physical, sexual, psychological, and economic violence are still a problem for women regardless of their income, age or education. Around one-third of women worldwide have experience any kind of violence in their lives (United Nations, 2015).

TABLE 11.9 Environmental Security, 2016.

	Male	**Female**	**Total**
Secure	71.0	47.5	59.1
Neither secured nor unsecured	13.8	15.6	14.7
Not secure	15.2	37.0	26.2

Source: TurkStat, Life Satisfaction Survey, 2016.

59.1% of people living in Turkey declared themselves as secure in 2016. 71% of male feels themselves as secure, whereas 47.5% of female declared themselves as secure. The proportion of female declaring themselves as secure is rather low when compared with male in Turkey (Table 11.9).

11.4.5 PROPORTION OF MALE AND FEMALE IN LOCAL GOVERNMENTS IN TURKEY

Men generally hold power and decision-making positions in societies worldwide and this could be evaluated as a severe and highly visible gender inequality. However, progress related to men position has been achieved in all regions. This progress seems to be slow and women continue to be underrepresented in many decision-making positions (United Nations, 2015).

The participation of women in a decision-making position in local government is an initial step in ensuring that needs, priorities, and perspectives of women are taken into account in local policies and budget allocations (United Nations, 2015).

TABLE 11.10 Proportion of Male and Female in Local Governments, 2014.

Mayor/mayoress	Total	Male proportion (%)	Female proportion (%)
City councilor	20,498	89.3	10.7
Member of provincial assembly	1251	95.2	4.8
Head of village	18,143	99.7	0.3
Member of village council	80,696	98.8	1.2
Head of quarter	31,635	98.0	2.0
Member of quarter council	137,781	97.5	2.5

Source: Ministry of Interior General Management of Local Authorities, 2014.

Table 11.10 shows that females participated less in local governments than males in Turkey. The proportion of female in city councilor is just 10.7% and the other representations are also quite low for females. Table 11.10 indicates that women are significantly underrepresented in the local governments in Turkey.

11.5 CONCLUDING REMARKS

The statistics and the findings related to women in labor force and social life summarizes the followings:

- Turkey has a young and growing population. Women population is slightly less than male population; however, women outnumber men in older age groups.
- Turkey has witnessed remarkable progress in participation in education since 1975. The rate of education completed has increased since then. There is still a gap between men and women's educational attainment. This indicates that contrary to developed countries, Turkey has not reached gender parity yet.
- The proportion of women in the labor force is not enough when compared with developed countries. Women employment rate is lower than men in Turkey.
- Unemployment rate for women seems relatively high in Turkey. Education is not always effective in protecting against unemployment rate of women in Turkey.
- Average annual core business earnings differ according to education level. However, women on average earn less than men no matter what education level they have. There is an obvious gender pay gap in Turkey.
- Higher educated people mostly find females' working appropriately.
- Half of women's first marriages were realized before age 20–24.
- More female than male declared themselves as happy in Turkey.
- Just half of the women in Turkey feel themselves secure in social environment in Turkey.
- Women are also significantly underrepresented in local government positions in Turkey.

It seems that the role and the position of women have been extensively changed through the passing years in Turkey. The demographic transition in women has a positive role in enhancing women's position in the society. The consequences of the demographic transition for women resulted in more women in the labor force and more educated women in society. This increase of women participation in education as well as in labor force seems parallel with the world tendency as female labor force participation rates have shown a tendency to increase slightly in the worldwide since 1980 (ILO, 2010). However, Turkey still has a long way to go in approaching the level of most developed countries. All the statistics reveal that Turkey still seems as a male-dominant society.

Analysis related to women in labor force and social life is essential indicators for the development of integrated strategies to promote nationwide fundamental principles. Turkey needs to set policies and strategy and then monitor the progress toward goals and to adjust policies where needed.

KEYWORDS

- **educational attainment**
- **employment rate**
- **population age pyramid**
- **unemployment rate**
- **women in labor force**
- **women in social life**

REFERENCES

Canning, D.; Raja, S.; Yazbeck, A. S. Africa's Demographic Transition: Dividend or Disaster? The World Bank: Washington DC, 2015. [Online] 2015, 10.1596/978-1-4648-0489-2 (accessed Feb 19, 2017).

Goujon, A. V. Demographic Transition and Education in Developing Countries, in Sustainable Human Development. In *Encyclopedia of Life Support System (EOLSS), Developed under the Auspices of the UNESCO*; Eolss Publishers: Oxford, UK, Laxenburg, 2003.

ILO. *Women in Labour Markets: Measuring Process and Identifying Challenges*; ILO: Geneva, 2010.

ILO. *Key Indicators of the Labour Market*, 9th ed., ILO: Geneva, 2016b.

Kırk, D. Demographic Transition Theory. *Popul. Stud.* 1996, *50*, 361–387.

Lee, R. D.; Reher, D. S. Introduction: The Landscape of Demographic Transition and Its Aftermath. In *Demographic Transition and Its Consequences*; Lee, R. D.; Reher, D. S., Eds.; Population Council: New York, 2011; Vol. 37; p 1.

Riyaza, F. Age and Sex Composition of the Population; Analytical Reports, Census of the Maldives. [Online], 2000. http://www.planning.gov.mv/publications/analytical_report/html/Source/ChapterIII.pdf (accessed Apr 20, 2017).

United Nations. *Beijing Declaration and Platform for Action: Beijing+5 Political Declaration and Outcome*; United Nations: New York, 1995.

United Nations. *The World's Women 2015: Trends and Statistics.* United Nations, Department of Economic and Social Affairs, Statistics Division: New York, 2015.

Verik, S. *Female Labor Force Participation in Developing Country*; IZA World of Labor: Germany, [Online] 2014. DOI: 10.15185/izawol.87 (accessed Mar 25, 2017).

Willekens, F. Demographic Transitions in Europe and the World. MPIDR Working Paper 2014-004; Rostock: Max-Planck Institute for Demographic Research, 2014.

World Bank. *Women are Less Likely Than Men to Participate in the Labor Market in Most Countries*; World Bank: Washington, DC, 2012.

World Bank. The Worldbank Database for all Countries and Economies. [Online] 2014, http://data.worldbank.org/indicator/SL.TLF.ACTI.FE.ZS on 03.04.2017, (accessed Apr 15, 2017).

Verik, S. Female Labor Force Participation in Developing Country. IZA World of Labor: Germany, 2014. DOI: 10.15185/izawol.87 (accessed Mar 25, 2017).

CHAPTER 12

Trajectories of Demographic Transition: The Challenges for the Elderly in India

SAHELI GUHA NEOGI GHATAK*

Department of Sociology School of Social Science, Adamas University, Kolkata, West Bengal, India

Corresponding author. E-mail: saheli.guhaneogi@gmail.com

ABSTRACT

A country's population size, structure, and composition have a precarious, while not deterministic—impact on development. India is undergoing rapid demographic change. India is in the process of demographic transition and has now moved into the phase of falling fertility thereby reducing natural growth and is predicted to reach low fertility mortality replacement level in the future (Agarwal, 2016). The changing nature of the composition of age is anticipated to create a paradigm of demographic dividend initially and will be converted to graying society in future. This paper contends the unique nature of demographic change of India along with the challenges faced by the elderly population. Without addressing these challenges seriously, demographic changes in the country may not return their foreseen benefits. In this context, the paper reviews trends in Indian population size and the drivers of population change. Furthermore, it endeavors the challenges for the elderly in India along with suggestive measure for coping with the situation.

12.1 INTRODUCTION

India, with its 17.64% share of the population in the world has been going through a dawdling even so stable demographic transition successively from the second half of the previous century. Demographic transition comprises various stages of population edifice, related to sluggishness and plunge in the birth rate and death rate, directing important impression on the progression rate of population in our country. At present, India is facing the 3rd phase of demographic transition (Haub and Gribble, 2011). This situation indicates the reciprocal decline of birth and death rate. The nature of conversion of age structure is going to produce demographic dividend first and later on will be converted to graying society in India.

Policy-makers in India started bringing various developmental reforms to the society for the last two decades. These reforms have a bizarre way of reducing the death rate faster than the birth rate. Hence, the net result is the population starts expanding fast, commonly known as the early expanding phase (Stage II). As the country progresses further, birth rates do start declining but not as good as death rates. So we enter a late expanding phase (Stage III), where India is currently in. However, India wanted low stationary phase (Stage IV). By more declining birth rate, another beautiful thing that awaits India in the very near future is thedemographic honeymoon. (Agarwal, 2016). India's base, the Gen-Y will soon be entering economically active age group, with a narrower base and India will have fair economics prosperity until this group enters the geriatric age group within next 30 years when the honeymoon will end. Usually, demographic transition provides only a 35–45 year demographic window of opportunity for the working age group. However, in India's case there will likely be only 30 years of demographic bonus due to India's accelerated fertility transition. According to current UN projections, India's old age dependency ratio will be far above 15% by the time its child dependency ratios will be below 30%. (James and Goli, 2016). Consequently, countries such as India and China will prematurely develop into aging societies. In contrast, countries such as South Korea, one of the East Asian Tigers which reaped the demographic dividend, became rich before getting old. Alternatively, countries such as Brazil have become rich without getting old because they have experienced a longer demographic bonus (Gribble and Bremner, 2012).

Twenty-first century is going to witness a transition from a younger society to elderly society. Population elderly is inescapable and we should try to understand the consequences and prepare accordingly. The concept "age quake" was used by Wallace, (1999), to refer the scenario of the aging society. If we comprehend the implication of graying society that is the "age quake" now, then only it will not spill down on us gruesomely alike an earthquake surrounded by death and extermination. Rather, we will formulate an age friendly society where elderly will get dignified life. The rate of population elderly, that is, 60+ age group is anticipating to get three times larger in number, whereas, the 0–14 age group, that is, child rate will endure stagnancy. The composition and distribution of the population in India will be similar to those countries presently which are facing high aging scenario. The countries like Russia and UK are now indicating the larger proportion of elderly people rather than child population. This kind of variations in the magnitude, composition, and dispersal of population predicted to have severe repercussions for community policies and market system. Although the scenario of aging population is a future challenge and not a current one, it is important that preparedness for it starts now.

12.2 DEMOGRAPHIC CHANGE AND ELDERLY PEOPLE

India is distinctively important in World, owing of absolute size of its population and the unconventional trajectory of its changing character. With a population of more than 1.3 billion, it contains nearly 18% of global population and it is anticipated to surpass China (population of 1.38 billion in 2015) within the next 6 years to convert into the most populous nation in the world (World Population Prospects, 2015). Interestingly, during the first census after India's independence in 1947, the population of India was only 68% that of China. It is now as high as 95%. According to 2015 estimates, the population of China is increasing by approximately 7 million people per year while the population of India increases by about 16 million per year. (James and Goli, 2016, World Population Prospects). Although, India has got the seventh leading position in the world economy, its rank acutely falls to 140 and 129 based on per capita income and per capita purchasing power parity figures, respectively (James and Goli, 2016). The percentage of the working population, that is, age group of 15–64, reached over 60% in 2015. This proportion is likely to go up further in the coming decade. It is projected to reach its apogee (64.8%) in 2030 and thereafter fall to a still-high 62.2% in 2050. The situation of demographic dividend for any nation sustains for a shorter tenure and gradually that nation will turn into a graying society. While not instant, transformation of the age composition from younger to older is supplemented by numerous social changes with substantial consequences on any nation.

The salient demographic concern for India in the 21st century is elderly population, an indicator assuring that India with 90 million elderly is anticipated to be 315 million, that is, 20% of the entire population by 2050 (UNFPA-BKPAI, 2011/2014; United Nations, 2013). The current and projected levels of the elderly population in India are unconventional compared to developed countries in the West and its Asian counterparts. This unprecedented aging expected in the country is going to create huge economic and health care challenges.

12.3 WORLDWIDE SCENARIO OF ELDERLY

The demographic transition commonly known as "*graying of nations*" is posing a challenge for the entire world (Bagchi, 1996), wheren both rich and poor countries have had a significant growth of life expectancy at birth. This "demographic time bomb" (Wallace, 1999) is impending explosion throughout the world. Globally, the proportion of elderly people enlarged from 9.2% in 1990 to 11.7% in 2013 and will continue to produce as a percentage of the world population, attaining 21.1% by 2050 (UN, 2013).

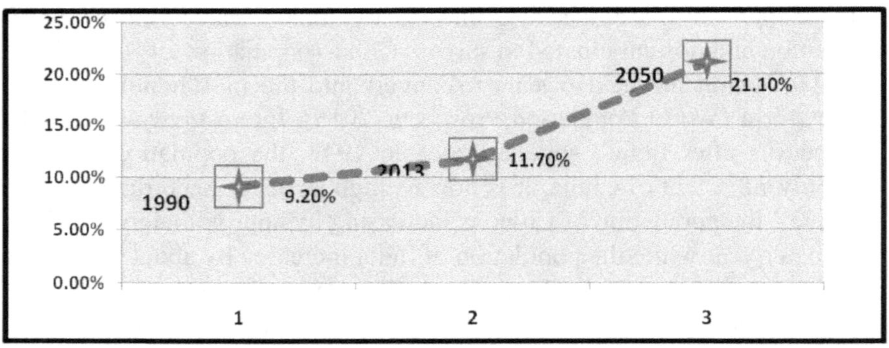

FIGURE 12.1 (See color insert.) Proportion of world population.
Source: UN 2013.

The World Health Organization (WHO) conveyed about the fact that in 2000, there were 600 million elderly persons, that is, 60 years and over and which will be 1.2 billion by 2025 and 2.0 billion by 2050. The WHO report mentioned that the developing countries will witness the aged society before they grow into rich, whereas the industrialized countries will turn into rich before they become elderly, accentuating that all the countries in the world need to be prepared for addressing the consequences of the elderly population. We should also note that with the passage of time, male and female life expectancy is predicted to escalate. The proportion of 60 and above age group of female will perpetually be greater than their male counterpart, which will be known as feminization of elderly. The UN projections, 2000, stated that the 60 and above age group of male population comprising 7.1% of the total male population, whereas for 60 and above female elderly it is 8.2% and by 2025, the male and female elderly proportions will be 11.9% and 13.4%, correspondingly, and by 2050, the comparable number will be 20.2% for males and 22.4% for females. (Bose and Sankardas, 2000)

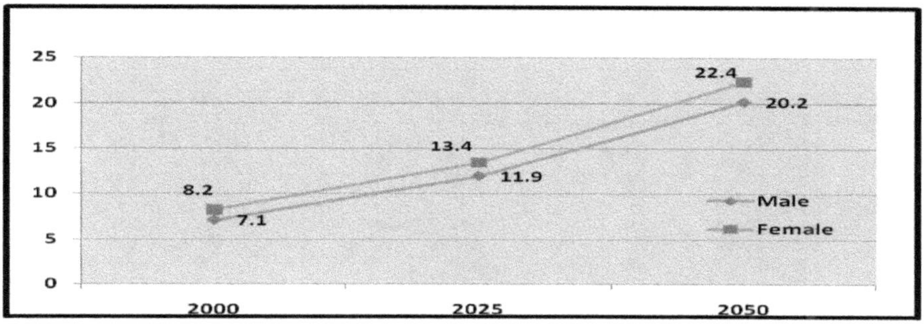

FIGURE 12.2 (See color insert.) Gender-wise proportion of the world population.
Source: UN 2013.

12.4 INDIAN SCENARIO OF ELDERLY

The changes in the demographic structure of the societies throughout previous few epochs, particularly in developing nations, have made the aged a socially more noticeable section of the population. Asia predicts for higher than 50% of the world's elderly (UN, 2007), with a comparatively young population, India is still composed to convert into the home to the second prevalent volume of elderly persons in the world, while China is impending closer to the initiation of a senile society (IFA, 1992).

With such a global phenomenon of population aging, India too is experiencing a dramatic modification in the population composition with phenomenal growth in the figure of elderly people and consequently, the demographers have identified India as a country that is entering the "age of ageing." In India, the amount of the elderly populace was 88 million, that is, 7% in 2009 and was projected to grow up to 315 million, that is, 20% by 2050 (BKPAI, 2011/2014), whereas it would be 11.10% in 2025 as per United Nations Department of Economic and Social Affairs (UNDESA,

2008) (Fig. 12.3), whereas, the total Indian population will grow up to 40% in between 2006 and 2050, and the aged population will grow by 270%.

According to United Nations World Population Projections to 2150, assessments stated that during 1995–2000, the life expectation of males hoisted at 62.3 years, whereas that of their female counterpart was 62.9 years, and it is 65.77 years for male and 67.95 years for female in 2011. Within the era 2020–2025, the numbers are 68.8 years for males and 72.1 years for females. And, for the era 2045–2050 the projections are 73 years for males and 76.9 years for females (Singh and Khan, 2012) (Fig. 12.4).

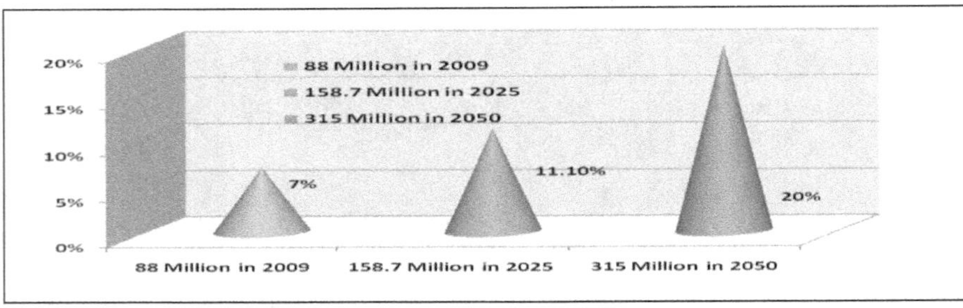

FIGURE 12.3 (See color insert.) Proportion of population increase.
Source: UNFPA-BKPAI, 2011/2014.

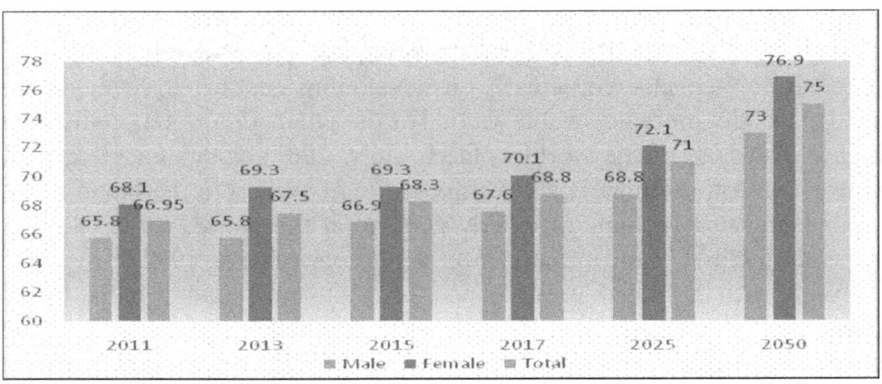

FIGURE 12.4 (See color insert.) Life expectancy pattern of India.
Source: Census, 2011; UN 2013; UN-World Population Projection 2150.

12.5 PERCEPTION OF ELDERLY IN INDIA

Accelerated population aging is a chief demographic concern in India in the 21 epoch, following its extensive effects on over-all socioeconomic condition. Immediate alterations of demographic parameters of previous few years, signifies, India will transfer into the graying country from youth country in coming decades. The Global Report on Ageing in 21st century, 2012 strengthens interpretations about India that multifarious discrimination occurs against elderly, predominantly with elderly women, comprising of opportunity for careers and health-related care, subjugation to various kind of exploitation, renunciation of basic human rights and deficiency of minutest income and social security (Help Age International, 2012 and UNFPA). Supplementary, extreme heterogeneity in the demographic transition across states, related to enormous dissimilarities in demographic set-up across socioeconomic–spatial groups (UNFPA, 2012).

The Madrid International Plan of Action on Ageing (MIPAA), 2002 is a most important source for showing the ways for future development of accomplishment through different nations. As signatory of MIPAA, Indian Govternment prepared NPOP, that is, National Policy on Older Persons and then it revised by National Policy for Senior Citizens (NPSC), though it is awaiting for cabinet sanction and NPHCE, that is, National Programme for Health Care for the Elderly Welfare to address various health-related challenges of elderly.

12.6 CHALLENGES CONFRONTED BY ELDERLY

The demographic transition accompanied by a social, cultural, and political shift with increased individualism destabilizing the status and role of the elderly in the family as well as in the society has altered living conditions for the elderly which create various challenges for elderly. The perception of the old as the repository of collective wisdom is also on the wane. The position of a large number of aged in the growing aged population portrays vulnerability due to various other reasons such as industrialization, western way of life, concept of privacy and space, consumerism, change of roles and expectation from women, urban space crisis, and attitude toward aged, enhancing the rapid and mushrooming growth of old age homes. Such are the predicaments of elderly from their increased longevity which with the passage of time leads to progressive generalized impairment of body functions, creating emotional insecurity, loneliness, lack of social contact and social support, and care (Lyyra and Heikkinen, 2006), giving space to acute sense of role loss, lack of adjustment, lack of decision-making power, lack of meaningful connectedness with a significant other, lack of sense of belongingness along with loneliness and a feeling of lack of accomplishment, and unhappiness due to nonrecognition and marginalization leading to loss of confidence, which significantly undermining the life-quality of elderly in India. With "Baby Boomer" generation transfers into aged people and instigate placing increased anxieties on the elderly care structure, it is vital for strategies to address numerous challenges of the elderly.

In recent time, the age old role of the family is being shared by institutions such as old age homes. Many of the elderly parents are obliged to leave their children and stay in old age homes that virtually compel the elderly to move out of the care of the family to such impersonal arrangements which lack warmth, sometimes create adjustment problems, loneliness, and depression (Joshi, 2012). With the fragmentation pattern of the traditional support base (the family), the nuclear family has been substituting the extended kinship structure leading to psychological isolation (Parsons, 1943, 1949, 1965). Correspondingly, the structuralist–functionalist and modernization theorists view of the breakdown of the traditional norms and increased dwindling of the bonds between adult children and their parents, (Goode, 1963; Parsons, 1943; Chattopadhyay and Marsh, 1999), age selective migration, and global employment opportunities have distanced the cohorts who were earlier living together. Thus, intimidating intergenerational interface, changing roles and expectation from the women, urban space crux and privacy, helped mushrooming growth of Old Age Homes, exposing the elderly to emotional, physical and financial insecurity. Such quandaries from increased longevity (Lyyra and Heikkinen, 2006) are generating a sense of loss of decision-making power, meaningful connectedness, belongingness, loneliness, unhappiness due to nonrecognition marginalization and loss of confidence significantly undermines life quality of elderly in India.

Many research studies have addressed the issues and problems of the elderly. Basu Mukherjee and Ray (2013), in their work on "The Problems & Potentialities of elderly in West Bengal" on 538 elderly from six districts including Kolkata, found that the problems of elderly in West Bengal were of five types—economic, physical, social, psychological, and legal knowledge and awareness. Various health problems, various disabilities affecting mobility and certain chronic diseases were also important. Joshi (1971) conducted a study on the medical problems of old age. He states that the trend of differential ageing both in form of physical and mental, emerged on the basis of environmental and social factors, for example, type of food, educational qualification, nature of job, adjustment with in family, professionalism and substance abuse. Major health problems of elderly include infective and parasitic diseases, diseases of respiratory systems, arthritis, rheumatism, hypertensions, congestion, heart, and diabetes mellitus. Srivastava et al. (2010) showed that the common psychosocial problems reported by older persons are impaired memory and intelligence, rigidity of outlook, anxiety, depression, dependency, and dissatisfaction with family members; occupation and earning have positive association with quality of life. Kumari (2007) in her study "The problems among people with special need (old age) in old age home of Delhi" mentioned that the elderly are facing various challenges in old age homes, such as physical helplessness, dependency, economic insecurity, changing pattern of living, making of new friends, development of new activities, becoming involved in community activities, and being victimized or taken advantage of by salesperson, etc. Pesic (2007) found that the elderly people are suffering from lack of social status and lack of power within present industrialized and modern societies. Chadha (1999) emphasized that psychological and environmental problems, feelings of neglect, loneliness, loss of power, and imbalance in the reciprocal relationship makes the aged feel unwanted and neglected. With current trends longevity, the problem of social isolation is getting importance in the context of high mobility and less number of children. Deshpande et al. (1998) asserted that depression, loneliness, and dementia were the main common psychological problem among elderly. The mood disorders were recognized as one of the important psychiatric concern among the elderly (Yvonneda et al., 2002). Bose (1990) in a study also pointed out that the elderly people feel neglected by their family members, friends, and society as nobody wanted to share their time to sit for a while with them. Intergenerational gap plays important role to boost this gulf and which, in turn, enhance the sense of despondency. Gupta (1988) found that aged are neglected a lot and leading a woeful life. This trend is mainly because of growing cost of living, widening of generation gap, erosion in moral values and desire of younger generation to live an independent life. Gangrade (1988) observed that the elderly people thought that the younger generation do not show any respect for them and care for them, which lead to neglect.

12.6.1 PROBLEMS FACED BY ELDERLY

Specific problems of elderly are mainly consists of physical problem, social problem, psychological problem, and economical problems. Major *physical health problems* are related to vision (visual problem of elderly were mainly cataract, distance power, reading power etc.), hearing, walking, dental problem, sleep/insomnia, and other physical ailments, such as joint/muscle pain, blood pressure, blood sugar, cardiac problem, acidity are all major physical health problems.

Social problems of elderly are mainly related to changing role, less power, less social status, feeling as burden, ignored, outdated and obsolete feeling, viewed as nonproductive, neglected, victimization, abuse, lacking care, participation in work, absent of social security, nonavailability of opportunity, etc. With the changing form of society, that is, more industrialize and globalize society, the structure of family changes. The joint and extended families are replaced by nuclear families. In present society, younger people are often move away from family because of various pull factors of migration and urbanization and the pattern of work of younger people itself consumes lots of time and energy that also relate to lack of time for looking after family members. The norms and values related to the care of elderly people are changing; various institutions are supplementing the role of family.

The globalization behemoth has added new dimensions to the susceptibility of India's elderly by exacerbating their social exclusion and making this segment of the aged vulnerable and marginalized. Traditional values attached with elderly, that is, authority, obedience, and self-discipline, are altering by the new concepts of equality, freedom, self-sufficiency, which, in turn, erode the traditional culture of India. Elderly are facing disrespect, ignoring attitudes from younger generation and isolation. The deficiency of social security and health care system unlike western society, make Indian elderly experience sheer reduction in their real role and status soon after their retirement from work (Help Age India, 2010; Mukherjee, 2011). The problem of elderly widow women are more because of the stigma attached to them.

Anxiety is one of the major *psychological problems*. Causes of anxiety among elderly were mainly related to their physical health, worry about routine activities of daily life, and relationship with family members, security, experience of losses, multiple medications, etc. According to Bruges (2006), fear of crime, that is, "a lack of a sense of security and feeling of vulnerability" was one of the most important reasons for anxiety of elderly. Psychological problems related to emotional disorders, which results in social maladjustments, various physiological and mental turbulence, jealousy, and bad temper were quite common and recurrent (Alemi et al., 2002) among elderly. Other psychological problems consisted of obsession, and excitements, loss of memory (dementia), depression are also confronted by elderly.

Economic problem of elderly related to the social structure, which make older people dependent, but in ways that lead some older people to be considerably more dependent than others (Estes, 1979; Walker, 1980; Townsend, 1981; Evans and Williamson, 1984). Lack of opportunities to work, poor pensions or no pension, and poor physical ability make older people powerless and dependent. Dependency of older people is in this perspective socially constructed by governments and markets that have acted in ways that have marginalized, dominated, and weakened older people. Reasons for such situation were the discriminations that elderly people experience often as a consequence of unjust resource distribution, mistreatment, abandonment, and restricted access to fundamental services.

12.6.2 ELDER ABUSE—VARIOUS DIMENSIONS

Elder abuse is social menaces in India; it is still not considered an act crime, though these are the base of humiliation and torment to aged persons, which goes unnoticed. Members of family, strangers, and formal caregivers entrust it. Elderly are abused because they are considered redundant, ineffective, and a liability.

According to WHO, older abuse is related to a solo or repetitive work or absence of proper act taking place inside any relationship, wherever, an anticipation of faith relates to impairment or strain to the elderly (WHO, 2012, p.1). The 2nd World Assembly of Aging, report of UN on Abuse of Older Persons (2002), mentioned that pure and compatible explanations of elder abuse in case of all developed and developing countries is lacking. Still, according to various studies conducted in the last two decades, the instances of elder abuse are prevalent all over the world. The documents of the aged opinion and experience of abuse includes—*violence, ill-treatment, neglect, abandonment, deprivation,* etc.

Mistreatment of elderly is referred to as " elder abuse." Older Abuse is made by family members, care givers or institution. Abuse can be done once or repeatedly, and various forms of abuse are like, *financial or economic abuse, physical abuse, sexual abuse, verbal abuse, political abuse, spiritual abuse, and neglect. Physical abuse* related to the deliberate acts which cause physical damage to the body of an old, for instance, hitting, beating, slapping, burning, pushing, etc. (Wolf, 2000; Iborra, 2009). *Verbal abuse* indicates the intended acts primarily in practice of words that could create mischief to an old people. This kind of abuse comprise with shouting, unnecessary blaming, using grimy language, insulting an elderly (Payne, 2005; Iborra, 2009). *Financial or economic abuse* related to planned work that involve illegal exercise of an aged people's financial and/or any other material resources relating harm to an elder person. This type of abuse related to willful exploitation of the elder people's financial and/or other material things, property, faking of signature through which the abuser can use the resources of elderly people ; forcing them to sign vital documents related to losing of their economic and material possessions. *Sexual abuse* comprises any deliberate act related to forceful sexual activities that create the sexual approaches of an elderly people without their own wishes or for the executor to achieve sexual pleasure. The type of sexual abuse contains rape, molestation, showing pornographic resources, compelling elderly to perform sexual activities amongst themselves, or kissing an elderly. *Neglect* related to deliberate failure to encounter one's own compulsions for caring of elderly people. The example of this kind of activities involve denying to attend to the requirements of elderly people although the resources are accessible, leaving them unaccompanied without anybody close by for helping them when they need help from others (WHO, 2012). *Spiritual abuse* concerned with any intended act to hurt the spirituality of elderly people for example—fake allegation to elderly people as witches, accusing them as misfortunes in family and society, accusing them for being demonized with others (Help Age International, 2011). *Political abuse* relates to any deliberate act to harm the political life of elderly people, which includes forcing the elderly people to vote for a specific political party in the election and intended utilization of elderly people's civic or political papers like ration card, voter's cards by individual or political parties to fulfill their interest (Help Age International, 2011*)*. In recent times, various other forms of abuse, such as *spousal abuse and medication abuse* are also getting importance (DWARF, 2002, 2003). *Loosing of respect, scapegoating, blaming elderly*, especially for women, for evils happening in the society. Elderly women, generally widows, are ostracized, tortured, injures, and even murdered in lieu of gaining regulation over their possessions (Puri, 2007). WHO, INPEA, Help Age International and researchers, documented various elder abuse as perceived by them (1) *Neglect along with isolation and social exclusion,* (2) *Infringement of social, legitimate and medicinal rights*, and (3) *Removal of choosing, judgments, position, funds and respect.* All of this type of abuses in any form or the other happens in both developing and developed countries across the continents. Presently, the marginalization of elderly are also recognized as cases of elder abuse based on various *political violence, displacement, disasters , armed conflict, and emergencies,* whereas needs of elderly are hardly conveyed for in release plans, *routine scolding, niggling, unawareness* about their requirements and diseases, material exploitation (Bose and Shankardass, 2004)

Various narratives of elderly person showed that *instances of burning, scalding, being pushed around, experience of rough handling* are also form of abuse. Elderly are *forced to eat unappetizing food, spit out* on them and frequently treated like a domestic helper and in some cases when aged parents are hospitalized, the children do not supply nutritious food and provide sufficient personal care. *Physiological assault, verbal embarrassment, material exploitation, putting elderly health at risk, lack of care, putting unjustified pressure on elderly and exposing them to humiliating behavior, send them in old age homes or other institutions* for the interest of family members.

12.7 CONCLUSION

This paper examined the current status of demographic transition of India which highlighted that the elderly people are growing at a disquieting rate and hence this century can be identified as the "Era of Population Ageing" (Ponnuswami 2005). With the current changing situation of demographic and social structure, the elderly people are confronting various challenges related to various social, physical, economic, psychological problems and more importantly they are facing various abuses (physical, verbal, psychological, legal, neglect, etc.) in our society. The trends of globalization, industrialization, and the economic policies of developed and developing countries growing volatility, and interconnectedness of economic development have changed social environments. Such change and challenges have facilitated the dynamics of social inequalities, more wide gaps in social strata which had impacted the transitional nature of Indian economy, social relationship, cultural norms, social structure, values and philosophical system, etc. Such changes and challenges have debilitated the family and kinship support for seniors across societies, thereby eroding the locale support traditionally available to seniors. More over due to loss of joint family system, illiteracy, inadequate social security the management of old age problem is a serious problems and it has been realized by all concerned that there is a immediate need for appropriate level of informational, emotional and instrumental support from the state and community for a vast elderly people.

Owing account of various challenges faced by elderly, following suggestions are put forward to decrease the problems of elderly and to improve quality of life of elderly (1) Though various welfare measures of India have taken initiative to improve the economic condition of elderly by providing pension to the elderly belonging to the BPL categories, the scope of pension for eligible persons are limited, so the Government should take proper steps to include all the eligible elderly within the scheme and should increase the amount and circulate those with in proper time. (2) Some of the steps like reverse mortgage and some of the pension schemes from the part of banks was introduced but there is requirement of spreading more awareness among elderly for the available schemes for elderly and also spread awareness regarding the savings from the young age so that financial quality of life can be maintained. (3) Various awareness programs related to the upcoming demographic structural change along with the challenges and various established schemes and laws related to welfare of elderly should be circulated through different types of medias, that is, radio, television, and other social networking system so that elderly can cope with the situation. (4) Inclusion of separate geriatric unit in hospitals and health care system can more helpful for elderly. (5) Various consultation services related to their health, financial, psychological, social and legal services should be offered to the elderly by diverse welfare centers. (6) Income generating proposal for elderly by using the potentialities of elderly skill and experience should be introduced to produce supplementary income for the elderly. (7) Various Self Help Groups (SHGs) should be formed for fulfilling various requirements of elderly and they should also be associated with innovative and developmental program. (8) Clubs for elderly and other organizations for facilitating various social, recreational, educational and cultural needs of elderly should be introduced. (9) Government should execute geriatric medical insurance schemes for elderly in India, which will extend the way to improve the problem of health care of elderly. (10) Various workshop for providing proper knowledge of health care practices and nutrition should be arranged and various indigenous health care approaches such as yoga, pranayam, etc. and traditional nutritional practices should also encouraged among elderly for achieving better quality of life. (11) As feminization of elderly is going to be the big concern for coming decades, specific focus on elderly women should be given to address and overcome their socioeconomic, psychological, and health care challenges and to provide them enriched quality of life.

KEYWORDS

- **trajectories**
- **demographic transition**
- **trajectories of demographic transition**
- **challenges**
- **elderly in India**

REFERENCES

Agarwal, A. K. A Challenge for the India to Take Advantage by Demographic Transition Dividends. *Indian J. Health Sci. Biomed. Res.* **2016,** *9*, 27–30. http://www.ijournalhs.org/text.asp?2016/9/1/27/183685.

Alam, M., et al. Report on the Status of Elderly in Select States of India, 2011: Building a Knowledge Base on Population Ageing in India. United Nations Population Fund, New Delhi, 2012.

Antony; et al. India: Opportunities and Challenges of Demographic Transition 2011. India Opportunities and Challenges of Demographic Transition. https://www.researchgate.net/publication/251119605 (accessed Jan 17, 2018).

Bagchi, K. Health Care of the Elderly is Homeward-Based. In *Aging: Indian Perspective and Global Scenario,* Kumar, V., Ed.; All India Institute of Medical Sciences: New Delhi, India, 1996.

Basu Mukherjee, S. Older Women's Empowerment: A Shifting Paradigm and Commitment. In *Older Women in India: The Context, Issues and Concerns,* Nayar, P. K. B, Ed.; The Women Press: Delhi, India, 2013.

Bloom, D. E.; Canning, D.; Malaney, P. N. Demographic Change and Economic Growth in Asia. Working Paper No. 15, 1998, Center for International Development.

Bloom, D. E.; Williamson, J. G. Demographic Transitions and Economic Miracles in Emerging Asia. *World Bank Econ. Rev.* **1999,** *12* (3), 419–455.

Bose, A.; Shankardass, M. K *Growing Old in India: Voices Reveal, Statistics Speak*; B. R. Publishing Corporation: New Delhi, India, 2000.

Bose, H. S. Rural Social Change and the Problems of the Aged. *Social Welfare* **1990,** *11*, 25–30.

Chadha, N. K The Other Side of Being Old. *Social Change* **1999,** *29*, 1–11.

Census 2011. http://censusindia.gov.in/Feb,3/2017/3.45p.m. Government of India: New Delhi.

Gangrade, K. D. Crisis of Values: A Sociological Study of the Old and the Young. In *Aging in India: Problems and Potentials;* Bose, A., Gangrade, K. D., Eds.; Abhinav Publications: New Delhi, India, 1988.

Goode, W. J. World Revolution and Family Patterns. Free Press: Glencoe, Scotland, 1963.

Gribble; Bremner, Achieving a Demographic Dividend. *Popu. Ref. Bureau* **2012,** *67* (2), 1–14.

Gupta, N. *Quality of Life of Older Women in Urban India.* Doctoral Dissertation, Department of Social Science, TISS, 2014. http://hdl.handle.net/10603/19504/ (assessed Sept 12, 2016).

Help Age India. Report on Elder Abuse in India. New Delhi, India, 2010.

Haub, C.; Gribble, J. The World at 7 billion. Population Reference Bureau. *Global J. Med. Public Health.* **2011,** *66*(2), 1–19. In Demographic Transition and Population Ageing in India: Implications on the Elderly of the Future Shradha Mathur, Navin Mathur.

Iborra, M. I. Elder Abuse in the Family in Spain. Queen Sofía Center: Madrid, 2009.

James, K. S.; Goli, S. Demographic Changes in India: Is the Country Prepared for the Challenge? *Brown J. World Affairs* **2016,** *xxiii* (i/21).

James, K. S. India's Demographic Change: Opportunities and Challenges. *Science* **2011,** *333*, 576. DOI: 10.1126/science.1207969.

Joshi, C. K. Aging in India in the 21st Century: A Research Agenda-Priority Areas and Methodological Issues, 2012. Report by Siva Raju, Mumbai, 2006.

Lyyra T. M.; Heikkinen, R. L. Perceived Social Support and Mortality in Older People. *J. Gerontol. Soc. Sci.* **2006,** *61B*, S147–S152.

Office of Registrar General of India and Census Commissioner, Census of India, 2011. Government of India: New Delhi, 2011.

Parsons, T. The Kinship System of the Contemporary United States. *Am. Anthropol.* **1943,** *45*, 22–38. As Cited in Goode William J. *World Revolution and Family Patterns*. Free Press: Glencoe, 1963.

Payne, B. K. *Crime and Elder Abuse: An Integrated Perspective*, 2nd ed.; Charles C Thomas Publisher: Springfield, 2005.

Paul, W. *Agequake: Riding the Demographic Rollercoaster Shaking Business, Finance and Our World*; Nicholas Brealey Publishing: London, 1999.

Pesic, L. Social and Health Problems of the Elderly. *Acta Medica Medianae.* **2007,** *46* (2), 45–48. www.medfak.ni.ac.yu (accessed Feb 21, 2017).

Ponnuswami, I. Aging: Worldwide Trends and Challenges for Care Giving. *Soc. Welf.* **2005,** *52*, 26–39.

ISEC, IEG, TISS, UNFPA. Population Ageing in India Report, *The Status of Elderly in West Bengal*, 2014.

Mupila, O. The Silent Cancer of Elder Abuse. Times of Zambia: Ndola, 2008

Singh, V. K; Khan, F. A. A Comparative Survey of Medical Disorders in the Elderly Persons of Rural and Urban Area of North India. *J. Clin. Diagnostic Res.* **2012,** 844–847. http://www.jcdr.net/Feb,14,2017/.

Srinivas, G.; Pandey, A. Is India Getting Old Before Getting Rich: Beyond Demographic Assessment. In *Challenges of the Twenty First Century: A Trans-Disciplinary Perspective*, Anand, S., Kumar, I, Srivastava, A.; Macmillan India: New Delhi, India, 2010.

Srinivas, G; Perianayagam, A. Trends in Health and Health Inequalities Among Major States of India: Assessing Progress Through Convergence Models. *Health Econ. Policy Law* **2013,** *9* (12), 143–168.

Srivastava, S. Family, Deviance and Delinquency. *Trends Soc. Sci. Res.* **1995,** *2* (1), 95–96.

UN, World Economic and Social Survey 2007. Development in an Ageing World. Department of Economic and Social Affairs (DESA). United Nations Publication, New York, 2007.

United Nations Department of Economic and Social Affairs, Population Division. World Population Prospects. 2008. http://esa.un.org/unpp/index.asp?panel=2.

United Nations. *World Population Prospects: The 2012 revision, Key findings and advanced tables*. Working Paper No, ESA/P/WP.227. 2013. http://esa.un.org/wpp/Documentation/pdf/WPP2012_ KEY FINDINGS.pdf.

United Nations World Population Projections to 2150: The Population Division of the United Nations Secretariat. *Population Develop. Rev.* **1998,** *24* (1), 183–189. Population Council Stable http://www.jstor.org/stable/2808146/ (accessed Mar 25, 2018)

UNFPA-BKPAI Survey for West Bengal. Building a Knowledge Base on Population Ageing in India, (2014).

Walker, A. (1980). The Social Creation of Poverty and Dependency in Old Age. *J. Soc. Policy* **1980,** *9*, 49–75.

WHO-World Health Organization. Good Health Adds Life to Years: Global Brief for World Health Day 2012. http://whqlibdoc.who.int/hq/2012/ (accessed Jan 14, 2013).

Wolf, R. S. The Nature and Scope of Elder Abuse: Changes in Perspective and Response over the Past 25 Years. *Generations* **2000,** *24* (2), 6–12.

World Data Bank Government of India, Economic Survey. 2014. World Data Bank, World Bank. http://databank.worldbank.org (accessed Mar 21, 2017).

PART 2
Economic Growth and Development

CHAPTER 13

Interrelation Between Economic Growth, Financial Development, Trade, Energy Consumption, and Environmental Degradation: A Multivariate Time Series Analysis on Turkey

KAMİL DEMİRBERK ÜNLÜ[1,*], NİHAN POTAS[2], and MEHMET YILMAZ[1]

[1]*Department of Statistics, Faculty of Science, Ankara University, Ankara 06100, Turkey*

[2]*Department of Health Care Management, Faculty of Economics and Administrative Sciences, Hacı Bayram Veli University, Ankara 06100, Turkey*

[*]*Corresponding author. E-mail: kdunlu@ankara.edu.tr*

ABSTRACT

The aim of this study is to explore the linkage between energy consumption, CO_2 emissions, economic growth, trade openness and financial development by utilizing multivariate time series techniques for the periods 1960 and 2013. In order to investigate the dependency between the variables vector error correction model is used. The cointegration test results indicates that these variables are related in the short run and in the long run. Granger causality tests indicates that there exists a bi-directional causality running from CO_2 to economic growth and trade to economic growth. In addition, this study enlights the unidirectional relation between economic growth and financial development.

13.1 INTRODUCTION

Economic development depends on macroeconomic factors such as investment rates, trade openness, unemployment rates, etc. The effects of these factors are widely studied in the literature such as Arestis and Demetriades (1997), Asafu-Adjaye (2000), Demurger (2001), and Christopoulos and Tsionas (2004). Since energy is very crucial part of the economy and the daily life, there are a variety of works done by Ebohon (1996), Soytas and Sari (2003), Zhang and Cheng (2009), and Soytas et al. (2007), which studies the association between economic activities and energy consumption. Willingly or unwillingly economic development and consumption of energy harm the nature. The relation between economic progress or energy consumption and environmental deterioration is also studied in the literature. Some of these studies are done by Chang (2010), Tiwari (2011), Soytas and Sarı (2009), Farhani and Ozturk (2015), and Dogan and Turkekul (2016).

In this study, we will try to synthesize the above literatures to explore the dependency between economic development, macroeconomic factor, and environmental degradation. We chose the following four macroeconomic variables. The first macroeconomic variable is the financial development. It is a factor that can influence the whole economy in many ways. A stable and strong financial structure will probably lower the cost of loans and create investment opportunist to corporations and individuals. Investment on manufacturing equipment would increase the efficiency in the production and lower the cost of production. In the same time, these technological advancements can decrease the CO_2 emission. Another effect of the low-cost loan can be the following. It may increase the consumption, which may lead to higher energy consumption and higher CO_2 emission. It can be seen that these variables are interrelated with each other. The second variable is trade openness. It is chosen because as the countries involve in open trade, they should develop some comparative advantages to overcome competition. These advancements and improvements in production or in technology would accelerate the economic growth and decrease the CO_2 emission. The last variable about the economy is the energy consumption. No doubt, energy is the key component of the production, and it is the main source of the CO_2 emission. Increase in production leads to higher income and increase in energy consumption. Since Turkey has scarce amount of energy resources, our aim here is to investigate the effect of energy on economic growth. If it causes economic growth, Turkey should invest in renewable energy sources or nuclear energy sources to maintain economic growth. Lastly, we introduce CO_2 emission, since we use carbon-based fuels to create energy, we should monitor the CO_2 levels to reduce the greenhouse effect to fight against the global warming. The fundamental contribution of this study is that, to best our knowledge, the first-time financial development variable is used to detect the relationship between economic growth, energy, trade, and CO_2 emission for the case of Turkey. The paper is organized as follows. Section 13.2 summarizes the literature review. The empirical analyzes are presented in Section 13.3. The last section concludes the paper.

13.2 LITERATURE REVIEW

We divided the literature review into two parts. In the first part, we review the literature about the causal relationship between financial and economic indicators and economic indicators such as trade openness, energy consumption, unemployment rate, foreign direct investment, export, etc. In the second part, we review the literature about the dependency between financial development and the environmental degradation. These works are mostly in developing countries.

Nair and Weinhold (2001) use two different panel data estimation method to enlighten the relation between foreign direct investment and economic advancement for 24 developing countries. Unidirectional casual relation between economic growth and energy consumption in India is showed by Paul and Bhattacharya (2004). Bootstrap panel Granger methodology is used by Hsues et al. (2003) to examine the linkage between financial stability and economic growth of Asian countries. Their results indicate that financial development leads to economic growth. Ahmad and Kwan (1991) use Granger causality test to identify the relationship between international trade and economic growth of 47 developing countries in Africa. Menyah et al. (2014) studied the relationship between trade openness, economic growth, and financial development for 21 African countries by employing panel data analysis. Their results show limited support for financial development economic growth nexus. Unlu and Selcuk-Kestel (2017) show the casual relationship between economic development and education in Turkey by using cointegration analysis. Wolde-Rufael (2014) uses bootstrap panel data analysis to show limited relation between economic advancement and energy consumption for 15 transition economies. The linkage between renewable energy and economic growth in new EU countries is revealed by Alper and Oguz (2016) by using autoregressive distributed lag. The causal relation between energy prices, energy consumption, and economic activities in Saudi Arabia is investigated by Alshehry and Belloumi (2015), and their results suggest unidirectional causality between economic development and energy consumption, while bidirectional causality between CO_2 and economic growth. Yıldırım et al. (2014) uses bootstrap autoregressive metric causality approach to identify the causal relationship between energy consumption and economic progress in next 11 countries. Samargandi et al. (2015) revisit the dependence between economic development and financial development for 52 middle-income countries; their findings show that the short run relation between these variables is insignificant.

Soytas and Sarı (2009) use multivariate time series analysis to identify the causal relation between CO_2 and economic development in Turkey. Kivyiro and Arminen (2014) discover that gross domestic product (GDP), energy consumption, and FDI cause CO_2 in sub-Saharan countries. Bozkurt and Akan (2014) analyzed the relation between economic growth and CO_2 in Turkey by using cointegration methodology; their results indicate that CO_2 emission negatively affects economic growth. Kasman and Duman (2015) use panel data approach on EU and EU candidate countries. Their results show that higher income cause more CO_2 emission. The linkage between economic growth and CO_2 emission in Israel is studied by Magazzino (2015) by forecast error variance decompositions, and their findings reveal the relation between economic growth and environmental deterioration. Hwang and Yao (2014) studied the relation between energy consumption, economic growth, and CO_2 emission. Their result shows bidirectional causality from energy consumption to CO_2 emission and unidirectional relation between economic growth and CO_2 emission. Yavuz (2014) studies the relation between income and CO_2 emission by utilizing the cointegration test with structural breaks. Their results show relation between CO_2 emission and economic growth in Turkey. The links between economic growth, CO_2, and coal consumption in India and China are studied by Govindaraju and Tang (2013) by using cointegration techniques. The results indicate causal relation from economic growth to CO_2 emission in China. Omri (2013) investigates the economic growth, energy consumption, and CO_2 emission of MENA countries. The result shows that there is no causal relation between CO_2 emission and energy consumption in the short run, but in the long run, there exists a cointegration between economic growth, energy consumption, and CO_2 emission. Economic growth and CO_2 emission of European Union is studied by Bengochea-Morancho et al. (2001) and their result differs from country to country. Hossain (2012) examines the linkage between CO_2 emission, economic growth, foreign trade, and urbanization. Their results enlight long-run relation between these variables.

13.3 DATA AND METHODOLOGY

The data set consists of annual GDP measured in US dollar (USD), annual trade openness (TRD) measured in USD, annual energy consumption (ENR) measured in kilograms of oil, annual credit to private sector (CRD) measured in USD, and carbon dioxide emission (CO_2) for the periods between 1960 and 2013. These time series are obtained from the World Bank Databank (http://databank.worldbank.org).

We use the cointegration procedure of Johansen (1991) and Johansen and Juselius (1990), which enables us to determine the linear association among the variables. If there exists any cointegration vector in the system, we can conclude that these variables are interrelated. Also by this methodology, we can identify the causal relation between the variables in the short and long run. In order to implement cointegration test, all variables should be at the same order of integration. We use unit root tests of Dickey and Fuller (1979) (ADF) and Phillips and Perron (1988) (PP) to detect the order of integration. The results of the unit root tests and the decision of integrations are given in Table 13.1.

TABLE 13.1 Unit Root Test Results.

	CO_2	GDP	ENR	TRD	CRD
ADF					
Level	−2.4440	−0.0424	−1.2806	−1.8836	1.0672
Δ	−7.9416*	−8.4494*	−6.9602*	−7.1979*	−6.1438*
PP					
Level	−2.5286	−0.0424	−1.3417	−1.8865	1.0403
Δ	−7.9299*	−8.4494*	−6.0545*	−7.2869*	−6.1381*
Decision	I(1)	I(1)	I(1)	I(1)	I(1)

*1% significant level.
Δ is the first difference of the data, and lag lengths are determined by minimizing Schwarz Information Criterion.

Since all variables are on the same order of integration, variables become stationary at the first difference, the next step in the analysis is to determine the appropriate number of lags, which is determined by minimizing Schwarz Information Criterion. The appropriate lag-length is determined as 1. In the next step, we apply the cointegration test of Johansen and Juselius (1990), and the results are given in Table 13.2.

TABLE 13.2 Cointegration Test Results (lag = 1).

H_0	λ_{trace}	5%	λ_{max}	5%	Eigenvalues
$r = 0^*$	102.3123*	69.8189*	44.7767*	33.8768*	0.5773*
$r \leq 1^*$	57.5356*	47.8561*	29.7571*	27.5843*	0.4357*
$r \leq 2$	27.7785	29.7971	15.6919	21.1316	0.2305
$r \leq 3$	12.0866	15.4947	11.39455	14.2646	0.1968
$r \leq 4$	0.6920	3.8415	0.6920	3.8415	0.0132

*5% significant level.

According to the maximum eigenvalue (λ_{max}) and the trace statistics (λ_{trace}), there exist two cointegration vectors. The existence of the cointegration among the variables implies Granger causality. In order to determine the direction of the causality, long and short run dependency between the variables we use the vector error correction model instead of vector autoregression model. We follow the work of Granger (1988) to estimate the Vector Error Correction (VEC) of the system, which is as follows:

$$ln\Delta CO2_t = \mu_1 + \beta_{1,1} ln\Delta GDP_{t-1} + \beta_{1,2} ln\Delta ENR_{t-1} + \beta_{1,3} ln\Delta TRD_{t-1} + \beta_{1,4} ln\Delta CRD_{t-1} + \theta_{1,1} ECT_{1,t-1} + \theta_{1,2} ECT_{2,t-1} + \varepsilon_{1,t}. \quad (13.1)$$

$$ln\Delta GDP_t = \mu_2 + \beta_{2,1} ln\Delta CO2_{t-1} + \beta_{2,2} ln\Delta ENR_{t-1} + \beta_{2,3} ln\Delta TRD_{t-1} + \beta_{2,4} ln\Delta CRD_{t-1} + \theta_{2,1} ECT_{1,t-1} + \theta_{2,2} ECT_{2,t-1} + \varepsilon_{2,t}. \quad (13.2)$$

$$ln\Delta ENR_t = \mu_3 + \beta_{3,1} ln\Delta GDP_{t-1} + \beta_{3,2} ln\Delta CO2_{t-1} + \beta_{3,3} ln\Delta TRD_{t-1} + \beta_{3,4} ln\Delta CRD_{t-1} + \theta_{3,1} ECT_{1,t-1} + \theta_{3,2} ECT_{2,t-1} + \varepsilon_{3,t}. \quad (13.3)$$

$$ln\Delta TRD_t = \mu_4 + \beta_{4,1} ln\Delta GDP_{t-1} + \beta_{4,2} ln\Delta ENR_{t-1} + \beta_{4,3} ln\Delta CO2_{t-1} + \beta_{4,4} ln\Delta CRD_{t-1} + \theta_{4,1} ECT_{1,t-1} + \theta_{4,2} ECT_{2,t-1} + \varepsilon_{4,t}. \quad (13.4)$$

$$ln\Delta CRD_t = \mu_5 + \beta_{5,1} ln\Delta GDP_{t-1} + \beta_{5,2} ln\Delta ENR_{t-1} + \beta_{5,3} ln\Delta TRD_{t-1} + \beta_{5,4} ln\Delta CO2_{t-1} + \theta_{5,1} ECT_{1,t-1} + \theta_{5,2} ECT_{2,t-1} + \varepsilon_{5,t}, \quad (13.5)$$

where *ln* is the natural logarithm of the variables, the symbol Δ represents the first difference of the variables, $\mu_1 \ldots \mu_5$ is the intercept of the equation. θ is the coefficient of the error correction term, which shows the speed of adjustment, and ECT terms represent the error correction vector, which represents the long run cointegration relation between the variables that is:

TABLE 13.3 Granger Causality Test Results.

	Short-Run					Error correction		Long-Run					
	F-statistics					Coefficients		Joint F-statistics					
Dependent variable	**ΔCO_2**	**ΔGDP**	**ΔENR**	**ΔTRD**	**ΔCRD**	**ECT_1**	**ECT_2**	**ΔCO_2**	**ΔGDP**	**ΔENR**	**ΔTRD**	**ΔCRD**	**Lag**
ΔCO_2	–	0.1602	0.5159	0.9599	0.7381	−0.1812**	0.01375*	–	6.1947*	5.3052*	5.3029*	5.2881*	1
ΔGDP	0.2723	–	1.5800	1.3742	0.4053	−0.5391**	−0.7413**	2.0299	–	2.7519**	2.3836***	2.4488***	2
ΔENR	1.0720	1.0286	–	0.1701	1.2984	0.01988	−0.0098	0.8698	0.5504	–	0.1618	0.6730	1
ΔTRD	0.2430	1.5218	0.6208	–	0.3382	0.04066	0.3291	0.5231	0.8452	0.5220	–	0.7232	1
ΔCRD	0.0004	3.7085**	0.0239	1.9018	–	0.03034***	0.4538***	1.9918	3.4727**	2.7303**	2.9655**	–	1

*1% significant level.

**5% significant level.

***10% significant level.

$$lnCO_2 = \lambda + \varphi_1 lnGDP + \varphi_2 lnENR + \varphi_3 lnTRD + \varphi_4 lnCRD + \varepsilon_t,$$

where λ is the intercept and ε_t is the residual. The models in eq 13.1–13.5 are estimated, and the results of the short-run and short-run Granger causality test statistics and the estimated coefficients of the ECT with significance levels are given in Table 13.3. Multiple diagnostics tests are performed on the models such as Jarque–Bera test for normality, CUSUM and CUSUM square test for stability of the parameters, LM test for serial correlation, and ARCH LM test for heteroscedasticity. The results of these tests are given in Table 13.4.

According to Table 13.3, Granger causality test results and cointegration results are summarized as follows

- In the first model, coefficients of the ECT are statistically significant; in the short run, neither of the variables causes CO_2 but in the long run, all of the variables cause CO_2, which confirms the results of Halicioglu (2009) and Soytas and Sari (2009).
- The ECT coefficients of the second model are also statistically significant, again in the short run neither of the variables triggers economic growth but in the long run energy consumption, trade openness and financial development cause economic growth.
- In the third and fourth models, all the coefficient of the ECT are insignificant and there are no short- and long -run relation between the variables.
- The coefficient of the ECT of the last model is statistically significant. In the short run, economic growth cause financial development, and in the long run, trade openness, energy consumption, and economic growth.

TABLE 13.4 Diagnostic Test Results.

Dependent variable	Serial cor. (F-Statistics)	Heteroscedasticity (F-statistics)	Normality (statistics)	CUSUM	CUSUM-SQ
ΔCO_2	1.3776	1.7364	2.4003	Stable	Stable
ΔGDP	0.7425	0.5765	6.5118**	Stable	Stable
ΔENR	1.1979	0.8374	4.9018***	Stable	Stable
ΔTRD	0.7535	1.0270	3.3049	Stable	Stable
ΔCRD	0.6368	0.9849	4.0602	Stable	Stable

**5% significant level.

***10% significant level.

CONCLUSION

In this work, we try to explore the relation between macroeconomic indicators, energy consumption and environmental degradation nexus for the periods of 1960 and 2013, which is the longest available data set. Our results show in the short run neither of the investigated variables cause CO_2 emission, but in the long run, we show that all variables cause CO_2. It is expected that economic growth cause CO_2 emission by increasing energy consumption but in our case economic development does not cause energy consumption in the short and long run. This result shows that Turkey does not have to forgo economic growth because of its scarce amount of energy reserves. In the long run, energy consumption causes CO_2 emission; this is not a surprising result since by 2016, Turkey (https://www.teias.gov.tr) produces only 8.39% of energy from renewable sources. Turkey produces almost 60% of its energy by burning carbon-based fuel so an increase in energy consumption naturally causes CO_2 emission. Trade openness and financial stability cause CO_2 in the long run. Financial development may affect CO_2 in two ways, firstly as the financial system become more stable and strong firms can access more amounts of funds. This may increase the efficiency of production by utilizing new technology and equipment, which probably decrease the CO_2 emission. Secondly, a strong financial system will decrease the cost of loans that increase the consumption. If households consume more, they will stimulate an increase in energy consumption. In the case of Turkey, second effect is superior to the first, so financial stability cause carbon emissions. As in the case of financial development, an increase in imports of goods reflects an increase in consumption so trade openness affects the CO_2 in the same manner. Thus, an increase in financial development and trade cause CO_2 emission. In order to protect the nature, policy-makers should consider these relationships. In the short run, neither of the variables does cause economic growth, but in the long run, trade and financial development cause economic growth. As the financial system gets stronger and stable, it enables more governmental and private investment, which leads to economic development. Energy consumption and trade openness also lead economic growth. When the industry produces more, it needs more energy for that reason increase in energy consumption cause economic growth. If the country involves in international trade, it must have some comparative advantages in order to cope with other countries. This competition among the countries leads development in the economy, for these reasons these macroeconomic variables cause economic growth. For the case of financial development, in the short run only economic growth causes development because as the country get higher wealth it will have more funds to distribute to the private and governmental sectors. As expected in the long run, all the macroeconomic variables cause financial development. Since these macroeconomic variables are highly interdependent and improvement in one of them would cause an increase in the financial development.

KEYWORDS

- **economic growth**
- **financial growth**
- **environmental degradation**
- **multivariate time series analysis**
- **cointegration**

REFERENCES

Ahmad, J.; Kwan, A. C. Causality Between Exports and Economic Growth: Empirical Evidence from Africa. *Econ. Lett.* **1991,** *37* (3), 243–248.

Alper, A.; Oguz, O. The Role of Renewable Energy Consumption in Economic Growth: Evidence from Asymmetric Causality. *Renew. Sustain. Energy Rev.* **2016,** *60*, 953–959.

Alshehry, A. S.; Belloumi, M. Energy Consumption, Carbon Dioxide Emissions and Economic Growth: The Case of Saudi Arabia. *Renew. Sustain. Energy Rev.* **2015,** *41*, 237–247.

Arestis, P.; Demetriades, P. Financial Development and Economic Growth: Assessing the Evidence. *Econ. J.* **1997,** *107* (442), 783–799.

Asafu-Adjaye, J. The Relationship Between Energy Consumption, Energy Prices and Economic Growth: Time Series Evidence from Asian Developing Countries. *Energy Econ.* **2000,** *22* (6), 615–625.

Bengochea-Morancho, A.; Higón-Tamarit, F.; Martínez-Zarzoso, I. Economic Growth and CO_2 Emissions in the European Union. *Environ. Res. Econ.* **2001,** *19* (2), 165–172.

Bozkurt, C.; Akan, Y. Economic Growth, CO_2 Emissions and Energy Consumption: The Turkish Case. *Int. J. Energy Econ. Policy* **2014,** *4* (3), 484.

Chang, C. C. A Multivariate Causality Test of Carbon Dioxide Emissions, Energy Consumption and Economic Growth in China. *Appl. Energy* **2010,** *87* (11), 3533–3537.

Christopoulos, D. K.; Tsionas, E. G. Financial Development and Economic Growth: Evidence from Panel Unit Root and Cointegration Tests. *J. Develop. Econ.* **2004,** *73* (1), 55–74.

Demurger, S. Infrastructure Development and Economic Growth: An Explanation for Regional Disparities in China? *J Comparative Econ.* **2001,** *29* (1), 95–117.

Dickey, D. A.; Fuller, W. A. Distribution of the Estimators for Autoregressive Time Series with a Unit Root. *J. Amer. statist. Assoc.* **1979,** *74* (366a), 427–431.

Dogan, E.; Turkekul, B. CO_2 Emissions, Real Output, Energy Consumption, Trade, Urbanization and Financial Development: Testing the EKC Hypothesis for the USA. *Environ. Sci. Pollution Res.* **2016,** *23* (2), 1203–1213.

Ebohon, O. J. Energy, Economic Growth and Causality in Developing Countries: A Case Study of Tanzania and Nigeria. *Energy Policy.* **1996,** *24* (5), 447–453.

Farhani, S.; Ozturk, I. Causal Relationship Between CO_2 Emissions, Real GDP, Energy Consumption, Financial Development, Trade Openness, and Urbanization in Tunisia. *Environ. Sci. Pollut. Res.* **2015,** *22* (20), 15663–15676.

Govindaraju, V. C.; Tang, C. F. The Dynamic Links Between CO_2 Emissions, Economic Growth and Coal Consumption in China and India. *Appl. Energy.* **2013,** *104*, 310–318.

Granger, C. W. Causality, Cointegration, and Control. *J. Econ. Dyn. Control.* **1988,** *12* (2–3), 551–559.

Halicioglu, F. An Econometric Study of CO_2 Emissions, Energy Consumption, Income and Foreign Trade in Turkey. *Energy Policy* **2009,** *37* (3), 1156–1164.

Hossain, S. An Econometric Analysis for CO_2 Emissions, Energy Consumption, Economic Growth, Foreign Trade and Urbanization of Japan. *Low Carbon Econ.* **2012,** *3* (3), 92–105.

Hsueh, S. J.; Hu, Y. H.; Tu, C. H. Economic Growth and Financial Development in Asian Countries: A Bootstrap Panel Granger Causality Analysis. *Econ. Model.* **2013,** *32*, 294–301.

Hwang, J. H.; Yoo, S. H. Energy Consumption, CO_2 Emissions, and Economic Growth: Evidence from Indonesia. *Qual. Quant.* **2014,** *48* (1), 63–73.

Johansen, S. Estimation and Hypothesis Testing of Cointegration Vectors in Gaussian Vector Autoregressive Models. *Econometrica: J. Econ. Soc.* **1991,** 1551–1580.

Johansen, S.; Juselius, K. Maximum Likelihood Estimation and Inference on Cointegration—with Applications to the Demand for Money. *Oxford Bull. Econ. Stat.* **1990,** *52* (2), 169–210.

Kasman, A.; Duman, Y. S. CO2 Emissions, Economic Growth, Energy Consumption, Trade and Urbanization in New EU Member and Candidate Countries: A Panel Data Analysis. *Econ. Model.* **2015,** *44*, 97–103.

Kivyiro, P.; Arminen, H. Carbon Dioxide Emissions, Energy Consumption, Economic Growth, and Foreign Direct Investment: Causality Analysis for Sub-Saharan Africa. *Energy* **2014,** *74*, 595–606.

Magazzino, C. Economic Growth, CO2 Emissions and Energy Use in Israel. *Int. J. Sustain. Develop. World Ecol.* **2015,** *22* (1), 89–97.

Menyah, K.; Nazlioglu, S.; Wolde-Rufael, Y. Financial Development, Trade Openness and Economic Growth in African Countries: New Insights from a Panel Causality Approach. *Econ. Model.* **2014,** *37*, 386–394.

Nair-Reichert, U.; Weinhold, D. Causality Tests for Cross-country Panels: A New Look at FDI and Economic Growth in Developing Countries. *Oxford Bull. Econ. Statist.* **2001,** *63* (2), 153–171.

Omri, A. CO2 Emissions, Energy Consumption and Economic Growth Nexus in MENA Countries: Evidence from Simultaneous Equations Models. *Energy Econ.* **2013,** *40*, 657–664.

Paul, S.; Bhattacharya, R. N. Causality Between Energy Consumption and Economic Growth in India: A Note on Conflicting Results. *Energy Econ.* **2004,** *26* (6), 977–983.

Phillips, P. C.; Perron, P. Testing for a Unit Root in Time Series Regression. *Biometrika.* **1988,** *75* (2), 335–346.

Samargandi, N.; Fidrmuc, J.; Ghosh, S. Is the Relationship Between Financial Development and Economic Growth Monotonic? Evidence from a Sample of Middle-income Countries. *World Develop.* **2015,** *68*, 66–81.

Soytas, U.; Sari, R. Energy Consumption and GDP: Causality Relationship in G-7 Countries and Emerging Markets. *Energy Econ.* **2003,** *25* (1), 33–37.

Soytas, U.; Sari, R.; Ewing, B. T. Energy Consumption, Income, and Carbon Emissions in the United States. *Ecol. Econ.* **2007,** *62* (3), 482–489.

Soytas, U.; Sari, R. Energy Consumption, Economic Growth, and Carbon Emissions: Challenges Faced by an EU Candidate Member. *Ecol. Econ.* **2009,** *68* (6), 1667–1675.

Tiwari, A. K. A Structural VAR Analysis of Renewable Energy Consumption, Real GDP and CO_2 Emissions: Evidence from India. *Econ. Bull.* **2011,** *31* (2), 1793–1806.

Unlu, K. D.; Selcuk-Kestel, A. S. Determination of Causality in Carbon Emission with Respect to Economic Growth and Education. *Gazi Univ. J. Sci.* **2017,** *30* (3), 161–179.

Zhang, X. P.; Cheng, X. M. Energy Consumption, Carbon Emissions, and Economic Growth in China. *Ecol. Econ.* **2009,** *68* (10), 2706–2712.

Wolde-Rufael, Y. Electricity Consumption and Economic Growth in Transition Countries: A Revisit Using Bootstrap Panel Granger Causality Analysis. *Energy Econ.* **2014,** *44*, 325–330.

Yavuz, N. Ç. CO2 Emission, Energy Consumption, and Economic Growth for Turkey: Evidence from a Cointegration Test with a Structural Break. *Energy Sour. Part B: Econ. Plann. Policy* **2014,** *9* (3), 229–235.

Yıldırım, E.; Sukruoglu, D.; Aslan, A. Energy Consumption and Economic Growth in the Next 11 Countries: The Bootstrapped Autoregressive Metric Causality Approach. *Energy Econ.* **2014,** *44*, 14–21.

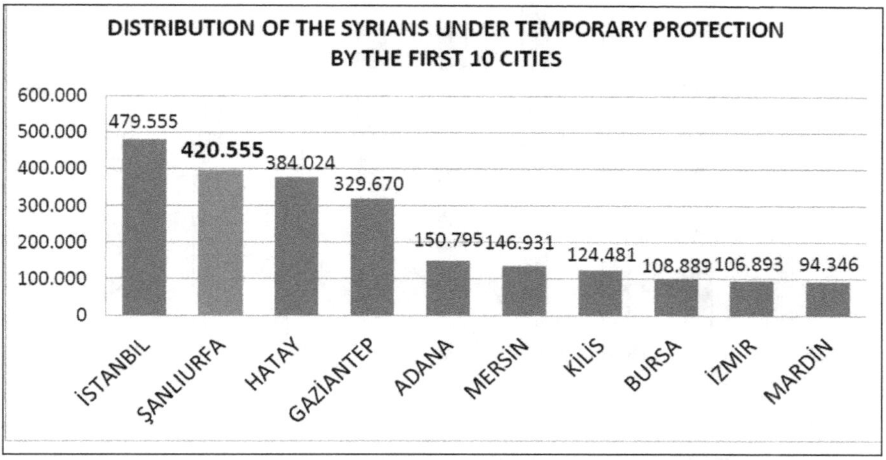

FIGURE 3.1 Distribution of the Syrians under temporary protection by the first 10 cities.
Source: http://www.goc.gov.tr/icerik6/gecici-koruma_363_378_4713_icerik

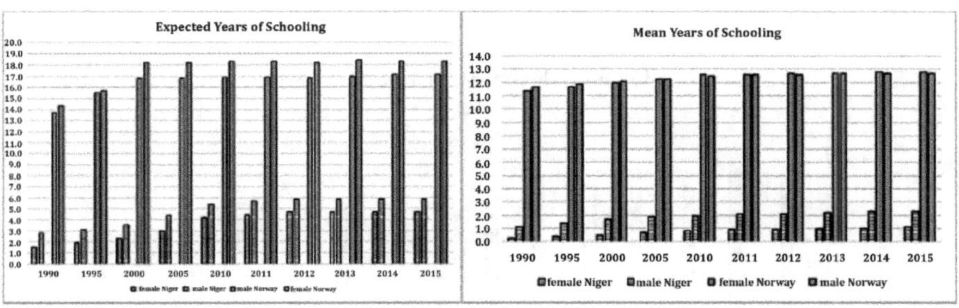

FIGURE 9.3 Expected years of schooling and mean years of schooling.

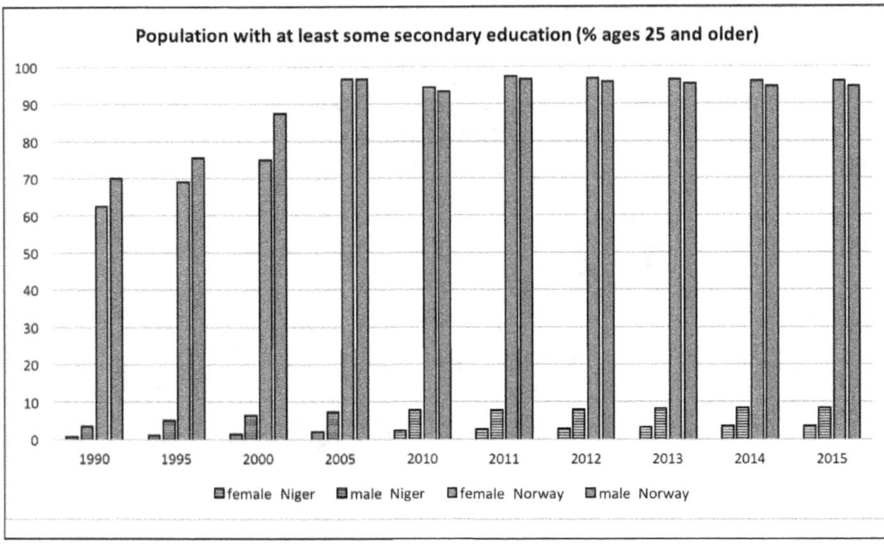

FIGURE 9.4 Proportion of population who completed at least secondary level in the country (% ages 25 and older).

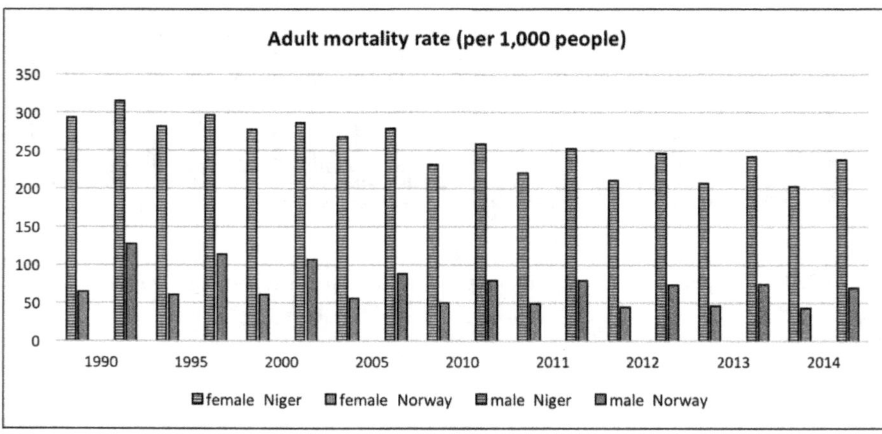

FIGURE 9.6 Adult mortality rate.

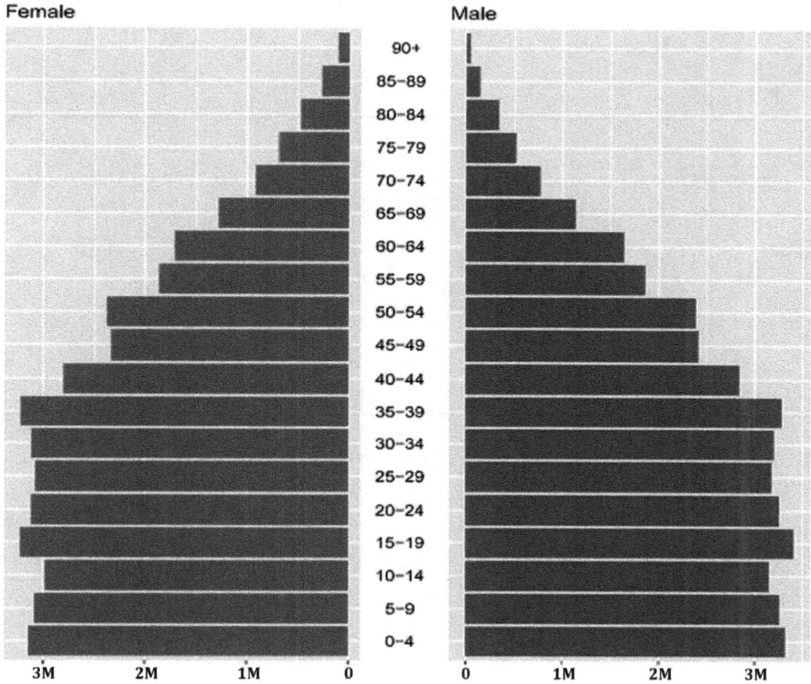

FIGURE 11.1 Population age pyramid for Turkey, 2016.
Source: TurkStat, Address Based Population Registration System (ABPRS), 2015, 2016.

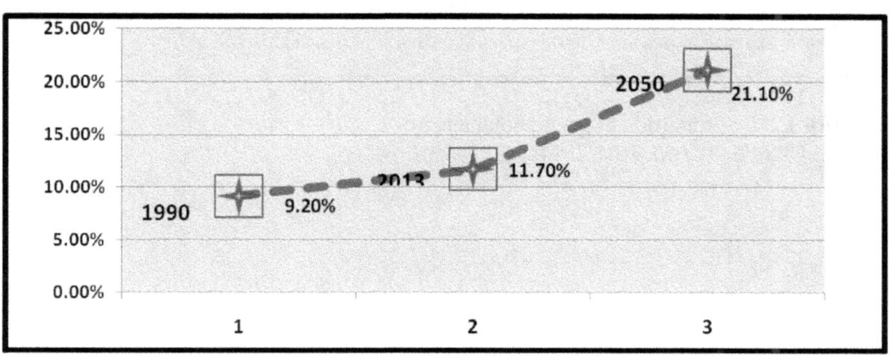

FIGURE 12.1 Proportion of world population.
Source: UN 2013.

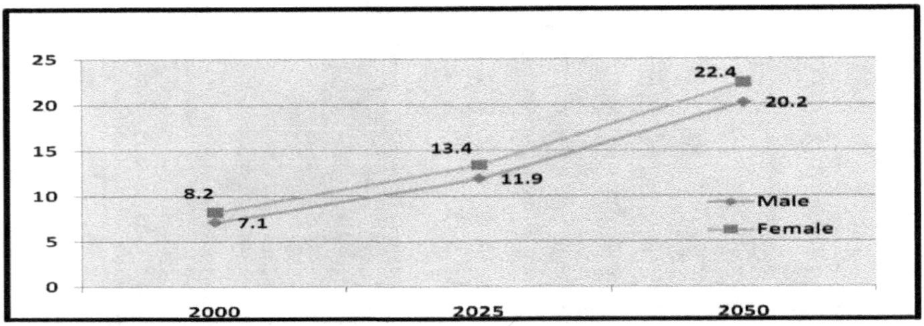

FIGURE 12.2 Gender-wise proportion of the world population.
Source: UN 2013.

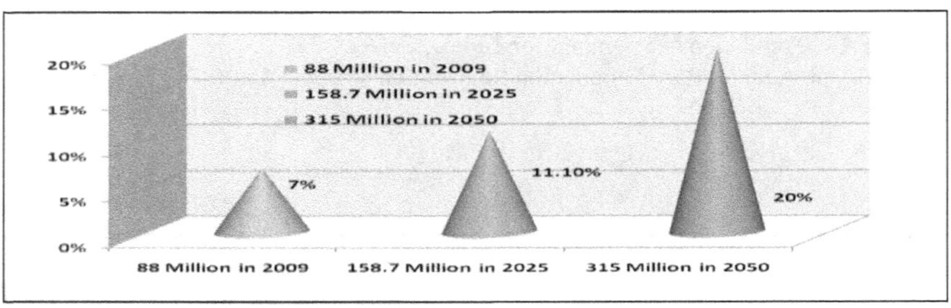

FIGURE 12.3 Proportion of population increase.
Source: UNFPA-BKPAI, 2011/2014.

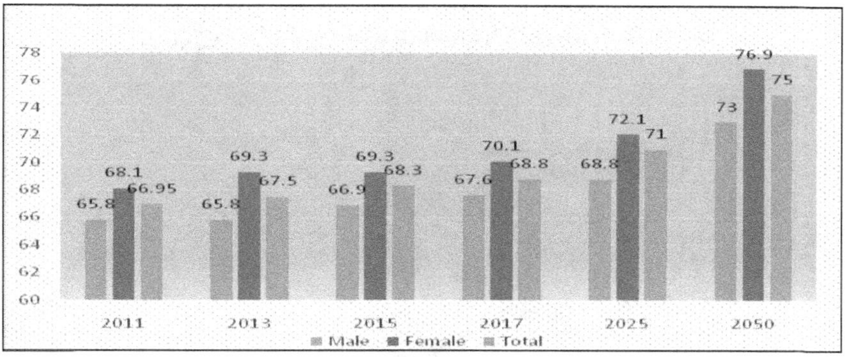

FIGURE 12.4 Life expectancy pattern of India.
Source: Census, 2011; UN 2013; UN-World Population Projection 2150.

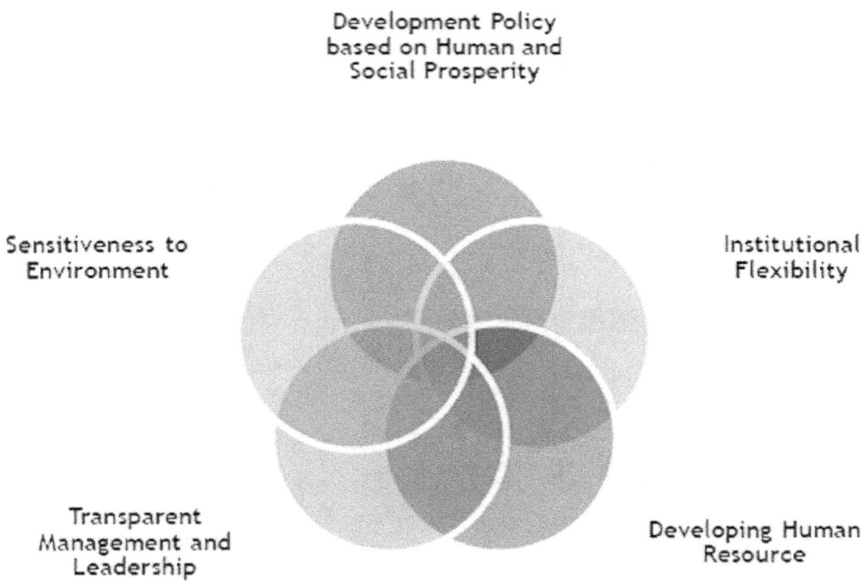

FIGURE 17.1 The possible dimensions of the challenges to economic growth and development.

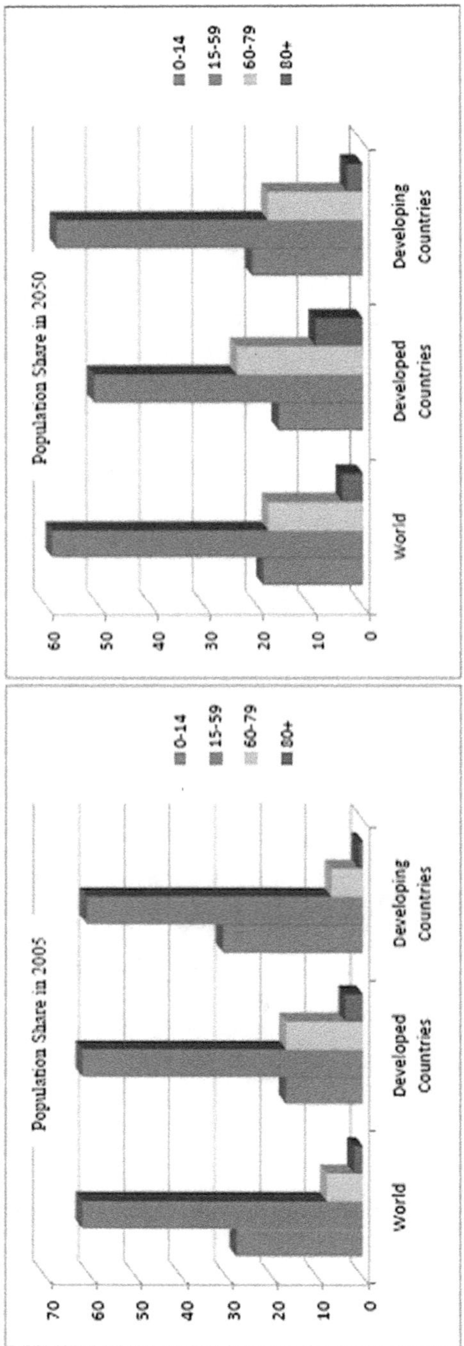

FIGURE 18.1 World and development-group age-structure changes.

Source: Adapted from UN (2009); Bloom et al. (2010).

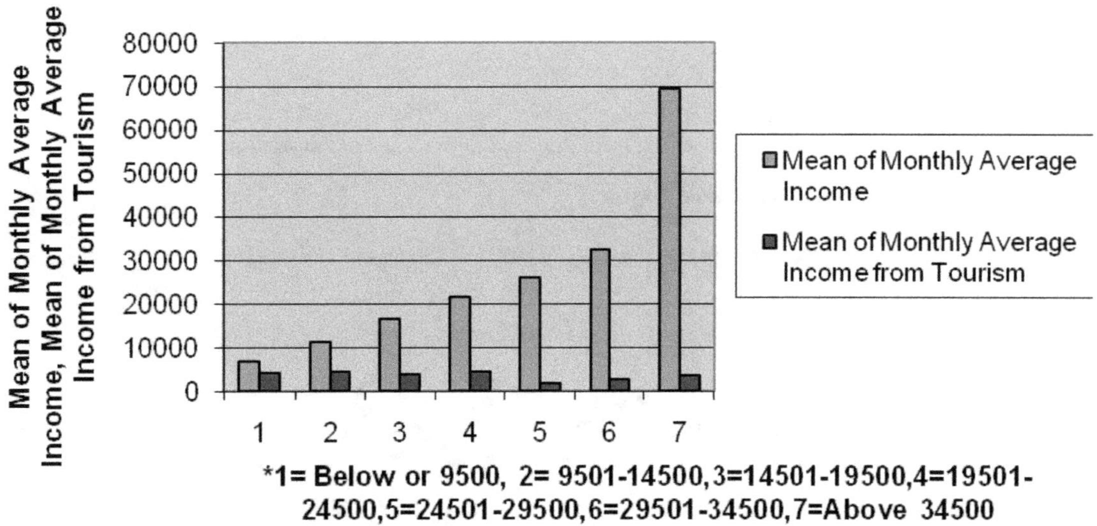

FIGURE 23.1 Mean of monthly average income and mean of monthly average income from tourism of the local people involved in tourism business in Sikkim.

Source: Field survey 2015–2016.

Note: Statistical analysis has been made using Excel.

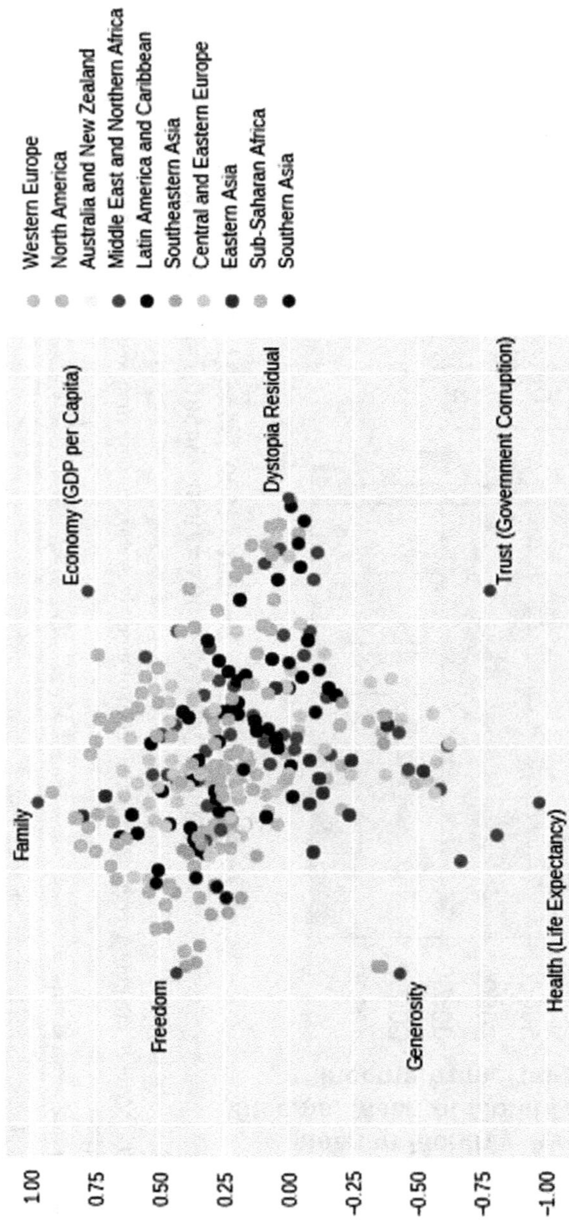

FIGURE 25.4 Scatter plot happiness by countries.

CHAPTER 14

Policy Designs for Economic Growth and Development

MUKUND DESHPANDE*

Department of Plant Science, Bharathidasan University, Tiruchirappalli–620024, India

**E-mail: mvdeshpande49@yahoo.co.in*

ABSTRACT

Public policy is an attempt by the Government to address a public issue. Government frequently designs policies to make laws, regulations, decisions, and actions. Public policy-making typically consists of three parts such as the problems, players, and the policy. Strategic tool has been recognized to endeavor economic prosperity that deliberately incorporates planning and priorities. Industry development that contributes greatly toward economic development has been considered as an imperative illustration which often seeks new approaches for addressing challenges owing to varying global environment. Third world nation is a nomenclature given to developing and poor countries. Such countries have the curiosity to know more about policies adopted by developed countries to create economic growth. They face challenges owing to chaos and complexity that has emerged due to globalization. Therefore, situation makes them difficult to implement plans of economic growth. The chapter aims to highlight method of addressing challenges and come out with a win-win situation. Further aim of this chapter is to acquire an improved understanding of the basis of designing policies and arrive at universal methodology and approach. Economic development is the prime area of interest amongst majority of countries and in that context the chapter discusses those emerging policies exhaustively.

14.1 INTRODUCTION

Designing a policy is an art that requires knowledge of background, critical dimensions, and the procedure. The policy is a document that remains in force for a long-term and displays declarations to address issues. The framing of policy initiates with the objectives and a designed path that is conceptualized using supporting information to produce the desired outcome. Knowledge inputs are sought from competent and conversant participants who set declarations after verifying the effect of the forecast. Policy statements are construed to achieve workable solutions. Developing a policy is fairly multifaceted as the determinants of the economy are yet to be completely recognized and understood amidst global challenges. Policy designing is an inclusive skill that needs overall and domain-specific knowledge. An appropriate instrument is indispensable for designing policy. Dimensions of economy and knowledge of issues are the further requirement. A line of thought perpetually depends upon processing relevant information and extending it to arrive at conclusion. Directing the perceptions to the goals of the policy finds first place whereas leading them to the required result comes out to be the next step in shaping it. As a result, policy-making is certainly an art and a serious affair that is essential to be performed on a strong footing with well-structured plan. Policies are being frequently designed for Financial, Social, and Political Institutions and therefore well known in the policy world. Nevertheless, there are hardly any approaches established when countries face challenges due to chaos and complexity as well as demographic transition due to globalization. As a result, the chapter is taken to identify the foundation of policy-making and to search economy development policies. The aim of this chapter is to highlight and reveal best practices of policy formulation particularly for economic development with reference to the background of third world nations and chaos and complexity theory. "policy designing" for all time aims at efficiency, and has continued to be an interesting arena for researchers especially in the light of "globalization." The effectiveness of the policy fundamentally is the other criterion emerging in policy circles which depends on the designed and the actual performance. When getting in this area, it was instantly evident that there is very little information is available on the steps that were being adopted by professionals in updating the tools for economic development. As a result of deficiency in the above-stated work, this chapter aims in developing a different line of thinking for a suitable and effective mechanism to establish an optimal approach in ensuring economic development.

14.2 METHODOLOGY

The aim of this chapter is to find opinions of experts and to compile them by surveying the literature for the knowledge of readers. Based on objectives, a systematic review of literature is planned and conducted with a stipulated questionnaire prepared to obtain responses. Prominence is placed on processes used to addressing challenges and designing economic policies through a multilateral approach. Data acquired through diverse sources have been organized in line with the objective and the leads to the designs of economic policy logically analyzed by an observation and interpretation approach.

14.3 LITERATURE REVIEW

With the advent of globalization designing a policy for economic development has become complex according to experts and researchers. Variety of dimensions, opinions, and situations has been observed in the literature. Following is a list of such discussions:

Farlex Inc. (2008) reports diverse types of policies developed according to the profession for the knowledge of readers. In general, a policy is said to be a map of activities accepted by an entity or community. Known is about a politician who keeps changing his policies for exhibiting dynamism in performing some activity; in that case, the core approach is on the political issues of focusing power in the main institution. Containment is another approach of evolving strategic treaty so as to control the growth of aggressiveness to compel it to bargain quietly. While modern-nationalism is the approach to acting with moderation, obscurantism is a terminology used to indicate disagreement to enlightenment or the spread of knowledge.

Paolo Savitti (2008) deliberated about the economic factors on diversity, format change, and development of the economy and pointed out that for growth to be effective variety is a necessary requirement for long-term economic development. "Variety growths, leading to new sectors, and productivity growth in preexisting sectors are complementary and not independent aspects of economic development."

GOC (2017) reports the features essential to accomplish high rates of sustainable development of the economy. If an administration is willing to adopt effective and visible public operation of the economy, to provide safe, stable, and attractive business and industry prospects, it will lead to strengthening of its ability to satisfy accountability for the welfare of the populace. When a business setting is such that it encourages new enterprise then it leads to competition and market expansion, imports outside knowledge, and rolls over investment prospects. A well-set and maintained infrastructure which includes transportation and communications systems allows good accessibility between marketplaces. The other factors highlighted are diverse, skilled, and qualified workforce which meets the demands of the economy. The social and economic environment is essential to create prospects for both women and men. Managing natural resources with due care of climate impact forms other factors to develop long-term sustainability of the economy.

Harberger (2005) reports scrutiny of the approach to the development of the economy is absolutely essential to identify that all economic growth takes place at the level of the productive enterprise. He further states that otherwise it is impossible to have a clear understanding of the growth mechanism. To explain it further, gross domestic product is measured as the sum of the goods manufactured in all identified economic activities of that country. Occasionally, it is measured at the level of final goods and services, but it evidently includes all the value added that went into making those final products. Even more often by collectively organizing national accounts, the value added by each activity is counted, thus catching all the various pieces that end up constituting the end product of the economy.

Wordpress.com (2011) has reported that the main strategy to develop an economy is to collectively gather domestic and foreign savings. Importance is placed on creating adequate savings to boost the growth of economy. The economic system through which more investment leads to further development is described by the Harrod–Domar model. The said model of economic growth considers that the rate of growth of gross national product is decided jointly by the national savings ratio and the national capital to output ratio. Therefore, the most prime strategy to grow the economy is to create savings and proportionately make an investment to boost Gross National Product (GNP) but the actual rate at which they can grow for any level of saving and investment relies on how much additional output can be had from an additional unit of investment. Capital adequacy is, therefore, the major obstacle to growth which became the ground for transfer of capital and technical assistance to the developing countries.

Santarelli and Ors (2007) have cited in the recent literature about the formation, survival, and growth of newly born firms. The intention is to specifically point out the entrepreneurial approach toward microeconomics of the industrial mechanism such as the entry into and exit from business and to point out the founder's prior characteristics and post business performance. The conclusion derived is that the entry of new firms has heterogeneous characteristics because innovative entrepreneurs find passive followers, over optimist gamblers and even the people escaped from unemployment. As the founders are found heterogeneous they are likely to make "entry mistakes." Therefore, policy incentives should be highly selective, favoring nascent entrepreneurs endowed with progressive motivation and promising predictors of better business performance. This would help in minimizing distortion in the post-entry market selection of efficient entrepreneurs.

Perrels (2001) has cited that Government Institute of Economic Research (VATT), points out that "Economic policy-makers and economists tend to focus on economic efficiency," as this perception mechanically implies that the measures taken are useful in terms of achieving reduction in inefficiencies. Policy Efficiency is measured by computing a ratio of actual performance to that with the planned outcome. The apprehension is that inefficient policies cause unnecessary loss of welfare or in the worst case to undermine the policy itself, due to relocation of manufacturing capacity or criticism in eagerness to invest and to innovate.

Shapiro (2013) has stated that economic growth could be promoted through innovation. Innovation and start-ups add to the economic growth. They are the eventual job creators who start with resourceful ideas, take risks, and create value for the consumer. The nation is built through encouragement to entrepreneurship and innovation. Particularly in the situation of economic slump in the United States, it is vital to follow national policies that encourage innovation to make certain that there will be adequate prosperity to carry on into the next generation.

Hendrickson (2012) points out that it is beneficial to live in a world that comprises of many countries rather than the one with the global government as talented people can compare how various policies work. Leaders of countries that lag behind in economic development can observe what wealthier countries have done to raise standards of living and then replicate those effective policies in their own countries. One famous example of this was that a number of national governments emulated Margaret Thatcher's hugely successful privatization policy and realized similar benefits.

Treasury (2015) highlights economic growth in developing countries. Benefits can be derived from global and regional trade which directly causes effects on poverty as an increase in trade volumes raises incomes. Nations which miss out on taking the benefits of world trade lose on the generating profit from international expertise, low-cost raw materials, and much-needed technology. Many prospects exist to open up global and regional trade further to benefit developing countries.

The government of Sweden (2010) states that on the whole, the aim for the growth of the economy is superior conditions for sustainable growth courses in poor and developing countries. The administration identifies three central areas in Swedish Development Cooperation as being particularly significant to boost economic growth. They are: (1) improved conditions for poor people to take part in growth mechanism, (2) encouragement toward new markets and entrepreneurship, and (3) ability to adapt to change, threats, and opportunities.

Hanushek (2013) has made a point in stating the importance of the quality of human resources which are directly accountable to create economic growth. The human quality in terms of cognitive skill is essential for economic growth and which can be improved through quality schools. Although developing countries have made considerable progress they were unable to reduce the gap with developed countries due to the issues of school quality. Without improving school quality, developing countries will find it difficult to improve their long-run economic performance.

Lall (2001) is concerned about the path of economic development and says that the suitable strategy for any country relies on collective factors considered for attaining its objectives on the economic situation which also include government policies and national views regarding the role of the state.

Curran (2000) describes that small business firms in the United Kingdom have performed a key role in the economic policies simultaneously causing huge development in support structures to promote them. Although good policy outcomes have been broadly claimed to develop a strong enterprise culture and promote the prosperity of the United Kingdom, the precise outcomes have been difficult to point out. The methodology has been pointed as the major problem for the evaluation of policy outcomes as policies are developed over the 20 years. To support and draw out some key implications for their future in the United Kingdom, problems of evaluating small business policies are examined. Even allowing for the problems of evaluation, poor take-up of the support offered is the clear finding. Although small businesses have become much more important in the UK economy, it is disagreed that this has been due to state involvement. Since there is well-settled agreement on the value of small business support in the United Kingdom, little attention has been given to whether the support represents good value for public money. Doubts prevail about whether policies are cost effective and also if they are needed in reality.

14.3.1 FREQUENTLY ADMIRED POLICIES

Some policies amongst many are by now recognized and established in the policy world as popular and having high utility value in variety of situations.

14.3.2 INCREMENTAL MODEL

Hayes (2013) has elaborated incrementalism. This terminology clearly points out the extension, addition, expansion in the continuation of previous policy, with minimal changes. The goal of this policy calls for a systematic periodic review so that increments could be introduced if so required. In the context of economic policy, the goal such a policy is to create in the increase of investments, payments, revenue, and savings. Disjointed incrementalism, as ironically called, is a policy-making process which leads to decisions that happen to be only marginally altered from past practice. Some analysts describe incrementalism as a muddling approach, in contrast to the ideal of the rational-comprehensive model of policy-making. The rational model makes an assumption of a great deal of information, clarity of goals and criteria, and it has the ability to define and analyze all possible alternatives, rendering a single clear solution. The real world is not so considerate. Given all the constraints examined, perhaps all that can be realistically expected is a small alteration, inertia, or business as usual. Policy users find this conclusion disappointing as incrementalism is regarded as a failure of the political system to come to grips with the underlying problems which put issues on the agenda. Risk-averters and some conservatives regard incrementalism as safe that has system-conserving behavior. However, incrementalism is considered the most likely outcome of the policy formulation process.

Forefront Identity Manager (2010), in their publication designing policy, described that while logical policy statements are written, in many cases, their compilation does not follow a specific format. This can make it hard for reviewers to provide meaningful feedback and is also a general source for possible misinterpretation of the meaning of a policy. To avoid vagueness and to improve the specificity, it is imperative to make use of a precise format to minimize on these freestyle policy statements. For a policy to be relevant, it is essential to secure the statements to conditions that can occur in the business environment and provide the appropriate response to them. When a condition occurs, the response needs to be stated. This document refers to policy statements that should follow this structure as logical policy statements.

National School of Government UK (1999) has revealed links to the practices of policy-making for the 21st century. While framing a policy a long-term and outward-looking view has been explained. Evidence of the success of policy in other countries has been considered as other option with due deliberation to recognizing regional variations. The other features of innovation wherein new and creative ideas and managing risk have been given priority while framing policy. The next aspect is that of flexibility have been given due weight age while framing a policy to make it applicable for changing situations is covered. The other feature is joined up meaning "looking beyond institutional boundaries; setting cross-cutting objectives; defining and communicating joint working arrangements across departments; ensuring that implementation is part of the policy process." The other aspect is "inclusive-consulting those responsible for implementation and those affected by the policy; carrying out an impact assessment." Another additional issue is "evidence-based-basing policy decisions and advice upon the best available evidence from a wide range of sources; ensuring that evidence is available in an accessible and meaningful form." Still, further aspect is, "evaluated-systematic evaluation of the effectiveness of policy is built into the policy-making process." It also covers, "reviews-existing/established policy is constantly reviewed to ensure it is really dealing with problems it was designed to solve." A further aspect is that of, "lessons learned-learning from the experience of what works and what does not."

14.3.3 GROUP MODEL

CSUlb (2002) has described highlights about the group model. The model is made to focus on the governing body, wherein the interest groups have the option to pressurize the executive. Groups may capture agencies that are meant to regulate, and in the background, the administrators become increasingly unable to make a distinction between policies that will help the general public and policies that will benefit the groups being regulated. The core of fact of politics is that the groups are made to interact between them. Individuals are vital in political affairs only when they act to serve the group interests. The group forms a link between the individual and the executive of the administration. The political system's role is to set up and negotiate between various conflicting interests in society.

14.3.4 ELITE MODEL

This model is made to focus on the interest of individuals within a society that has most power than the demands of the masses. Elite who makes policy acts (CSUlb, 2002) in a background distinguished by apathy and information distortion governs a largely passive mass. In this model, therefore, policy flows downward from the elite to the mass. Society is split into two divisions of power born individuals and those without power. Elites, who have the power, share values that distinguish them from the mass. Due to the intervention of elites in the policies, the prevailing public policies reflect elite values, which generally preserve the status quo. The power born, elites, have higher income, more education, and higher status than the mass.

14.3.5 INSTITUTIONAL MODEL

Public policy is launched by the government of that country wherein it is determined by political institutions, which give policy legitimacy. The government makes the policy to apply universally to all citizens of society and since it has the absolute monopoly uses force while applying that policy, (CSU, 2002) which has the focus on the traditional organization of government. The policy describes the duties and arrangements of bureaus and departments. While framing policy, consideration is given to constitutional provisions, administrative and common law, and judicial decisions. The policy has its focus on formal arrangements such as federalism executive reorganizations, presidential commission, etc.

14.3.6 PROCESS MODEL

Hahn (N.D.) has highlighted the process model of framing policy. Policy creation is a process that consists of multiple steps such as identification of a problem and demand for government action. Policy legitimating is known as formulation of policy proposals by various parties, selection, and enactment of policy. The process model is a combination of steps and therefore the evaluation of policy has been criticized for being overly linear and simplistic. This model has some weakness since the stages of the policy process may overlap or never happen in reality. Also, this model is unable to take multiple actors attempting the process itself as well as act with each other, and therefore complexity prevails.

14.3.7 RATIONAL MODEL

Hahn (N.D.) has also described the rational model. The policy is generally designed to achieve the greatest social gain. Best policy is the one that maximizes benefits at low costs. This reflects part of the rational choice theory. This is known as step by step mode of analysis. It has its own limitations. The thought process implied is linear and can face hitches in extraordinary problems or chaotic problems which have no sequences of happenings.

14.3.8 BEST POLICY

Research has revealed that a simple answer to this question of best practice or approach is just impossible or unavailable. References on decision-making reveal the options that hardly give any clear conformity between them. Some obvious weaknesses always carry on. A vision-led approach, for example, is one which vitally relies on the individual with the vision. When that individual leaves office, it may demonstrate very hard to keep away from completely abandoning the strategy. While a plan-led approach for that matter can become excessively reliant on the vision of professional planners, who may lose sight of the wants of stakeholders as well as politicians. A consensus-led approach is the one, which may be quickly reached and sustained unless agreed and may lead to delay and inaction. Therefore, most cities invariably adopt a mixed approach. UK practice, for example, is of a cyclical approach, wherein objectives and problem specification are decided through discussions. A strategy is then developed and reviewed in the light of experience with execution. Best suggestions therefore happen to be to choose the blend of approaches which best suits a city's conditions but, having done so, they need to be maintained, and that leads to the future development of the strategy.

14.4 TERMS USED IN POLICY DOMAIN

14.4.1 POLICY

It means and includes a definite plan of action opted by the government, institution, group, enterprise, or individual which describes options and alternatives amongst and in the light of given conditions to guide and, generally, to resolve on current and future conclusion. Policy is a comparatively cohesive set of responses to solving the problem that has arisen while performing work. While developing activities, governments have policies, plans, programs, and projects in succession which may be short-term, more specific in place or with more timing than the previous and each successively more executive rather than legislative. Further explanation on policy is that the definite plan of action is a compilation of statements written to deal with the problematic situation that is arisen while executing activities to bring in positive transformation with a view to benefit the firm or country so as to keep it away from losses. The policy of government is known as public policy.

Web definition on policy (2008) states that a policy is a statement describing an action stated to achieve a rational outcome. The term is normally referred to as either procedure or protocol.

14.4.2 POLICY OBJECTIVE

Objective designates aim or goal of the policy and the purpose for which policy is designed. There can be single or multiple objectives or may be in a combination of main and supplementary objectives. Objectives need to be clear, that is without vagueness, so that approach could be clearly designed to attain the desired outcome. In the present context of the chapter, the actual outcome is quantifiable in terms of the economic growth of the country. Enhancement in yearly revenue, employment, investment, and steadiness remain immensely significant factors for measurement of growth. As a result, while testing the outcome of the policy these factors become extremely accountable.

14.4.3 POLICY RESOURCES

Policy resources are the assets accessible to the enterprise (Deshpande, 2011) or institution which may be available in the forms like financial, materials, services, staff, or other possessions which are utilizable for processing in order to produce benefit and in the process may be exhausted. Resources of all kinds are essential to act upon the policy. Therefore, they may be materials, monies, machines, manpower and the like. With the help of resources, policy processes could be worked out to reach the objective and arrive at the outcome.

14.4.4 POLICY DELIVERY

Policy delivery has concern with the effectiveness, and that is determined not only by the policy choices but also by policy delivery. Turning into this latter issue it is observed that there are several grounds for favoring simplicity and consistency in the policy delivery.

14.4.5 POLICY DECISION

Lindblom (2000) reported that administrators often make policy and advice in the making of decisions. In that context, decision-making and policy-making are to be treated as synonymous.

14.4.6 POLICY DESIGN

An institution, government, or firm when initiates policy it has to set objective and identify the course of action using appropriate resources and processes to accomplish the desired outcome by eliminating its undesired impact if any. Good decision-making requires clarity of judgment and firmness (Deshpande, 2011) when they are to be implemented effectively.

14.4.7 GOOD POLICY

The overall objective of the good policy is to ensure that it fits well for the purpose it is designed. It should exhibit an efficiency of resources and should be sustainable. It should as well be flexible and easily adaptable and attractively formulated. In that context, a good design is a process which essentially calls for effective leadership that also properly engages with all key stakeholders throughout.

14.4.8 CHALLENGES

Challenges are evolved due to complex situations in social, economic, and business circles when an Institution or firm foresees barriers in setting its plans. Until they are reduced or eliminated, the growth is unable to proceed. Policies are devised (Deshpande, 2011) after studying challenges and identifying an appropriate course of action to pacify them and to balance the intricate situation. For example, the course of action in connection with development or enhancement of economy will be decided with narration of the process as to how it will be attempted. The challenges of globalization are many and they create barriers in designing a policy. In that connection, conceptual design considered for course of action might just counteract with the desired outcome thus posing a challenge of designing.

14.4.9 CHALLENGE GAUGING

Gauging is making an assessment of the challenge and measuring it in terms of the monetary value. Managing the accountability arising out of the challenge, into an opportunity, is through the right decision. This is an all-time desired action by the authorities. For example, the IT service industry has brought in a massive revolution by introducing a tool like Enterprise Resource Planning (ERP). Need is to accept such changes as they offer multiple advantages of speed, accuracy, and efficiency and if uncared for would pose breakdown of the organization sooner or later, by losing competitive edge. Successful organizations continuously monitor the challenges by classifying them either under technical, financial, economic, or management head. This helps them in finding clarity in complexity that is essential in creating a robust policy. Challenge due to the environment would be grouped as technical since it is related to the preservation of the environment. Another challenge such as recruitment is classified as administrative. Likewise, the third would be maintaining organizational profitability that is classified under the financial challenge.

14.4.10 DEBATE ON THE TRADITIONAL APPROACHES

Gauging is making an assessment of the challenge and measuring it in terms of the monetary value. Managing the accountability arising out of the challenge, into an opportunity, is through a right decision. This is an all-time desired action by the authorities. For example, the IT service industry has brought in a massive revolution by introducing a tool like ERP. Need is to accept such changes as they offer multiple advantages of speed, accuracy and efficiency and if uncared for would pose breakdown of the organization sooner or later, by losing competitive edge. Successful organizations continuously monitor the challenges by classifying them either under technical, financial, economical or management head. This helps them in finding clarity in complexity that is essential in creating a robust policy. Challenge due to environment would be grouped as technical since it is related to preservation of the environment. Another challenge such as the recruitment is classified as administrative. Likewise, the third would be maintaining organizational profitability that is classified under the financial challenge.

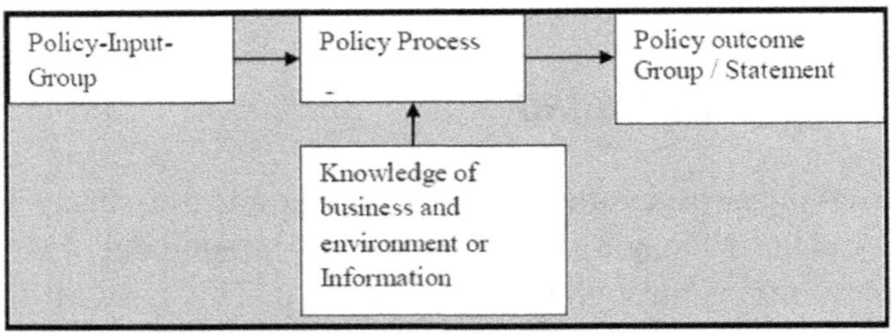

FIGURE 14.1 A trilateral instrument.
Source: Compiled by author.

14.4.11 POLICY-DESIGNING ASPECTS

Diverse terms have been suggested by researchers and experts. Use of policy philosophy and policy target is invariably meant to start framing the policy. Other term like policy means is used to indicate the resources required to satisfy the objectives. Further terms such as policy object group, policy executing principal, person and authorities in charge of policy, regulation, and institution are the people who are involved in designing the policy while social, economic and cultural dimensions, as well as the willingness of the government and enterprises become the next consideration while arriving at policy formulation. The scope covers the features of the accountability of the designer or else the policy may lose grounds of sustenance. The methodology of designing the policy consists of three main steps. Each step has been identified and linked with its variables under following heads. Figure 14.1 is a diagram that exhibits the sequence of activities undertaken to design policy.

1. Policy input group, conceptual factors (objectives, target, and issues)
2. Policy process group (policy mapping, procedure, moral principles, history and past lessons learned and statement drafting)
3. Policy outcome analysis group (features and effectiveness).

Logical analysis of the above three sides of "policy designing" program reveals that it is a "trilateral instrument." The above logic also goes according to the following well-known formula:

Input + Process = Outcome

14.4.11.1 TRILATERAL INSTRUMENT

In the first step amongst three of this trilateral instrument, policy input group formation is initiated between experts who are accountable to decide objectives, target and highlight issues evolved therewith. Emphasis is given on the consensus between the group members to finalize the statements of objectives. Objectives are broadly fixed, for example, to develop economy while targets are with specific figures meaning that they need to be in units of economy planned to be achieved within the desired time period. If a country is to grow its economy to 25 billion dollars, then such figures are specified to plan the approach to reach.

While framing policy, the policy-makers usually seek clarity of objectives in order to decide the priorities and then make choices of instruments or approaches which most efficiently promote to achieve the goals. Based on priorities the sequence of objectives is decided by choices. Next, the choice about which instruments are to be engaged to pursue the best of those depends upon the philosophy of the authorities. Therefore, new instrument has emerged called the philosophical approach. Since the policy needs to survive and succeed, policy designers set to rationalize on reaching to their goals and going beyond the ordinary leads to risks. The risks, in that case, need to be balanced by the achievement of extraordinary gains. Responsible people engaged in policy-making, through their philosophy create influences that make competitive choices among available processes which invariably stand for either one or many of them. Research study in this field has revealed various traits useful while designing policy. They are passion for growth, analytic mind, resources and capability, and achiever personality. Philosophy positively influences the performance of the organization as it is a choice among the main and supportive approaches. Statistical analysis on various approaches has revealed that the main approach is primarily used to achieve the objective while supportive approaches are meant to address challenges. Philosophical basis dominates the orientation of the authorities which is responsible for policy success.

The second side of designing comes up with policy process group that is the one formed between experts who are responsible for policy mapping, and having the knowledge and expertise in drawing procedure, consider moral principles, know the history and past lessons learned and draft policy statements, and make use of available and required additional resources which could be transformed to desired outcome using a particular path or procedure or approach. Policy mapping is nothing but drawing a diagram that is planned to be used by aiming the objectives and reaching toward the desired outcome.

The third part amongst the three sides of the trilateral instrument proceeds with policy outcome analysis group that is formed between experts who have the knowledge and ability to analyze the outcome of a policy and if found fit the policy statement is finalized or else it is modified according to the requirements. Efforts through a deep thought process are put in to eliminate any undesired outcome and path or the route or the approach is altered if needed.

Policies are being frequently designed for financial, social, and political institutions and therefore well known in the policy world. Nevertheless, there are hardly any common approaches established for economic development particularly when facing the narrated challenges due to changing environment or "globalization." As a result, the study is taken to identify the foundation of policy-making and to search economy policies utilized in the globe.

While formulating a basis for designing policies for economic development policy designing for all time aims at efficiency, and has continued to be an interesting arena for researchers especially in the light of "globalization." The effectiveness of the policy fundamentally depends on the actual outcome, based on the designed outcome.

14.5 MAIN FOCUS OF THE CHAPTER

The chapter focuses on designing policies for economic development. A steady increment and equilibrium between dimensions of economy such as investment, income, payments, and savings is imperative for economic growth to come about. These dimensions of economy are applicable to all countries irrespective of their class whether rich, medium, or poor. The government that is run by people has the power to make increment in the said dimensions; as a result, people only happen to be the catalyst for economic growth. Design for economic growth, therefore, relies on the people who plan and govern the desired outcome. Policy of investment in projects is well known nevertheless investment in people has greater significance, as it generates entrepreneurship and leadership that is always responsible to create economic growth. Peoples' quality improvement is another goal of such an investment. Quality of people is further vital issue cited by experts in the economic growth of the countries. Public good quality is said to accrue from a large percentage of highly educated, talented, and work skilled citizens who have the intent and urge to develop economy. Countries poor with resources have the alternative to opt for prudent collaboration policies. Development of policy initiates with the fixation of its objective. The framing of main objectives happens to be the vital task of the designing authority. For developing economy, understanding its definition becomes the starting point to bring in clarity in the process of designing. Once the dimensions of economy are defined and background explained, the path to policy roles over resources to transform them through appropriate processes to reach the desired outcome. With this background, policy designing has come out as a trilateral instrument that considers resources, processes, and outcome analysis as its three inherent and essential dimensions.

14.5.1 ISSUES, CONTROVERSIES, AND PROBLEMS

14.5.1.1 BACKGROUND OF CHAOS AND COMPLEXITY

Tuomioja (2001) reports that the democracy is dispersing across the globe, but so is the disappointment about its workings. The key to this paradox is said to be "globalization," according to the Finnish Foreign Minister. "Globalization" is said to create positive results due to its potential, but to make the best of globalization, a common approach is essential since it creates "them and us" politics between the countries. Chaos and complexity theory has been playing its role in organizational science. Moreover, globalization has brought in a variety of customs and cultures because of which there is a rise in complexity. Taking the views affirmatively, globalization has reflected open access to all countries and citizens of the world by which an enormous exchange of people, products, funding, technology, and culture between countries is proceeding at pace. Development in one country rapidly spreads to other countries due to the emergence of internet communication technology. Many countries try preserving their own culture; however, it is changing with the advent of new customs of the foreign culture. The occurrence of changes is gradual and no country is able to defy them. Postglobalization, countries own culture is steadily getting transformed while accepting global changes. Without contradicting, the developing countries invariably found accepting the culture of developed countries for having proven their worth. Activity forces perform in linear, parallel, and cross-functional directions resulting in chaos and complexity. As a result, there is a negative effect that leads to destructive impact on the life of country. The chaotic situation reflects disorder in the social circle while complexity a complication. Assorted thinking of the people and their obdurate attitude is often pointed out by experts to lead to chaotic and complex situation. Peoples' choices on utility, priority, and viability on diverse issues of life differ and so evolve confrontational, dissimilar, and peculiar circumstances. Advancement of global customs, cultures, and competition in the home country add fuel to the fire thus aggravating the situations. Complexity is the result of several processes, customs, or cultures acting in chorus that is presumed to create messy outcomes and without a firm direction attaining the desired goals is a complicated task. Therefore, public, in general, remain rigid and mostly contest as they find them complicated for the development of the nation.

14.5.1.2 THIRD WORLD ECONOMY

Researchers and experts have mixed views on third world countries. This nomenclature, as observed from literature, is assigned to the developing and poor countries in the world. First world countries generally include developed countries and their allies, whereas second world countries include developing countries and their allies. By and large, this nomenclature for the nations in the globe is classified into three groups based on political and economic situations. The third world was normally seen to include countries which are mostly poor countries. In the present context, major indicators of economy include public dimensions and their values or in other words the demography of the country. Per capita income of a country, for example, can be compared with others to judge the position of a country in assessing its economy. Third world countries have been considered extremely poor and nonindustrialized, therefore, nomenclature that came out for poor countries is "third world countries." History cites few exceptions that, some of the nonaligned countries and a few of these nonindustrialized countries came up to be very prosperous.

14.5.1.3 DEMOGRAPHIC TRANSITION

Saucedo (2017) has highlighted on demographic transition theory. Transition indicates changeover and therefore demography is the basis of this theory. Size, density, and distribution of the human population have been primary consideration in the transition theory. Transition in birth and death rates, age distribution, and other factors that influence the size and growth of a population is the domain of this theory. Statistical data related to the people, products, places, prices, processes, and promotions of a country continue to change year by year. Industrialized and nonindustrialized countries as well have extensive variation in the demography. Transition is a state of change in progress. Global demography has been changing continuously due to development and emerging of new customs with pace and this is a challenge for designing sustainable policy. Continuous change is observed in the social, economical, political, environmental, and business-related surroundings due to emerging novel customs and technologies that are posing further additional challenges. Dealing with this continuously changing situation has become a regular concern to all countries in the globe.

14.6 SOLUTIONS AND RECOMMENDATIONS

Recent decades have shown that the countries in the world have grown increasingly interdependent since they need to address challenges. The challenges which have evolved due to globalization, formulating economic growth policy, and making them take notice of at the countrywide and worldwide are a matter of skill. Rapid changes are taking place in the society's expectations, the role of people and technology as well as and business. In addition, the changes cover everything from the environment to community relations. Therefore, to address these new demands created out of the evolution, society, and business has to consider variable input to major initiatives, due to global impact. Policy designing extensively demands to make right decisions. Although theory has given good methodology to arrive at decision, it is equally important that designer focuses the attention on the quality of a decision. The level of confidence will be enhanced if a set of steps and rules is made to provide an assurance through thoroughness and rigor. Splitting the objectives into small independent objectives and deciding the issues step by step at a time aids in working out decisions. With a proper process or framework, the mechanism is to set to secure the quality of decisions. The framework is normally used for breaking down objectives and working decisions of virtually any size. The complexity begins with two large ideas such as declaring a decision and working decision. Each of those larger elements needs to be then broken down into sub-components, and the decision-making path is followed.

14.6.1 ADDRESSING THE CHALLENGES

Cole (2003) points out that Globalization is the ground for chaos and complexity. Macnamara (2003) also points out that the change conversation that is repeatedly extended these days is around the chaos theory or even the science of complexity. Observation reveals that countries which have taken it in negative sense are facing challenges. People perceived that chaos and complexity have been evolved owing to the advent of differed customs and cultures amongst the countries by which is resulted unrest amidst public. Human behavior when turns abnormal cause agony to all the public in a country resulting in a negative impact on the life of people. Therefore, to address such challenges human behavior needs to be essentially guided and influenced. Nevertheless, by understanding the grounds of challenge, issues are able to be resolved through positive and participative approach. People are habituated in doing certain things in a certain way. The habits which are deeply rooted in human beings make them prefer those practices since they are duly established and considered most efficient. However, the perceptions can be challenged by bringing new ways of doing things into their conscious attention. By designing such policies, the individuals can find larger incentives to adopt more efficient practices, instead of continuing with less efficient habits. The other side of globalization reveals that many countries have taken its benefits in real terms since countries have come closer to each other in the domain of business, cooperation, and cultural exchange, employment opportunities are widely expanded. When public knows about the benefits, they admire it, however, the government has the pivotal role in guiding and directing them. People generally go for optimal outcome without engaging in a detailed search. People go to certain extent to accept changes, however beyond their capacities, they do not keep searching. Behavior-driven policy is the one which considers fairness to the user populace, without degrading their status, balanced social norms, and with maintained personal identity in motivating behavior instead of focusing exclusively based on low cost or financial incentives. It is hard to change social norms, however, balance could be maintained using people with certain influences. Benefits could be derived by encouraging greater personal involvement, instead of focusing solely on the complexities evolved. New Zealand Ministry in (2006), advocated that "policies are effectively designed on the basis of theories from economics as well as philosophy, while psychology, cognitive science, and sociology are equally valued and therefore accountable." Policy-makers, therefore, have to keep abreast with the latest developments and make sure that their involvement in policy-making reflects these mechanisms. Policy analysts should, therefore, enquire if an issue is driven by the behavior exhibited by people or decision-making. So also find if the intervention adequately tackles the reasons for behavior. A ready reference guide supports the policy analysts about the "behavior and decision-making." Since perfect information is not available with the people most decisions are made under doubt. Care should be exercised when assessing the outcome of a decision. If an outcome turns out bad for a person it does not necessitate it was a bad decision. It is normal to consider that some people make mistakes and that some decisions will be wrong. Stakeholders in policy-making should keep away from relying too heavily on viewpoints of others as it is really hard to interpret what someone else's beliefs are while using only their behavior is an indication. The department needs to concentrate its academic attention in three fields such as knowledge management, socioinformatics, and decision-making.

In the end therefore, two important approaches have come out to address globalization challenges

a) By influencing people's behavior
b) By implementing innovative and people friendly policies

Remedial measures therefore would be to make changes by influencing people's behavior and people friendly policies. For improvement in people's behavior they need to be educated with added skills and capabilities that should form part of policy toward economic development. Entrepreneurship and Leadership building becomes the subsequent policy to implement innovative projects to furthering economic development.

14.6.2 DESIGNING POLICY TO DEVELOP ECONOMY

Basis of design means and includes the approach, instrument, process, tools, elements, and philosophy used for arriving at solutions to the problems. Philosophy means and includes the thought process, values, attitude engaged in designing the policy. Third world and other countries that are aspiring to make economic growth essentially need to think of designing a robust policy wherein all the dimensions of economy need to be engaged by designing a suitable structure (Savitti, 2008) and setting specific targets. Model to economic growth needs to cover six Ps meaning products, prices, people, places, processes, promotions and in view of that fundamentally need to employ combination of core and complementary policies (Deshpande, 2011) to complete policy development. Economic growth is, as well, a function of four steps such as plan, entry, survival that finally leads to growth. Also, economic development needs to know about the background situation and situation such as plan, entry, survival, and growth (progressive policy). Therefore, the structure of policy for economic development turns out to the one shown in Figure 14.2.

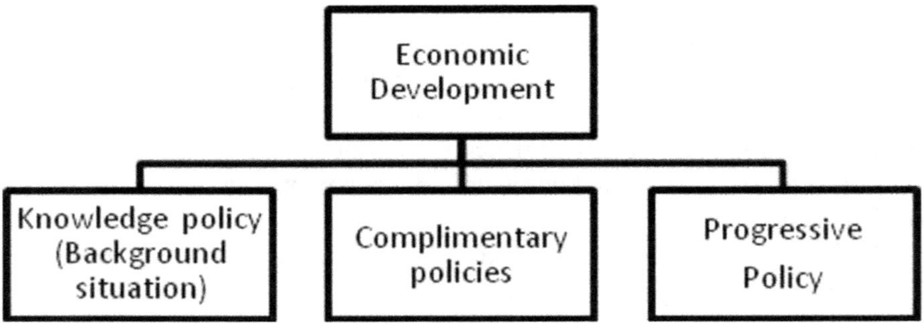

FIGURE 14.2 Development of economy.
Source: Compiled by author.

Globalization has opened doors to all the countries to enhance economy through business or trade between them. By trading internationally, many countries have derived large benefits due to globalization and this policy happens to be the complementary policy to the major economic policy of the country. Economy by its definition considers complementary policies as important to deal with the core policy. Authorities craving to fix goal of economic development has to devise those policies so that a full-fledged and integrated economic policy is productively drawn. First and foremost work of the policy authorities is to create a basis of economy by setting targets on the said policies. This step essentially results in clarity of objectives so that initiatives to produce the desired outcome are able to set. Authorities have to set a target as increase in investment, revenue, and reserves. Once set, the approach designed to achieve the target is able to operate. Enhancing the value of production may be another objective. Another area of economic policy relates to taxes and interest rates. The level of interest rates is determined by expert in the line, that is, by the monetary policy committee which convenes meetings at regular intervals. A rise in interest rates raises the costs to business of borrowing money, and also causes consumers to reduce expenditure, leading to a fall in business sales. The government spending policy also influences business. For example, if the government spends more on schools, this will increase the income of businesses that supply schools with books, equipment, etc. The government also provides subsidies for some business activity, for example, an employment subsidy to take on the long-term unemployed.

1. Economy dimension-driven policy: To enhance macro- and micro-economy dimensions is a core policy which uses following policies as the complementary to the core policy.

a. Capital raising policy (equity and borrowing policy)
b. Investment policy (projects, schemes, infrastructure, etc.)
c. Revenue policy (direct and indirect taxes, interest, sale of goods and services)
d. Payments (expenditure) policy (CAPEX and OPEX)
e. Reserves (savings/profit) policy

a. Capital-driven policy: Capital is essential for investment in promising projects and is an important path to enhance economy. Capital raising policy for a needy nation is essential to seek capital from other countries. Organization for funding relates to the global capital flows (Michel Camdessus, 1997). In this context, many countries have gone in to make various structural economic reforms such as removing barriers to make improvement in trade, streamlining investment procedures to enhance investments, and liberalizing their own financial markets to make their economies much more attractive to investors.

b. Investment-driven policy: Echandi (2016) states that attracting; promoting and retaining foreign investment is covered under this policy while developing economy. However, potential benefits investment is not automatic. Therefore, investment needs to be planned and implemented. This policy is also meant for the national security (OECD, 2009) of the country. These policies are formed and guided by the principles of nondiscrimination, transparency and predictability of outcomes, proportionality of measures, and accountability of implementing authorities. The other approach to investment policy is that each country should identify the portfolio and projects that need to be developed and make assessment of the size of investment essential in them.

c. **Revenue-driven policy:** This is by and large related to the mobilization of revenue through domestic and international trades. Raising revenue (Carlo Cottarelli (IMF), 2011) is the central aim of any tax system, but it is not the only anxiety. Developing countries spend more monies and therefore their needs are extensive. Foreign funding assistance can be obtained by such countries because their needs are greater and ultimately more sustained. Developing economies often face challenges of effective tax structure and its collection. It is the government's inability to efficiently collect the stipulated tax from the wealthy people and therefore very little funds become available to finance essential public sector venture in services the social benefits of which are much high. The effective tax collection system is therefore essential. Developing economies may face constraint in manufacturing of basic products and lose comparative advantage. Due to volatile prices, low-income elasticity of demand, and finite nature, there is a limitation to the economic development. For that reason, the development of economy may need government encouragement to set up new industries such as manufacturing in different sectors. Lewis model of a dual economy is popular and although it is argued and known for shifting labor to manufacturing, it may be launched for a temporary commitment to gain tariffs. Diversifying attempts from agriculture may produce mixed results. At times with a poor basic level of infrastructure, countries struggle to make effective use of capital investment in the manufacturing sector. It is also argued by some that attempts of government to persuade the manufacturing industry is mislaid because of lack of knowledge for promoting best kinds of industries. Better is the stand of government to leave the decisions to free market as to which industries to invest in. Encouragement in the private sector and deregulation is the other important path to rapid economic development, and this decision will lead to the development of a mixed economy. In many countries, state-owned industries have diluted their shareholding to encourage privatization. This way it helps giving firms a profit motivation to cut costs and aim for greater efficiency that results in the rise of the number of private firms. State-owned monopolies may face competition if deregulation is implemented. Greater pressure due to competition can help to create incentives to cost-cutting. Competition may be increased through liberalized trade and opening markets to international firms. Privatization may pose a potential problem such that it can aggravate disparity in society.

d. **Payment-driven policy:** Spending the money for a right cause at right time is considered important underpayment policy. This policy relates to the payment to vendors (UOW, 2015). This involves working with the various project authorities and heads of risk management segment to comply with domestic and foreign country regulations regarding payment to vendors. Minimizing losses from currency conversion and complying with domestic and host-country banking regulations when wiring money is the aim. Making payments from home-country bank accounts should be with the minimum number of transactions and reduction in the need to replenish host-country bank accounts is essential. On another front, Mitchell (2005) has highlighted the impact of government spending on the growth of economy. According to him, a negative economic impact is often observed due to government spending. He further states that the deficit is not the critical variable but the size of funds spent rather than the size of government. Growing evidence cites that due to high government spending, economy could not grow faster until the burden of government was reduced to a great extent. Tejvan (2016) has interestingly cited about the policies on expenditure switching. Suggestions are cited on making changes in the goods that people buy. Making domestic good relatively cheaper is possible through devaluation, for instance, and foreign imports more expensive. This move will lead the consumers to switch from importing foreign goods to domestic goods. The other example cited is that the supply side policies make British goods relatively more attractive. He also highlighted the policies on reduction in expenditure and policies to reduce overall spending on imports as well. Involvement of tight fiscal or monetary policy is cited as an example.

e. **Reserves-driven policy:** This policy (Amadeo, 2016) relates to the generation of savings from domestic and foreign sources. While domestic savings are accrued from the positive cash flows of national revenue, the foreign currencies held by a country's central bank and its member banks in terms of foreign currency or exchange reserves. A policy of reserve creation is useful to countries to honor commitments on balance of payments and to save the country from emergencies arising out of natural catastrophes or unintended erroneous work done by people.

2. Policy to create all-round growth: By overall growth is meant the development of economy dimensions such as products, people, places, prices, processes, and promotions. These dimensions support the economic growth of the country. Valentina Pasquali (2015) has reported that globalization has brought in a positive view of creating all-round growth. This comes from inclusive policies. The goal of economy generation has multiple routes to reach to. Each country has its own resource routes nevertheless, agriculture industry, services industry, manufacturing industries, industrial and consumer trades, etc. are common amongst most of them. The government policy should include these routes to bring in all the resources under the coverage of economic policy to generate multiple employment and revenue for the country.

3. Policy to create leaders: Alvord (2015) has highlighted the role of leadership in economic development. Developed countries have generated strong leadership in almost all segments of trade and business. Amongst the variety of advantages that globalization has brought in, the modern management practices have essentially spread across the world. To address the challenges of education in management practices is considered indispensable. In that context, leadership creation has been thought out important since many issues could be sorted out by talented leaders. Government policy has the power to eradicate a variety of livid issues to create comfort and development. Good policy can be favorably and gracefully launched by strong leadership in the ruling government of the country to effectively address the challenges. Strong leadership has the power to convince and direct people to act upon good cause by implementing viable schemes that are required to generate employment and developing economy of the country. Leaders in every field of movement become the potential citizens who can enhance the growth of the country via multiple routes. Leadership training, therefore, becomes an inherent part of education.

4. Policy to augment higher education to masses: Developed countries in the globe have introduced a variety of skills and capabilities for employment in the organizations that are managing businesses in a modern way. Therefore, adoption of those practices is a matter of prudent policy. Analysis of the public dimensions of the countries reveals literacy rate and percentage count of the highly educated citizens of a country as two vital measures that can bring about a transformation of the chaotic conditions into better social life and economic independence. As a result, this policy carries prominence.

Market failure is often observed in areas such as education, health care and transport (Pettinger, 2012) as the free market is unable to provide levels of education that are required. Increasing literacy and the number is a key factor in improving economic development. It is very difficult for economy to develop into higher value-added industries without gaining basic levels of education and training. Mixed returns on investment in education are observed. Investment in such areas often takes a long time to feed through and result in higher rates of economic growth. However, literacy should an aim for the development of economy.

5. Policy to breed entrepreneurship: Wiens (2014) has highlighted entrepreneurship role in economic development. Global practices have revealed that to improve the economy of the nation, it is really essential to produce entrepreneurs. Entrepreneurs are known to be the economic developers. More the entrepreneurs, more the businesses and more are the sources of revenue. Small and Medium Enterprises (SMEs) as well as multinationals are the enablers of generating new products, and they scale up revenue building. Entrepreneurs are the real wealth of nations as they are the pillars of economy. Creating an environment in which people can develop their full potential is the real human development since this only can lead creative lives in accord with their interests and needs. Expanding the choices people have to lead lives that they value is, therefore, an important feature of development. Making a startup enterprise survive amidst global challenges is a highly admired skill while extending it to sustainability is intelligence and a matter of patience. Enterprises staying in the business indefinitely are the real leaders advocating sustainability instrument.

6. Resource-based policy: Globalization has exposed the fact that a variety of resources is required by various countries due to demand–supply imbalance. Every country has its own potential natural and man-made resources (Patil, 2016) like minerals and metals, other materials or machines or skilled manpower. As a result, the country has the opportunity to promote those resources to third countries to spawn revenue. A country like India has the potential in supplying talented and skilled manpower to various countries. Italy has the potential in designing machines. There are many countries which are rich in oil and gas, minerals, and metals. Such material resources can be sought by developing countries to grow, reduce poverty, and manage without going for international finance. Countries such as Arab are rich in petroleum oil have therefore opted for business of petroleum resources. Jordan is known to produce minerals, whereas the United States is known to conduct basic and applied research on a large scale in all fields. Similarly,

Japan is known to produce high-class automobiles. Resource-based policy ensures high revenue growth forming a major part of economics of the country.

7. Repeated iteration policy: Kolko (2015) has reported that design thinking comes of age and is a continuous process. Repeated iteration is a new approach of continually revisiting the environment and making changes accordingly that is in large part a response to address the increasing complexity of modern technology and modern business and is a need of time. Complexity evolves through a variety of forms. That could be dealt comparatively easily by using the software. This software could be sometimes placed at the center of a product and then integrated with hardware and made intuitive and simple. This is essential from the user's point of view. At times the problem may be multifaceted to be tackled. Sometimes the business environment is very volatile so that a firm must experiment using multiple approaches to survive and grow.

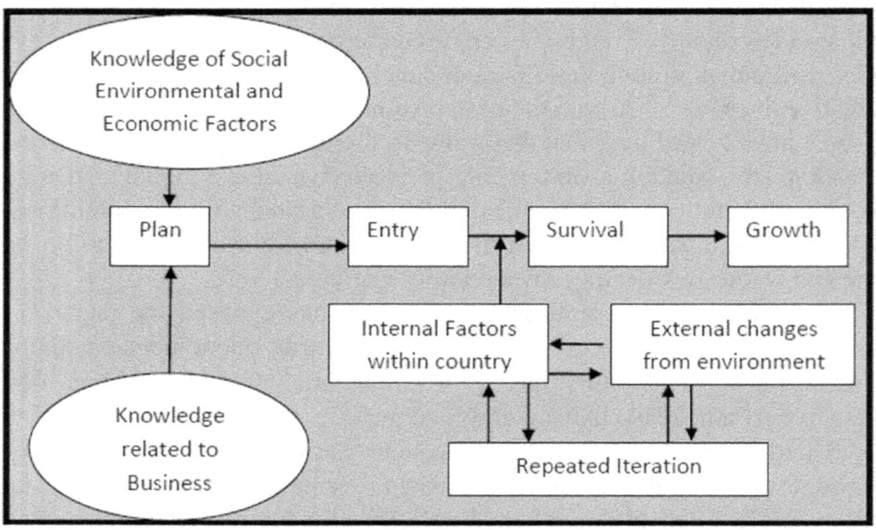

FIGURE 14.3 Repeated iteration policy.
Source: Compiled by author.

Globalization has produced an impact that there would be continuous change in the demography due to development in the world and the changes are formidable and cannot be stopped. To capture and update the changes in global environment this inward-looking and forward-looking policy is implemented. Capability to analyze data on demographic and various dimensions of economy from the global environment is the need of the time and this policy needs to be continuous forever. On evaluating various dimensions, changes need to be identified and corrective measures to adopt them could be planned and carried out based on viability. Figure 14.3 exhibits the flow diagram that explains the development steps starting from inputs to growth, and especially demonstrates the iteration that needs to be conducted continuously to assess changes from external environment happening due to improvements due to globalization.

8. Policy to promote geographical growth: Gallup et al. (2016) have reported some effects of geography on economic development. As far as location and climate are concerned, they have been observed to create large effects on income levels and growth due to the consequence on transport costs, disease burdens, and agricultural productivity, amongst other aspects. Impact on economic policy is also observed owing to choices of geography. Geographic regions where population densities are high and rapid rise in population have faced problems of economic growth as those regions have not been conducive to modern economic growth. Regions which are located far from coasts and ocean-navigable rivers, for which the transport costs of international trade, are high and due to tropical regions, which bear a heavy burden of disease have resulted into a disadvantage. In addition to this, a large portion of population growth is expected during next 30 years turning these geographical regions into disadvantage. Therefore, designing policy considering the basis of geography needs to be undertaken carefully. Full geographical coverage of the economic policy ensures growth of urban and rural resources since they are unleashed to development and all population gets the benefit of this policy. This policy has the power to reduce imbalance between the urban and rural populace and stop the migration of job-hunting rural population to urban places.

9. Policy to create knowledge banks: Globalization has rewarded a variety of benefits as they are improving economies through knowledge management. The knowledge economy is swiftly and incredibly spreading among nations. Enhancing the economy through knowledge (UNIDO, 2008) of issues, resources, processes, and outcomes quickly ensures growth as processing accuracy is gained at pace in dealing with the issues. Managing the knowledge is the prime requirement to go with such a policy. In the modern world, these are the requisites majority of nations have accepted this custom and become ready to go for this policy.

10. Policy to make reforms: World Bank (2008) reported political economy of policy reforms for social development. The government makes reforms creative conducive policies with a greater level of transparency to boost private investors' confidence. Public–private partnerships, popularly known as the PPP model, is an example of reform and the indication of the interest shown by private players in building national infrastructure. Diverse economic programs are being suggested whilst infrastructure has been recommended in the policy reforms to furthering industrial growth.

11. Overhauling policy: This policy is a thorough examination of a system or existing policy that is pointing out for changes, for instance, "a major overhaul of environmental policies." When policies become ineffective and start creating negative outcome time is right to revise or overhaul them. Considering overhaul, the government needs to fix up fresh and appropriate objectives with the background of globalization and a path is identified to reach the desired goal. Many countries have fallen prey to the old policies for many years and when they realized, altered them in a new way. Once a fresh policy is framed, improved outcome follows. Incremental policy has inherent limits, as there is a deep-rooted bias toward the status quo. When circumstances change this tendency of status quo gives way to a situation that creates departure from prior practice and turning to a situation called overhaul. Still, political institutions continue to run with incrementalism approach since it finds them more pragmatic. The inherent tendency of continuing status quo, however, helps resisting critical inquiry and comprehensive revision. Faux, (2008) reported a case wherein the basic economic assumptions in the agreement of North American free trade clearly went wrong. The promoters of the organization those offered stable jobs to American workers told them not to fear of losing jobs as Mexican labors were available at much lower wages in their comparison. The promoters felt that because it is a free trade by itself, it would create booming Mexican economy and huge middle-class market for US goods. Due to surplus, they predicted that trade with Mexico would grow, generating new good jobs in the United States, and at the same time, help reducing illegal labor immigration from Mexico. But Mexican boom or vast middle-class market never materialized. The gap between wages in the two countries actually widened. Therefore, policy-makers had to go for overhaul of the policy.

12. Synoptic policy: Jha (2007) has presented a synoptic view on fiscal policy in developing countries. This is an instrument that constitutes a synopsis; presenting a summary of the principal parts or a general view of the whole. It may be an economic growth policy that is presented in a summary format for the purpose of conveniently spreading knowledge to the masses of people.

13. Policy to trade internationally: International trade has emerged due to globalization in a very big way and that will continue to rise in future. It has become the prime attention of many countries since it helps increase revenue. Care is nevertheless crucial in abiding the laws of the country to which goods are supplied by the local supplier. These days, variety of products is supplied on large scale by countries across the world.

14. Opportunity-driven policy: Countries keep studying and examining the sources from where opportunities are generated. Opportunities generally arise when the world market demand, supply, and pricing position changes. They capture such occasion to avail business exchange that generates good revenue. International entrepreneurship through opportunity based approach has emerged due to globalization. It is an integrative approach to small firm internationalization. Greater engagement of small firms with international markets has made to attract this approach. Existing internationalization models and approaches have exhibited certain weaknesses while seeking to internationalize. Opportunity-based approach argues that at the heart of firm internationalization is the formation while it has the ability to exploit entrepreneurial opportunities. The approach also views firm internationalization as dynamic, nondeterministic, and nonlinear, and seeks to integrate current International Entrepreneurship approaches which include stage theories, network, resources, and capabilities. It also addresses the literature on international new ventures and born-again global in order to develop a holistic perspective of small firm economy development through internationalization. Interesting empirical findings have been cited on the internationalization processes of eight small, high-technology firms in New Zealand. Using case studies and an opportunity to map these firms were tracked through time and their internationalization processes, as well as the entrepreneurial processes of opportunity formation, decision-making, and opportunity exploitation. The study also revealed the principle of opportunity interconnectedness which shows how firm internationalization typically involves multiple opportunity elements such as knowledge, resources, networks, capabilities, globalization, technologies, etc. Combination of different opportunity elements has been said to typically facilitate foreign market entry. However, it makes prediction of stereotypical internationalization patterns difficult.

15. Evidence-based policy: Policy implemented by some country becomes the example of success and hence it so the evidence. Therefore, other countries equally become inquisitive and study that policy since that policy is proven. Countries may adopt policy of that successful country in toto or else may make nominal changes. Asian Development Bank (2006) emphasized the need of evidence-based policy-making. Responsibility and accountability for policy-makers, problem-solving, and decision-making are the skills needed. Problem-solving is integral to decision-making that is particularly central to management and leadership. Processes and techniques have been developed to improve decision-making and the quality of decisions. For certain personalities, decision-making is more natural. Essentially these people should focus more on improving the quality of their decisions. Less natural decision-makers are often able to make quality assessments, and once assessment is made they need to be more decisive in acting upon. Problem-solving and decision-making and each requires creativity in identifying and developing options. Brainstorming technique is particularly useful for developing options. Good decision-making requires a mixture of skills such as creative development and identification of options, clarity of judgment and firmness of decision.

16. Clear, simple, and easy policies: Decker (2015) states that policies are to be framed easy to understand. Policy design must go with clarity of objectives, simple processes, and easy path. This trilogy is considered indispensable since knowledge, nature, and life flow in easy path. Complicated processes do not survive as they are frequently declined by the public and also because they create chaos and do not serve the purpose of reaching to the masses.

17. Capability- and strength-driven policy: Business improvement methodologies and strategic planning within large corporate are continuously developed for business excellence. In the United States, Europe, and Middle-East, strength-based policies, lean thinking, and six sigma methodologies are in vogue. Management tools such as performance metrics, key performance indicators, and scorecards are widely adopted in the organizations across the world. Tool such as strength-based approach is widely appreciated as that is used to exhibit strength of the organization in particular areas. On the other hand, traditional approaches have their focus primarily on problems, needs, and weaknesses. There is a thought on what works in the strength-based approach and therefore, it focuses on what is working. This approach aligns with research on resiliency, positive psychology, and asset-based thinking. Since hidden assets, strengths, and resources exist within every individual, team, and organization they are highlighted and exhibited to the customers. The firm is able to achieve the desired outcomes when attention is focused on achieving these collectively. Concentration is given to building on "what's working" which creates collective resiliency. Firms can always identify something that is working well. Therefore, solutions already exist within teams and organizations which need to be discovered. Strengths-based approach builds on the best aspects of the company and creates new positive images of the future, images of possibility and solutions. The approach strengthens teamwork and relationships, and creates the energy and momentum for change. Smallwood and Ulrich, (2004) have very usefully reported that the capabilities, the collective skills of the employees and their abilities are covered under the banner of this approach. Expertise of an organization is the strength and that is the outcome of investments in staffing, training, compensation, communication, and other human resources areas. All of them collectively represent the ways that people and resources are engaged to accomplish work. Identity and personality of the organization are created by defining what it is good at doing. Competitors find them more difficult to copy than capital market access, product strategy, or technology. The capabilities give investors confidence in future earnings by knowing the strengths.

18. Holistic approach: Policy-makers have to emphasize the importance of whole and their interdependent part that is "holistic approach" (Davis, 2008). People in world have fragmented thinking (Tejomayananda, 1999) and it is not holistic and so incomplete. When firms face problems, they focus on particular aspects arriving at solutions without taking the totality into account. Therefore, the solution to their problem is also partial and works only for short time. Many countries govern and often run on an ad hoc basis. Therefore, holistic approach is the one which counts for total thinking while working on any solution.

14.6.3 GLOBAL CASES ON ECONOMIC POLICY

14.6.3.1 CASE OF THE US ECONOMIC POLICY

Prestowitz (2016) has stated that economic policy-makers in America have some of the hardest jobs to work in the world. Well known is the fact that the United States is the richest and most powerful economy in the world. It is further known to be the engine of economic growth across the world. Therefore maintaining a healthy economy happens to be a prerequisite for policy-makers. So also, improving the living standards and quality of life of Americans and individuals the world over is their accountability. While maintaining a healthy economy, study and solution over the challenges is very important. Opportunities and changes have been created by globalization.

Policy-makers, therefore, have to combat unfair trade practices from abroad and rising protectionism at home to ensure that the global economy continues to run smoothly.

14.6.3.2 CASE OF UNIDO (PUBLIC–PRIVATE PARTNERSHIP APPROACH)

Industry has been recognized as a major segment of economic growth since it produces a variety of products which supports economy. Industrial policy and strategy development Program has been designed and presented by this organization. Designing policy essentially considers economic analysis to cover activities such as competitiveness analyses as well as the setting up of knowledge banks. The design counts for best practices and benchmarking systems and all vital elements in the struggle to boost productivity. United Nations Industrial Development Organization (UNIDO) has been active in helping to set up mechanisms such as PPP at local and national level. Policy monitoring and assessment is also the task of policy-makers. Learning by monitoring is always exercised in order to fine-tune and improves strategies. Impact assessment helps policy-makers to think through the consequences of policy proposals. Quality of policy could be improved by advice to government and by encouraging more informed public debate on policy issues.

14.6.3.3 CASE OF NEW ZEALAND

Sound understanding is that education and the quality of teaching is a priority for economic growth and living standards. The Treasury (2013) has been on the public record numerous times stating about the quality of education and in particular. A skilled workforce is affirmed to be crucial to raising growth and productivity. By lifting student achievement in terms of skill there happened to be growth impact. As overall student achievement was lifted to among the top in the Organization for Economic Cooperation and Development (OECD), GDP is said to rise by 3–15% higher.

14.6.3.4 CASE OF CHINA

Tang (2013) has highlighted policy on the Chinese economy in the 21st century. China has made a choice in adopting recycling economy approach. The focus is on both a stable economic growth and an economical use of energy and resources. Explicit indices on energy and waste discharge have been settled and considered to reduce energy consumption and cut release of major pollutants in the coming five years. Development of the recycling economy, for that matter, is of vital importance. Attention is given on the most efficient use of energy and resources with the least release of wastes, pollutants, and toxic substances during industrial production. Scientific knowledge and due respect from policy-makers for that knowledge are considered vital in carrying out environmental policies. Experts in multidisciplines, including politics, economy, society, environment, and science are considered to be pooled in making major policies. Failure in this respect came to be the source of many problems China is facing.

14.6.3.5 CASE OF JAPAN POLICY (RATIONAL APPROACH)

Prof. Akira Suehiro (2006) provides useful information on industrial policies in Japan. Information revealed that the fiscal investment and loan program promoted trade and industry until early 1960s. Small loans were granted by Japan Development Bank and Exim Bank but it impacted on two important fronts. First one is catalyst for larger commercial bank loans while the second front is that, information sharing happened between business and government. The rational approach to policy-making initially sets goal and opts for a path to achieve it. Decision-makers are able to specifically state the ends, analyze the means to attain them and they want least possible input of scarce resources per unit of valued output, utility value, and cost–benefit analysis.

14.6.3.6 CASE OF UK'S MONETARY POLICY (MARKET APPROACH)

Biz/Ed (2010) cites an example of market-related approach to UK's monetary policy which was the development over time. Setting the target of monetary policy is considered as the essence of policy design. Policy player obviously became the UK government. Policy approach was then identified to reach the set targets. This has been the practice followed. In the 1950s and 60s, for example, monetary policy relied mainly on direct controls over Banks and other financial institutions. Then in the 1970s, a lot more attention was directed to monetary policy. In 1971, the United Kingdom moved to a much more market-related approach to monetary policy and distinctly away from the direct controls of the previous decade. Later in the late 1970s, the government had begun to set targets for the growth of the money supply. The 1990s saw the adoption for the first time of an explicit inflation target. In 2003, the official measure of inflation changed from the Retail Prices Index (RPI) to the Committe on Public Information (CPI) and the target rate was set at 2%.

CONCLUSION

Policy designing is a skill and requires knowledge and the ability to grasp the issues to address them effectively. Policy designing is a trilateral instrument that is convenient to remember for the users in the policy world. Economic policy development is a combination of core and complementary policies that cares for the overall economic growth of the nation. Countries in the globe have adopted various policy models for economic development depending upon their views and significance they attached to. Nevertheless, resource-based policy has been widely accepted and adopted since it is a major revenue generating instrument. Issues on third world nations and chaos and complexity could be addressed and are resolved by sorting them out by duly studying and eliminating the grounds they are generated from. Repeated iteration policy and introducing timely corrections and changes are indispensable to avoid debacle from globalization since they remain in line with the trends in the globe.

KEYWORDS

- **policy designs**
- **government**
- **economic growth**
- **economic development**
- **third world nation**

REFERENCES

Alvord, D. Leadership, 2015 https://www.eda.gov. (accessed Feb 18, 2017).

Amadeo, K. Seven Ways Central Banks Use Foreign Exchange Reserves, 2016 https://www.thebalance.com.

Biz/Ed. Market Related Approach to UK's Monetary Policy, 2010 https://www.ecb.europa.eu.
China. Circular economy approach, 2008 http://www.indigodev.com.
Cole, K. Globalization: understanding complexity, 2003 http://journals.sagepub.com.
CSUlb. Models of Public Policy Making, 2002 http://web.csulb.edu.
Davis, R. E. *Absolute Knowledge*; Full Circle Publishing: New Delhi, 2008.
Decker, L. Make Your Policy Easy to Understand, 2015. http://www.urbanbound.com.
Deshpande, M. Designing Policies for Business. PhD Thesis. University of Pune, India, 2011.
Echandi, R. Three Key Ideas for Creating Effective Investment Policies, 2016. http://blogs.worldbank.org.
Faux J. Overhauling NAFTA, 2008. http://www.epi.org.
Gallup, J. L. et al. Geography and Economic Development, 2016. http://journals.sagepub.com.
GOC. Development Sustainable Economic Growth Strategy, 2017. http://www.international.gc.ca.
Government of Sweden. Policy for Economic Growth, 2010. http://www.government.se.
Hahn, A. J. Policy Making Models and Their Role In Policy Education, N.D. http://ageconsearch.tind.io.
Hanushek, E. Economic Growth in Developing Countries: The Role of Human Capital, 2013. https://hanushek.stanford.edu.
Harberger, A.C. On the Process of Growth and Economic Policies in Developing Countries, 2005. http://www.econ.ucla.edu.
Hayes, M. T. Incrementalism, 2013 https://www.britannica.com.
Hendrickson, M. W. Country Economic Policies: What the U.S. Could Learn from Other Countries, 2012. https://www.forbes.com.
Jha, R. A Synoptic View on Fiscal Policy in Economic Development, 2007. https://crawford.anu.edu.au.
Macnamara, D. Change, Chaos, Globalization and other Windmills, 2003. http://www.banffexeclead.com.
Mitchell, D. Impact of Government Spending on Economic Growth, 2005. http://www.heritage.org.
Patil J. Resourced Based Policies, 2016. http://www.jayendrapatil.com.
Perrels, A. Workshop on Good Practices in Policies and Measures (Oct 8–10, 2001, Copenhagen).
Prestowitz, C. US Economic Policy, 2016. http://www.econstrat.org.
Santarelli; Or. Entrepreneurship and the Process of Firms' Entry, Survival and Growth, 2007. https://www.academic.oup.com.
S.hapiro, G. Six Ways to Create Economic Growth, 2013. https://www.forbes.com.
Smallwood; Ulrich. Capitalizing on Capabilities. Harvard Business Review, 2004 https://hbr.org.
Saucedo, J. Demographic Transition Theory & Overpopulation, 2017. http://www.brighthub.com.
Savitti P.P. Variety, Structural Change and Economic Development, 2008, pp 1–14. http://www.ideas.repec.org.
Tang R. Chinese Economic Policy in the 21st Century: Growth, Imbalance, and Considerations for Australia, 2013.
Tejomayananda Address to the Delhi CHYKS. *Chinmay Jagruti* **1999,** *3* (5).

Treasury. Education Key to Economic Growth and Equity, 2013. http://www.treasury.govt.nz.

Treasury Gov. Uk Government Policy, Economic Growth in Developing Countries, 2015. https://www.gov.uk.

Tejvan (2016). Policies to Reduce a Current Account Deficit. http://www.economicshelp.org

UOW, 2015, https://f2.washington.edu.

Valentina Pasquali. Highest GDP Growth, 2015. https://www.gfmag.com.

Wordspress.com Theories of Economic Development, 2011. https://erikkrantz.wordpress.com.

World Bank. The Political Economy of Policy Reforms, 2008. http://siteresources.worldbank.org.

Wiens, J. Entrepreneurship Role in Economic Development. http://www.kauffman.org.

CHAPTER 15

Evaluation of Development of Teachers' Performance in Terms of Economic Growth

NURSEL YARDİBİ*

Assistant Professor, Independent researcher.

**E-mail: dryardibi@gmail.com*

ABSTRACT

In terms of organizational and individual sense, in-service training is expected to provide many benefits. The most fundamental benefit of in-service training in terms of economic sense is to make a positive difference between the costs of the organization and the outputs. The planning, programming, implementation and evaluation of the training programs to be provided for this benefit have a certain cost to the organizations. In educational organizations, teachers adaptation to changes in terms of knowledge by supporting them with in service trainings, reflect students' growth according to the age and directly affect economies of countries, because today's students will be tomorrow's employees and employers. Managing chaos in educational organizations depends on effective relationships of all stakeholders in organizations such as students, parents, teachers, managers and also depends on adaptation to changes of world. So, effective trainings of teachers in terms of global changes can be solution to overtake rapid changes in economy. In this chapter, we aimed to make evaluation of development of teachers' performance in terms of economic growth.

15.1 INTRODUCTION

The realization of organizational performance is possible by attaining organizational goals, which depends on the performance of the employee.

Supporting the development of the performance of their employees in their work by managers to reach organizational goals is very important for organizations. In this sense, things can be done in a managerial sense in support of occupational development of employee can be listed as follows (Bozkurt Bostancı, 2004):

- clear definitions of individual and organizational goals,
- joint planning,
- linking organizational strategy with business objectives,
- performance agreements, recognition of the training, and development needs of the employees,
- making a development plan for employees, ensuring to use the gained knowledge in performance behavior change, and
- establishment of internal and external communication networks, evaluation of employee performance according to performance criteria, and use of evaluation results.

When educational organizations are dealt with; it is clear that the organization should support teachers in a managerial way in order to increase teacher performance. In order for teachers' professional development and school goals and objectives to be more easily accessible by supporting teachers' increasing performance, it is important to eliminate the individual needs of teachers in the professional sense, guidance, and orientation, and in-service training activities. In addition, as the teacher performance increases, such arrangements as teacher counseling and supervision studies to be done in line with the needs of the teachers, solution-focused commission activities, seminars and courses to support the teachers, and increase the knowledge level by reading the professional publications will increase the teacher performance. In this chapter, we aimed to make an evaluation of the development of teachers' performance in terms of economic growth.

According to Amstrong (1998) and Trethowan (1991), learning that can be done in order to improve occupational performance, learning by working as a manager, team leader, mentoring, job rotation, job enrichment, and cooperative learning, learning on the project, learning freely out of work. Eraut (1997) reveals that teachers can learn with three methods: self-learning (self-learning about what they will do at work), supported learning (learning from the supervisor and teachers), and performance management (goal setting, regular evaluation, and development). In addition, teachers' professional counseling and mentoring and small teacher groups' focus on professional development and innovations are also activities that support teacher performance.

15.2 IN-SERVICE TRAINING

In-service training is the continual training or training of a professional person for his/her profession from the first day she/he started his career to the day she/he left his job (Başaran, 1960). Given the fact that the prisoners from overseas and the workers in the country are trained at work as foremen, it can be said that the history of in-service training of labor is as much as the ancient Egyptians (Sims, 1990). Today, employee education is organized in the form of formal education and lifelong education. The concept of learning organizations has come to the agenda in the 1980s. The basic idea of this movement is that the ability of organizations to make their own transformations according to environmental changes depends on training and occupational development (Pearn et al., 1995).

Developments in science and technology necessitate learning new knowledge and technologies in every profession and educating employees in these fields. Organizations in the information society have been now looking for competencies that can comprehend complex intraorganizational and interorganizational skills and have ability to comply with effective teamwork. One of the common problem areas in almost every organization is that the quality of personnel does not meet the business requirements.

In the social and economic areas where technology has been rapidly developing, the qualifications sought for personnel have been changing in this period, when innovation and therefore knowledge have been increasing and the concept of personnel management in institutions has been being replaced by the management of human resources. In such an environment, the importance of in-service training in order to earn the required competencies of the staff working in organizations is increasing day by day (Taymaz et al., 1997).

In addition, it is increasingly necessary to train in-service personnel who have difficulties in meeting the rapidly changing society and the conditions of the profession with the education they had previously seen. The necessities, requirements, and obligations are essential when defining in-service training objectives. In order to determine the aims of in-service training in an organization, it is necessary to first determine the reasons for training.

In terms of organizational and individual sense, in-service training is expected to provide many benefits. The most fundamental benefit of in-service training in terms of economic sense is to make a positive difference between the costs of the organization and the outputs. The planning, programming, implementation, and evaluation of the training programs to be provided for this benefit have a certain cost to the organizations. On the other hand, as a result of training programs, increasing the performance of the employees, gaining more positive business attitudes, achieving satisfaction, and increasing the power of the heart are the outputs of education (Kearsley, 1994).

According to Sims (1990), the main benefits provided by in-service training are as follows:

- In-service training increases the profitability and profit of organizations.
- The knowledge and skills related to work develop at all levels in the organization.
- Morale in the organizational environment increases.
- Employees' recognition and adoption of organizational goals are facilitated.
- Assistant–upper management relationships develop.
- Openness and trust develop within the organization.
- Organizational development is achieved.
- Knitting feedbacks are provided by the employees.
- Understanding and monitoring of organizational policies are ensured.
- Preparation of business directives is ensured.
- The effectiveness of organizational decisions and problem-solving processes increases.
- Increased organizational reputation.
- Leadership skills, motivation, loyalty, and positive attitudes toward mature develops.
- The quality of production and productivity increase.
- Misuse of the means of production and organizational waste is reduced.
- Creates an atmosphere suitable for communication and development in an organization.
- Adapting to change for employees get easier.
- Helps in reducing stress and tension and resolve conflicts.
- Helps employees to make better decisions and develop effective problem solving skills.
- Develops motivational variables which are success, development, accountability, and progress.

- Encourages self-development and self-confidence.
- Increases the ability of employees to cope with stress, tension, frustration, and conflict.
- Increase job satisfaction and recognition.
- Education enhances the person's self-directed approach while increasing the person's interaction and communication skills.
- Provides satisfying personal needs.
- Prepare a personal environment to individuals for self-determination and development.
- Employees develop their writing, speaking, and listening skills.
- It removes fear of passing or rising to new tasks.

Today, knowledge, using knowledge in working life and education is different from yesterday because of the global changes. When people's abilities improve as a member of societies that they live in, education and training in that society increases social cohesion, reduces crime, and improves the distribution of income (World Bank, 2003).

A country's economic growth can change as age populations; because different age groups have different production capacity, so economic changes of countries are to assume constant age-specific behavior with respect to employment, savings, and so on (Bloom et al., 2010). On the other hand, another factor that effects the economic growth of the societies is social developments.

Two major tendencies that cause intense debate, marked social developments over the past three decades, these are (Malitza, 2005);

- The first of these is developments in lifelong learning concept. The traditional education approach has been replaced by a more flexible approach that extends over increasingly long periods.
- The focus has gradually shifted to learning in work life.

In this content, various types of in-service training activities are applied in order to overcome the training deficiencies and requirements of the employee or to adapt to new developments. In the most general sense these types of training are on-the-job training and off-job training (Aydın, 2008);

- *On-job training*: It is the basic approach to acquiring skills. It happens in the real environment. For this reason, when the on-the-job training programs are well organized, the level of effectiveness is high. We can give practical training during the candidate teacher period if we give an example from education.

- *Off-job training*: Programs which are outside the work and employees benefit from the training opportunities in outside the organization.

In-service training programs can be classified as follows according to application purposes (Taymaz, 1992):

1. *Orientation training*: These are trainings designed for new employees who come to the organization to define the purpose and policy of the organization and to define their own duties, authorities, and responsibilities.
2. *Fundamental education*: This is done in order to gain the knowledge, skills, and attitudes required for the work to be started in an institution. It can also be described as preparation, adaptation, job orientation, internship, and candidacy training.
3. *Development training*: It is the training applied for the development of the employees who are working at the institution in the field of development and innovation and the development of their abilities. This training is programmed to raise employee morale, increase commitment to the organization, and gain sensitivity.
4. *Completion training*: These are the programs that are applied to the employees who need to change their duties to acquire competencies required by the new task.
5. *Upgrading training*: Programs designed to meet the promoting needs of employees and staffing in the organization. We can give as an example to trainings for the school principals and assistant manager exam preparation.
6. *Private field training*: These are the programs that are applied to train employees in various fields for special services. As an example, we can say that class teachers can work in the special education field by participating in special education seminars.

15.3 CONCLUSION

Actually, chaos is a source of life and it is a reality for people and organizations as self-organizing complex systems (Erçetin et al., 2014). So education is inextricably interwoven with chaos and complexity (Erçetin et al., 2015) which renders it largely unpredictable especially at a time when contemporary paradigms seem unstable and ever-changing with varying patterns (Erçetin and Bisaso, 2016). Managing chaos in educational organizations depends on effective relationships of all stakeholders in organizations such as students, parents, teachers, managers, and also depends on adaptation to changes of the world. So, effective trainings of teachers in terms of global changes can be a solution to overtake rapid changes in the economy.

The knowledge economy is a global concept; however, the countries' responses to such a phenomenon are different from each other (Singai et al., 2016). So, better appreciation and recognition of the value of experienced employees should be combined with occupational retraining programs (Luchsinger, 2016).

Being successful in the knowledge economy requires mastering new competencies like basic academic skills, the ability to use information, communication technology, and the ability to use these skills effectively (World Bank, 2003). In educational organizations, teachers' adaptation to changes in terms of knowledge by supporting them with in-service trainings, reflects students' growth according to the age, and directly affects economies of countries, because today's students will be tomorrow's employees and employers.

KEYWORDS

- **teacher**
- **education**
- **economic growth**
- **in-service training**
- **teachers' performance**

REFERENCES

Armstrong, M. *Managing People*; Kogan Page Limited: London, 1998.
Aydın, İ. *Öğretimde Denetim*; Pegem A Yayıncılık: Ankara, 2008.
Başaran, İ. E. *Hizmet içi Eğitim Araştırması*; MEB Eğitim Birimi Müdürlüğü Yayınları: Ankara, 1960.
Bloom, D. E.; Canning, D.; Fink, G. Implications of Population Ageing for Economic Growth. *Oxf. Rev. Econ. Policy* **2010,** *26* (4), 583–612.

Bozkurt Bostancı, A. *Türkiye'deki Resmi ve özel Ilköğretim Okullarında Öğretmen Performans Yönetimi*. Yayınlanmamış Doktora Tezi, Ankara Üniversitesi Eğitim Bilimleri Enstitüsü, Ankara, 2004.

Eraut, M. Perspectives on Defining "The Learning Society." *J. Educ. Policy* **1997,** *12* (6), 551–558.

Erçetin, Ş. Ş.; Bisaso, S. M. The Influence of Changing Paradigms on Educational Management and School Administration. In *Applied Chaos and Complexity Theory in Education*; Erçetin, S. S., Ed.; Information Science Reference: Hershey, PA, 2016; pp 11–28 DOI: 10.4018/978-1-5225-0460-3.ch002.

Erçetin , Ş. Ş.; Bisaso, S. M.; Saeed, F. Understanding Chaos and Complexity in Education Systems Through Conceptualization of Fractal Properties (Chapter 12). In ŞŞ & S *Chaos, Complexity and Leadership 2013*; Erçetin, S. S.; Banerjee, S., Eds.; Springer Proceedings in Complexity, Switzerland, 2015; pp 147–161.

Erçetin, Ş. Ş.; Hamedoğlu, M. A.; Açıkalın, Ş. N. Can Managerial Narcissism Which Likely Leads to a Chaotic Process be Measured? Managerial Narcissism Scale (MNS) (Chapter 7). In *Chaos and Complexity in World Politics*; Ercetin, S. S.; Banerjee, S., Eds.; IGI Global: USA, 2014.

Kearsley, G. *Cost, Benefits, and Productivity in Training Systems*; Addison Wesley Company: California, 1994.

Luchsinger, G. *Shaping the Future: How Changing Demographics can Power Human Development*; Asia-Pacific Human Development Report, UNDP United Nations Development Programme: New York, 2016.

Malitza, M. The Double Helix of Learning and Work. In *Health, Ageing and Work, the Counter-ageing Society*, European Papers on the New Welfare 1; The Risk Institute: Trieste; Geneva, 2005; pp 123–127.

Pearn, M.; Roderick, C.; Mulrooeny, C. *Learning Organizations in Practice;* McGraw Hill Company: London, 1995.

Sims, R. *An Experiential Learning Approach to Employee Training Systems*; International Personnel Management Association: Chicago, 1990.

Taymaz, H. *Hizmet içi eğitim*; Pegem A Yayıncılık: Ankara, 1992.

Taymaz, A. H.; Sunay, Y.; Aytaç, T. Hizmet içi Eğitimde Koordinasyon Sağlanması Toplantısı. *Milli Eğitim* **1997,** *133,* 13–17.

Trethowan, D. M. Managing With Appraisal Achieving Quality Schools Through Performance Management. Management Education; Paul Chapman Publishing Ltd: London, 1991.

World Bank. *Lifelong Learning in the Global Knowledge Economy: Challenges for Developing Countries*; World Bank: Washington, DC, 2003.

CHAPTER 16

Population, Age Structure, and Economic Growth in India: Evidence From the Toda–Yamamoto Noncausality Approach

P. K. MISHRA[1,*] and S. K. MISHRA[2]

[1] *Deparment of Economics, Central University of Punjab, Bathinda, Punjab, India*

[2] *Deparment of Economics, Siksha 'O' Anusandhan University, Bhubaneswar, Odisha, India*

[*]*Corresponding author. E-mail: pkmishra1974@gmail.com*

ABSTRACT

In India, the recent fall in fertility rates and increases in life expectancy combined with the past changing patterns of birth and death rates have been producing a noteworthy shift in the population age structure. Particularly, the share of working-age population in the total population has been increasing with the growing evidence that they are healthier than before. This has implications for earnings, employment, and savings thereby indicating maximization of the demographic dividend of the nation subject to the availability of harmonious policy environments. In this context, this paper examines the dynamic nexus between the growth rate of total population and economic growth, and also investigates the relationship between the growth rate of the share of the working-age population and economic growth in India during 1961 to 2015 using Toda-Yamamoto noncausality approach. The results provide the evidence of the feedback causality between the former, and unidirectional causality between the latter in favor of economic growth. This finding, on the one hand, implies that population growth in India is both the cause and consequence of economic growth, and on the other hand, implies that growth of the share of working-age population has positive impacts on the economic growth of the country. It is, therefore, essential to have complementary policies and facilitating institutions to capitalize this window of opportunity in the country to boost the macroeconomic performance.

16.1 INTRODUCTION

The process of economic growth and development involves the utilization of physical resources of a nation by the labor force so that the productive potential of that country is realized (Datt and Mahajan, 2012). It means labor force, broadly the population size, exerts a crucial influence on the economic development of a country (Shen, 2015). In other words, the economic performance of a country is to a substantial degree influenced by the country's demographic situation (Furuoka, 2010; Mahmud, 2015). Thus, the size of the population and its growth can be an important determinant of the economic growth of a nation (Sethy and Shaoo, 2015). Therefore, the nexus between population and economic growth is a topic of keen interest among demographers, economists, and policy-makers as well (Yao et al., 2013; Chang et al., 2014).

In the literature on development economics, there exist at least three paradigms concerning the relationship between population and economic growth, namely, the Malthusian, Boserupian, and Kelleyian paradigms (Malthus, 1798; Boserup, 1965; Kelley, 1968, 1988). The Malthusian "pessimistic version" states that rapid population growth leads to lower economic growth and causes poverty. It proposes that the rapid growth of population may lead to massive increase in demand for food, living space, and natural resources, which would result in a decrease of productivity and thus, hinder economic development (Yamaguchi, 2001; Shen, 2015). This pessimistic theory has been supported by the findings of Coale and Hoover (1958), Ehrlich (1968), Meadows et al. (1972), and Barney (1980). Second, the Boserupian "optimistic version" states that higher rate of population growth positively contributes to economic growth by augmenting the desire to consume and invest, inducing technological change, increasing the stock of human capital, creating markets, and by increasing the governmental investment on infrastructure. This optimistic theory has been supported by the findings of Keynes (1937), Hansen (1939), Kuznets (1966, 1973), Yamaguchi (1973), Birdsall (1977), Simon (1977, 1981), Simon and Kahn (1984), and Kremer (1993). Third, the Kelleyian "neutralist version" states that the population growth has no impact on economic growth. This neutralist theory has been supported by the findings of Dawson and Tiffin (1998) and Thornton (2001). Recently, however, a new dimension of demography is being discussed in the literature which has challenged the neutralist version, and puts in place the argument that the age structure of a population can play a key role in promoting the economic growth through an increase in the share of the working-age population (Bloom and

Williamson, 1998; Bloom et al., 2003; Safdari et al., 2011; Thakur, 2012; Joe et al., 2015).

In support of the abovementioned paradigms, the empirical journey marched ahead since the publication of the pioneering work of Darrat and Al-Yousif (1999). The studies significant in finding the positive effect of population growth on economic growth include Darrat and Al-Yousif (1999), Kothare (1999), Furuoka (2009), Gideon et al. (2010), Furuoka and Munir (2011), Adediran (2012), Furuoka (2013), and Garza-Rodriguez et al. (2016). The studies important in finding the negative effect of population growth on economic growth include Darrat and Al-Yousif (1999), Dao (2013), and Yao et al. (2013). The studies which found no evidence of any relationship between population growth and economic growth are Dawson and Tiffin (1998), Thornton (2001), and Singha and Jaman (2012). It is, thus, inferred that the general consensus whether population growth is beneficial or detrimental to economic growth of a country is confusing. This justifies for furthering the research on the issue.

In view of the complex and ambiguous impact of population growth on economic development, Becker et al. (1999) with the help of a simple theoretical model demonstrated that the population growth has both positive and negative impacts on economic development. On the one hand, a rapid growth of population promotes an increase in investment and facilitates the further division of labor, and thus, positively contributes to the economic growth. On the other hand, population growth has been argued to reduce productivity and thus, negatively affect economic development. The empirical findings of Shen (2015) also support this view. Shen concluded that in high populous countries such as China, India, and Indonesia, the population growth and increase in the number of skilled labor fails to promote economic growth. In relatively developed countries such as Japan, South Korea, and the Philippines, population growth boosts economic growth. And, in countries such as Pakistan and Vietnam, population growth does not explain economic growth. In case of India, few studies including Bhat (1995), Haldar (2009), Bloom et al. (2010), Eberstad (2010), and Da Vanzo et al. (2010) have concluded the existence of both positive and negative impacts of population on economic growth.

Amidst such perplexity, another strand of research on the relationship between population and economic growth concluded any of the following four hypotheses: economic growth-induced population growth hypothesis, population-led economic growth hypothesis, feedback hypothesis, and neutrality hypothesis. In the Malthusian sense, the relationship between population growth and economic growth is bidirectional (Malthus, 1798).

On the one side, higher economic growth increases population by stimulating early marriage and high birth rates, and by reducing mortality. On the other side, higher population pushes down the economic growth through diminishing returns to labor. Tsen and Furuoka (2005) found the empirical evidence of the existence of bidirectional causality between population and economic growth for Japan, Korea, and Thailand; population-led economic growth hypothesis for China, Singapore, and the Philippines; economic growth-induced population growth hypothesis for Hong Kong and Malaysia; and the neutrality hypothesis for Taiwan and Indonesia. These findings clearly indicate that population growth could be beneficial or detrimental to economic growth and economic growth could have an impact on population growth. In a recent study, Chang et al. (2014) found the empirical evidence in favor of economic growth-induced population growth hypothesis for Canada, Germany, Japan, Norway, and Switzerland; population-led economic growth hypothesis for Finland, France, Portugal, and Sweden; bidirectional hypothesis for Austria and Italy; and neutrality hypothesis for Belgium, Brazil, Denmark, Netherlands, New Zealand, Spain, Sri Lanka, the United Kingdom, the United States, and Uruguay. Garza-Rodriguez et al. (2016) provides empirical evidence in favor of mutually reinforcing or feedback relationship between population growth and economic growth for Mexico. The empirical findings of Furuoka (2010) support the economic growth-induced population growth hypothesis for the Philippines. Sethy and Sahoo (2015) provides the evidence that population growth spurs economic growth in India, whereas Mahmud (2015) found the evidence in favor of economic growth-induced population growth for India.

The contribution of Bloom and Williamson (1998) opened up another dimension of thinking that the growth rate of the share of the working-age population (15–64 age group) can have meaningful implications for the economic growth. It has been argued that as countries experience various phases of demographic transition from high fertility and high mortality to low mortality and low fertility, the age composition of population changes (Bloom, 2011). If this change in age composition is such that the working-age population grows larger than the growth of total population then the country is likely to benefit from it, often called "demographic dividend" (Bloom and Canning, 2008). Bloom and Williamson (1998) and Bloom et al. (2003) through cross-country studies, argued that an increase in working-age population tends to increase productivity, labor supply, stock of human capital, savings, and thus, tends to provide ample capital for investment. Therefore, there can have a positive impact on economic growth (Bloom and Canning, 2008; Kumar, 2010). However, this is contingent upon the presence of complementary policies,[1] and facilitating institutions[2] (Kumar, 2010). Aiyar and Mody (2012–2013) found a large and significant growth impact of both the level and growth rate of the working-age ratio in Indian states. Song (2013) also found the positive effect of the growth in the working-age population on rapid economic growth in Asia. Zhang et al. (2015) provides evidence of the significant contribution of working-age population to the provincial economic growth rates in China.

It is inferred from the foregoing discussion that the relationship between population and economic growth is rather confusing and conflicting, both in terms of impacts and causalities. Although the extant studies have been conducted using different methods and resulted in diverse conclusions having unique characteristics with respect to differences in sample time and space, these provide a theoretical as well as an empirical base for further investigation of the issue. It is with this backdrop, this chapter is an attempt to reinvestigate the causal relationship between population and economic growth in the context of the second most populous country, that is, India. Furthermore, this chapter investigates the relationship between growth in the share of the working-age population and economic growth in the country. Thus, the rest of the chapter is organized as follows: Section 2 gives a brief picture of the status of the population growth in India; Section 3 describes the data and methodology used in the study; Section 4 makes the empirical analysis and discusses the findings; and Section 5 concludes.

16.2 POPULATION GROWTH IN INDIA

The dynamics of India's population that we notice today and that was prevailing in the pre-independence era are quite different, and also have implications for the economic growth of the country. Datt and Mahajan (2012) placed the growth rate of India's population in four groups—stagnant population (1891–1921), steady growth (1921–1951), rapid high growth (1951–1981), and high growth with a definite sign of slowing down (1981–2011). During the first phase of 30 years (1891–1921), the total population of India grew from 236 million to 251.32 million with the compound annual growth rate of 0.19% per annum. This minimal growth rate for the period was

[1] The complementary policies include health, population and family planning, labor markets, macroeconomic, financial, and education policies.

[2] The facilitating institutions include retirement policy, pension, and healthcare finance, the efficiency of labor and capital markets, and the structure of national and global economic systems.

mainly driven by the high average annual birth rate (48 per thousand) and high average annual death rate (45 per thousand) thereby putting India in the first phase of demographic transition. During the second phase of 30 years (1921–1951), the population grew from 251.32 million to 361.09 million with the compound annual growth rate of 1.22% per annum. This moderate growth rate for the period was mainly due to a very minimal fall in average annual birth rate to 44 per thousand, and substantial decline in the average annual death rate to 32 per thousand thereby making India move into the second stage of demographic transition. During the third phase of 30 years (1951–1981), the population grew from 361.09 million to 683.33 million with the compound annual growth rate of 2.14% per annum. This alarming growth of population for the period was mainly due to a fall in average annual birth rate to 39.5 per thousand and a sharp decline in the average annual death rate to 17.4 per thousand thereby exploding the population bomb in the country. During the fourth phase of 30 years (1981–2011), the population grew from 683.33 million to 1210.19 million with the compound annual growth rate of 1.89% per annum. Specifically, from 2001 to 2011, the annual average growth rate of the population declined to 1.64%. This phase of high population growth in India with a definite sign of slowing down was mainly driven by fall in birth rate to 21.8 per thousand as well as death rate to 7.11 per thousand.

It is seen from Figure 16.1 that the population of India which was 238.4 million in 1901 has been more than quadrupled over a period of 100 years, and became 1210.19 million in 2011. The pace of population growth in the country was slow up to 1921, with some decline during the decade of 1911–1921 due to rapid and frequent occurrence of epidemics such as cholera, plague, influenza, and famines, but picked up after 1921 and accelerated after 1951, with decadal growth rising above 20% after 1961 through 2001.

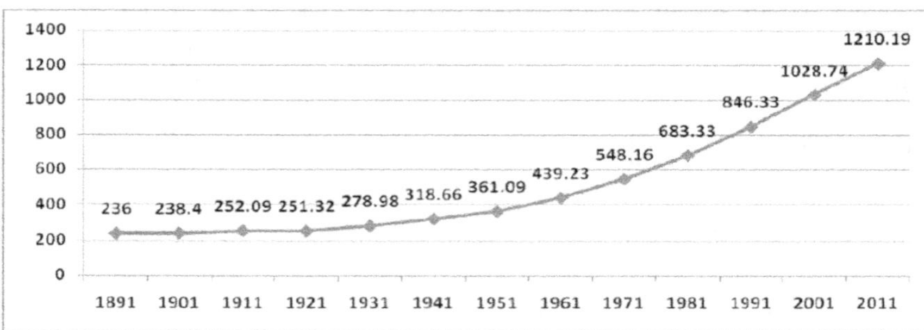

FIGURE 16.1 Census population in India, 1891–2011 (million).
Source: Census of India and Authors' Own Plot.

However, it is sanguine to mention that the population in India has grown at a decreasing rate particularly after 1971 (Fig. 16.2). The latest census has shown a clear fall in decadal growth of population in India to 17.64% in 2011. The reason may be the narrowed down gap between crude birth rate (CBR) and crude death rate (CDR) during this period (Fig. 16.3).

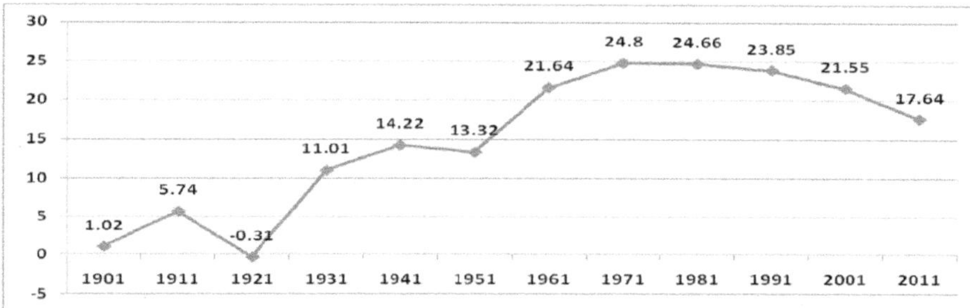

FIGURE 16.2 Decadal growth rate population in India, 1901–2011(%)

Source: Authors' Own Plot.

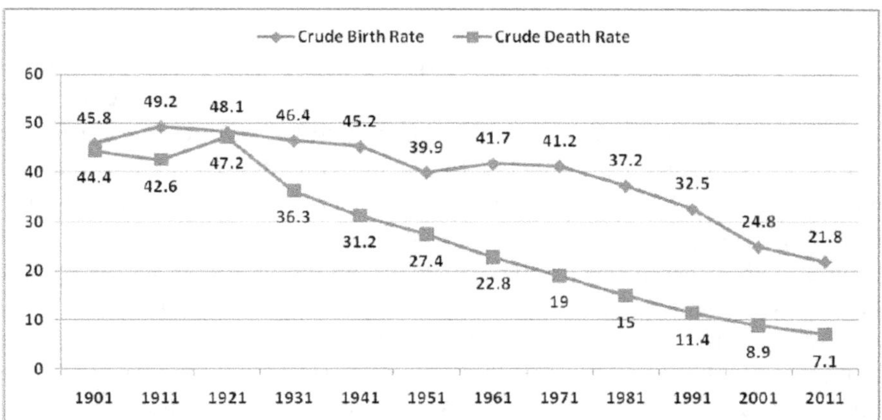

FIGURE 16.3 Crude birth rate and crude death rate in India (per 1000).

Source: Census of India.

In accordance with the changing structure of CBR and CDR over years, the age structure of India's population has been undergoing some gradual changes during the last few decades having implications for the growth of the country. In broad terms, the age distribution has shifted toward middle ages, with fall in the share of the child population and a corresponding rise in the share of adults (Fig. 16.4).

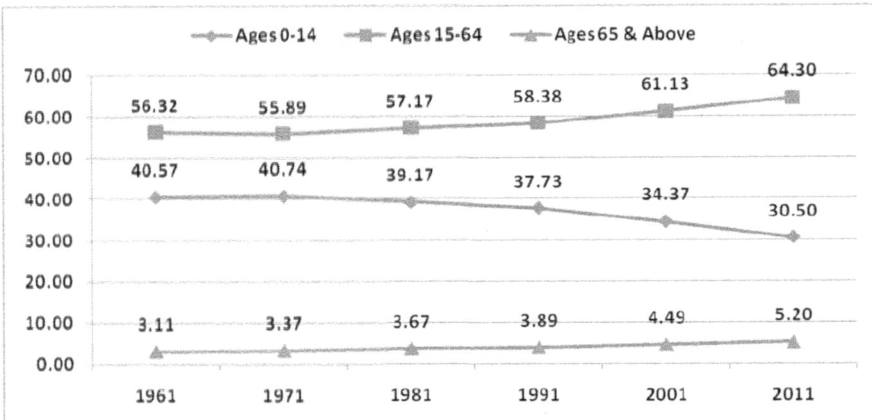

FIGURE 16.4 Age structure of India's population (as % of total population).

Source: World Development Indicators; World Bank.

Specifically, the proportion of the younger population in the age groups of 0–14 years has been showing a declining trend with a decrease from 40.57% in 1971 to 30.50% in 2011. The share of the population in the working age group 15–64 years has been showing an increasing trend with a rise from 56.32% in 1971 to 64.30% in 2011. The proportion of people in the age group 65 and above years has also increased marginally from 3.11% in 1971 to 5.20% in 2011. A positive consequence of the changes in the age distribution is that the dependency ratio has fallen, particularly on account of the drop in the young age dependency ratio. Clearly, the requirement of supporting children has come down as couples now have fewer children than in the past. Thus, with an increase in the share of the working-age population, India's window of demographic opportunity has opened in recent decades. This huge opportunity brings with it a positive socioeconomic consequence for the nation possibly in the form of higher levels of productivity, income, savings, and higher levels of standard of living (Kumar, 2010). Hence, it is essential to examine the nexus between the growth of total population and economic growth, and also imperative to investigate the relationship between the growth in the share of the working-age population and the economic growth.

16.3 DATA AND METHODOLOGY

On the basis of the aforesaid discussions, this study puts forth two different but well-connected hypotheses that (1) the growth of the total population and the growth of the Indian economy have causal relationships between them and (2) the growth of the share of the working-age population causes the economic growth of India. For this purpose, we have used the annual growth rate of population (*pop*), annual growth of the share of the working-age population in the total population (*wap*), and annual growth rate of GDP per capita (*gdp*) to test the aforesaid hypotheses. The annual data on these variables for the sample period 1961 to 2015 have been compiled from the database on World Development Indicators published by the World Bank. The specifications (1) and (2) are proposed to be estimated:

$$gdp_t = \alpha_1 + \beta_1 pop_t + \varepsilon_{1t} \tag{16.1}$$

$$gdp_t = \alpha_2 + \beta_2 wap_t + \varepsilon_{2t} \tag{16.2}$$

The specification (1) presumes the relationship between the growth of total population and the economic growth of India, while specification (2) assumes the relationship between the growth in the share of the working-age population and the economic growth of India. The estimation results would help us in predicting the causal relationships between the specified indicators. Depending on the stationary properties of the key variables of interest (Table 16.1), this study performs the Granger noncausality test in a multivariate framework proposed by Toda and Yamamoto (1995). This method is relatively more efficient and appropriate for time series for which the order of integration is either not known, or may not be necessarily the same, or the order of integration is two or more. In our case, the variable *gdp* is integrated of order zero, *pop* is integrated of order two, and *wap* is integrated of order one. Thus, the selection of the estimation technique is justified.

The Toda and Yamamoto (1995) methodology of Granger causality test by directly performing the test on the coefficients of the levels Vector autoregression (VAR), minimizes the risk associated with possibly wrongly identifying the orders of integration of the series and the presence of cointegration relationship (Galies, 1997; Mavrotas and Kelly, 2001). The basic idea in the Toda and Yamamoto (1995) procedure is artificially augmenting the correct VAR order, *k* with *d* extra lags, where *d* is the maximum likely order of integration of the time series in the empirical system. Thus, at the outset, it is required to determine the maximum order of integration of time series, say, d_{max}. Then the optimal lag length of the VAR model is determined using Akaike information criteria (AIC), say, k. In the third step, the ($p = k + d_{max}$) th order of VAR is estimated with seemingly unrelated regression (SUR). At last, the null hypothesis of no causality is tested using a standard Wald statistic, say, W. The implementation of the Toda and Yamamoto (1995) noncausality approach necessitates linking the two variables of the study in a bivariate framework which can be stated as:

$$Y_t = A_0 + A_1 Y_{t-1} + ... + A_k Y_{t-k} + \varepsilon_t \tag{16.3}$$

Here, $Y_t = \begin{bmatrix} Y_{1t} \\ Y_{2t} \end{bmatrix} = \begin{bmatrix} gdp_t \\ pop_t \end{bmatrix}$ and s; and A's are 2 × 2 matrices of coefficients.

The augmented level $VAR(p = k + d_{max})$ is estimated to test the null hypothesis of no causality and this VAR is specified as:

$$Y_t = \alpha + A_1 Y_{t-1} + ... + A_k Y_{t-k} + A_{k+1} Y_{t-k-1} + ... + A_p Y_{t-p} + \varepsilon_t \tag{16.4}$$

This augmented VAR system is estimated using SUR technique of regression. And, the null hypotheses of the study are H_{01} and H_{02} and these are tested by the Wald test.

H_{01}: Y_{2t} does not cause Y_{1t}, that is, $a_{12}^1 = a_{12}^2 = ... = a_{12}^p = 0$

H_{02}: Y_{1t} does not cause Y_{2t}, that is, $a_{21}^1 = a_{21}^2 = ... = a_{21}^p = 0$

Let $e_1 = \begin{bmatrix} 1 \\ 0 \end{bmatrix}$, $e_2 = \begin{bmatrix} 0 \\ 1 \end{bmatrix}$, and $D = I_k \otimes e_2$ with I_k being the $k \times k$ identity matrix. Let $vec(A)$ be the column vector obtained by stacking the rows of the matrix A. Then the Wald test statistic is given by: $W = T\left((e_1^{'} \otimes D^{'})vec(\hat{A})\right)$ $\left((e_1^{'} \otimes D^{'})\hat{\Sigma}(e_1^{'} \otimes D^{'})^{'}\right)^{-1} (e_1^{'} \otimes D^{'})vec(\hat{A})$, where $\hat{\Sigma}$ is a consistent estimator of the asymptotic variance matrix of $\sqrt{T}vec(\hat{A} - A)$. The Wald test statistic (W) has an asymptotic χ^2 distribution with k degrees of freedom. The reason for ignoring the remaining d_{max} autoregressive parameters in testing for Granger causality is that it helps overcoming the problem of nonstandard asymptotic properties associated with standard Wald test for integrated variables. It is established in the empirical literature that the Wald test experiences efficiency improvement when SUR model is used in the estimation (Rambaldi and Doran, 1996). The similar estimation methodology is also assumed for the specification (2).

16.4 ESTIMATION RESULTS AND DISCUSSION

In the first step of this noncausality analysis, the order of integration of each of the variables *gdp*, *pop*, and *wap* has been determined using the augmented Dickey–Fuller (ADF) unit root test, and the results are reported in Table 16.1.

TABLE 16.1 Results of ADF Unit Root Test (with Trend and Intercept).

Variables	ADF statistics at level	*p*-value	ADF statistics at first difference	*p*-value	ADF statistics at second difference	*p*-value	Decision
gdp	−8.069[a]	0.000	NA	NA	NA	NA	I(0)
pop	−2.292	0.429	−0.375	0.986	−4.647[a]	0.002	I(2)
wap	−0.821	0.956	−4.268	0.008[a]	NA	NA	I(1)

ADF, augmented Dickey–Fuller.

[a]Significant at 1% level.

Source: Authors' Own Estimation.

It is evident that the null hypothesis of no unit roots for *gdp* is rejected at its level since the ADF test statistic value is less than the critical value at 1% level of significance. Thus, it is level stationary and integrated of order zero, that is, I(0). But the variable *pop* is integrated of order two, that is, I(2) as the ADF test statistic at the second difference is less than the critical value at 1% level of significance. Similarly, the variable *wap* is integrated of order one, that is, I(1) as the ADF test statistic at the first difference is less than the critical value at 1% level of significance. Thus, the results obtained from the ADF test suggest that the maximum order of integration of the time series in this study is two, that is, $d_{max} = 2$. Therefore, the Toda–Yamamoto test involves the addition of two extra lag of each of the variables to control for potential cointegration in specification (1). But in specification (2), this test involves the addition of one extra lag of each of the variables. Then it is required to select the appropriate lag length for the VAR in order to perform a causality test. In this study, the AIC and final prediction error (FPE) techniques have been used to determine the optimal lag length for both the specifications (1) and (2). In a study having $n < 60$, AIC and FPE are superior to other information criteria (Lutkepohl, 1991; Liew, 2004). The results of such test are presented in Table 16.2. The optimal lag length for the estimation of specification (1), thus, selected is $k = 1$, and that of for specification (2) is $k = 2$.

TABLE 16.2 Selection of Lag Length.

Lag length	Final prediction error	Akaike information criterion
	Specification (1)	
0	0.959	5.634
1	0.005[a]	0.320[a]
	Specification (2)	
0	76.610	10.014
1	0.090	3.269
2	0.033[a]	2.266[a]

[a]Lag order selected by the criterion at 5% level.

Source: Authors' Own Estimation.

In the next step, the augmented VAR of order 3 ($p = k + d_{max}$) is estimated for both the specifications (1) and (2). And, the Wald test is carried out using standard chi-square distribution for both the specifications. The results of this Toda and Yamamoto Ganger noncausality test are reported in Table 16.3.

TABLE 16.3 Results of Toda and Yamamoto Granger Noncausality Test.

Null hypothesis	Chi-square statistic	*p*-value	Decision
	Specification (1)		
gdp does not Granger Cause *pop*	6.321	0.0970	Reject
pop does not Granger Cause *gdp*	20.638[a]	0.0001	Reject
	Specification (2)		
gdp does not Granger Cause *wap*	1.320	0.724	Accept
wap does not Granger Cause *gdp*	18.325[a]	0.000	Reject

[a]Significant at 5% level of significance.

Source: Authors' Own Estimation

Specification (1): The results in the upper part of Table 16.3 show that the null hypotheses of "*gdp does not Granger Cause pop*" and "*pop does not Granger Cause gdp*" are rejected at 5% level of significance. This means *gdp* and *pop* cause each other, that is, a feedback causal relationship is predicted between population growth and economic growth in India over the sample period. In other words, population growth can be considered as the cause as well as the consequence of economic growth in India. In the country, population growth has the ability to drive the economic growth, and also the economic growth has the capability to influence the growth of population. This finding corroborates to the findings of Tsen and Furuoka (2005) for Japan, Korea, and Thailand; Chang et al. (2014) for Austria and Italy; and Garza-Rodriguez et al. (2016) for Mexico.

Specification (2): The results in the lower part of Table 16.3 show that the null hypothesis of "*gdp* does not Granger Cause *wap*" could not be rejected at 5% level of significance. This means that no causal relationship runs from economic growth to the growth of the share of working-age population in total population in India. However, the null hypothesis that "*wap* does not Granger Cause *gdp*" is rejected at 5% level of significance. This indicates that the causal relationship runs from the growth of the share of working-age population in total population to the growth of India economy. This finding is of key importance because the age structure and its dynamics are critical in understanding the impact of population growth on the growth prospects of a high populous developing country like India (Kumar, 2010).

16.5 CONCLUSION

In recent years, India, the second most populous country of the World next to China, has become one of the fastest growing economies. Amongst the principal reasons, the growth of India's population is also noteworthy. The population can play an important role in influencing the economic growth of a nation through population size, labor force, and human capital. Additionally, the changes in the age structure create the demographic window of opportunity which has been claimed to have a positive impact on the long-run economic growth of a country possibly through productivity gains and employment growth. In this background, this chapter examined the nexus between population growth and economic growth, and the relationship between the growth of the share of the working-age population in total population and economic growth in the context of India. The results provide the evidence that the population growth is both the cause and consequence of economic growth in India. In addition, the results also predict the unidirectional causality that runs from the growth of the share of the working-age population to economic growth in the country. Precisely, there exists the window of opportunity or demographic dividend in India, and to get benefits of such opportunity, it is essential to accelerate investments in human capital (education, training, and skills) and promote research and innovation. In other words, special focus is warranted on the investment of education, training, and skill development. It is also required to bring the women in the working age into labor force in order to increase per capita output in the country. Since India is having lots of low-skill workers, it is important to enhance medium-skill occupations such as operating machinery, driving vehicles, doing clerical work, etc. Infrastructure development will also go a long way in utilizing the window of opportunity in the country. It is required that the policy environment and institutional arrangements of the country complement to propel economic growth. The last but not the least is the requirement of good governance for reaping demographic dividends in the long run.

KEYWORDS

- **age structure**
- **causality**
- **economic growth**
- **India**
- **population growth**

REFERENCES

Adediran, O. A. Effect of Population on Economic Development in Nigeria: A Quantitative Assessment. *Int. J. Phys. Social Sci.* **2012,** *2* (5), 1–14.

Aiyar, S.; Mody, A. The Demographic Dividend: Evidence from the Indian States. *India Policy Forum* **2012,** *13*, 105–137.

Barney, G. O. *The Global 2000 Report to the President of the US: Entering the 21st Century*; Pergamon Press: New York, 1980.

Becker, G.; Glaeser, E.; Murphy, K. Population and Economic Growth. *Am. Econ. Rev.* **1999,** *89* (2), 145–49.

Bhat, Mari P. N. Age Misreporting and Its Impact on Adult Mortality Estimates in South Asia. *Demogr. India* **1995,** *24* (1), 59–80.

Birdsall, N. Analytical Approaches to the Relationship of Population Growth and Development. *Popul. Develop. Rev.* **1977,** *3* (1/2), 63–102.

Bloom, D. E. Population Dynamics in India and Implications for Economic Growth. *PGDA Working Paper No. 65*; Program on the Global Demography of Aging, Harvard University, 2011.

Bloom, D. E.; Williamson, J. G. Demographic Transitions and Economic Miracles in Emerging Asia. *World Bank Econ. Rev.* **1998,** *12* (3), 419–456.

Bloom, D. E.; Canning, D. Global Demographic Change: Dimensions and Economic Significance. In *A supplement to Population and Development Review*; Prskawetz, A.; Bloom, D. E.; Lutz, W., Eds.;, Population Council: New York, 2008; Vol. 33.

Bloom, D. E.; Canning, D.; Sevilla, J. The Demographic Dividend: A New Perspective on the Economic Consequences of Population Change; RAND, MR-1274: Santa Monica, California, 2003.

Bloom, D. E.; Canning, D.; Hu, L.; Liu, Y.; Mahal, A.; Yip, W. The Contribution of Population Health and Demographic Change to Economic Growth in China and India. *J. Comp. Econ.* **2010,** *38* (1), 17–33.

Boserup, E. *The Conditions of Agricultural Growth: The Economics of Agrarian Change Under Population Pressure*; Earthscan Publications: London, 1965.

Chang, T.; Chu, H-P.; Deale, F. W.; Gupta, R. The Relationship Between Population Growth and Economic Growth over 1870–2013: Evidence from a Boots-trapped Panel-granger Causality Test. *Department of Economics Working Paper Series No. 2014-31*; University of Pretoria, South Africa: Pretoria, 2014; p 1–30.

Coale, A. J.; Hoover, E. M. *Population Growth and Economic Development in Low-income Countries: A Case Study of India's Prospects*; Princeton University Press: Princeton, NJ, 1958.

Dao, M. Q. Population and Economic Growth in Africa. *Econ. Internazionale* **2013,** *66* (2), 215-230.

Darrat, A. F.; Al-Yousif, Y. On the Long-run Relationship Between Population and Economic Growth: Some Time Series Evidence for Developing Countries. *Eastern Econ. J.* **1999,** *25* (3), 301–313.

Datt, G.; Mahajan, A. *Indian Economy*, 67th ed.; S. Chand & Company Pvt. Ltd.: New Delhi, India, 2012; p 59.

DaVanzo, J.; Dogo, H.; Grammich, C. In *Demographic Dividend or Demographic Drag? A Net Assessment of Population Trends in China and India, 2020–2025, and Their implications*, Proceedings of the International Union for the Scientific Study of the Population (IUSSP) Seminar on Demographics and Macroeconomic Performance, Paris, France, June 4–5, 2010.

Dawson, P. J.; Tiffin, R. Is There a Long-run Relationship Between Population Growth and Living Standards? The Case of India. *J. Dev. Stud.* **1998,** *34* (5), 149–156.

Eberstadt, N. Asia-Pacific Demographics in 2010-2040: Implications for Strategic Balance. In *Asia's Rising Power and America's Continued Purpose;* Tellis, Ashley J., Andrew Marble, Travis Tanner, Eds.; The National Bureau of Asian Research: Seattle and Washington, 2020; pp 236–2040.

Ehrlich, P. R. *The Population Bomb*; Ballantine Books: New York, 1968.

Furuoka, F. Population Growth and Economic Development: New Empirical Evidence from Thailand. *Econ. Bull.* **2009,** *29* (1), 1–14.

Furuoka, F. Population Growth and Economic Development: Empirical Evidence from the Philippines. *Philippine J. Dev.* **2010,** *37* (1), 81–93.

Furuoka, F.; Munir, Q. Can Population Growth Contribute to Economic Development? New Evidence from Singapore. *Econ. Bull.* **2011,** *31* (4), 3226–3239.

Furuoka, F. Population and Economic Development in Indonesia: A Revisit with New Data and Methods. *Acta Oecon.* **2013,** *63* (4), 451–467.

Galies, D. Causality Between the Measured and Underground Economies in New Zealand. *Appl. Econ. Lett.* **1997,** *4* (1), 63–67.

Garza-Rodriguez, J.; Andrade-Velasco, C. I.; Martinez-Silva, K. D.; Renteria-Rodriguez, F. D.; Vallejo-Castillo, P. A. The Relationship between Population Growth and Economic Growth in Mexico. *Econ. Bull.* **2016,** *36* (1), 97–107.

Gideon, K.; Thuku, G.; Obere, A. Impact of Population Change on Economic Growth in Kenya. *Int. J. Econ. Manag. Sci.* **2013,** *2* (6), 43–60.

Haldar, S. K. Economic Growth in India Revisited: An Application of Cointegration and Error Correction Mechanism. *South Asia Econ. J.* **2009,** *10* (1), 105–126.

Hansen, A. H. Economic Progress and Declining Population Growth. *Am. Econ. Rev.* **1939,** *29* (1), 1–15.

Joe, W.; Dash, A. K.; Agrawal, P.. Demographic Transition, Savings and Economic Growth in China and India, *IEG Working Paper No. 351*, Delhi, India, 2015.

Kelley, A. C. Demographic Change and Economic Growth: Australia 1861–1911. *Explor. Entrep. History* **1968,** *5* (3), 207–277.

Kelley, A. C. Economic Consequences of Population Change in the Third World. *J. Econ. Liter.* **1988,** *26* (4), 1685–1728.

Keynes, J. M. Some Economic Consequences of a Declining Population. *Eugen. Rev.* **1937,** *29* (1), 13–17.

Kothare, R. Does India's Population has a Positive Effect on Economic Growth? *Social Science Series No. 410*, 1990. http://pages.cs.wisc.edu/~dluu/data/papers/rkothare99.pdf. access date 15.02.2017

Kremer, M. Population Growth and Technological Change: One Million B.C. to 1990. *Q. J. Econ.* **1993,** *108* (3), 681–716.

Kumar, U. India's Demographic Transition: Boon or Bane? A State-level Perspective. *Economics Program Working Paper Series No. 10-03*; The Conference Board, Economic Program: New York, 2010.

Kuznets, S. S. *Modern Economic Growth: Rate, Structure and Spread*; Yale University Press: New Haven, 1966.

Kuznets, S. S. *Population, Capital, and Growth: Selected Essays*; Norton Press: New York, 1973.

Mahmud, M. A. In *Econometric Model on Population Growth and Economic Development in India: An Empirical Analysis*, Proceedings of the International Symposium on Emerging Trends in Social Science Research, Chennai, India, April 3–5, 2015.

Malthus, T. *An Essay of the Principle of Population;* Pickering: London, 1798.

Mavrotas, G.; Kelly, R. Old Wine in New Bottles: Testing Causality between Savings and Growth. *Manch. School* **2001,** *69* (s1), 97–105.

Meadows, D. H.; Meadows, D. L.; Randers, J.; Behrens III, W. W. *The Limits to Growth*; Universe Books: New York, 1972.

Rambaldi, A. N.; Doran, T. E. Testing for Granger Non-Causality in Cointegrated System Made Easy. *Working Papers in Econometrics and Applied Statistics No. 88*, Department of Econometrics, University of New England, 1996.

Safdari, M.; Mehrizi, M. A.; Elahi, M. The Effect of Population Age Structure on Economic Growth in Iran. *Int. Res. J. Finan. Econ.* **2011,** *72*, 62–69.

Sethy, S. K.; Sahoo, H. Investigating the Relationship Between Population and Economic Growth: An Analytical Study of India. *Ind. J. Econ. Bus.* **2015,** *14* (2), 269–288.

Shen, Z. Analysis of the Correlation Between Population Growth and Economic Development in Asian Countries. *Cross-Cult. Commun.* **2015,** *11* (11), 6–11.

Simon, J. L. *The Economics of Population Growth*; Princeton University Press; Princeton, NJ, 1977.

Simon, J. L. *The Ultimate Resources*; Princeton University Press: Princeton, NJ, 1981.

Simon, J. L.; Kahn, H. *The Resourceful Earth: A Response to Global 2000*; Basil Blackwell Publisher Limited: Oxford, 1984.

Singha, K.; Jaman, M. S. Nexus Between Population and Economic Growth in India: A Co-integration Analysis. *Sci. J. Pure Appl. Sci.* **2012,** *1* (3), 90–96.

Song, S. Demographic Changes and Economic Growth: Empirical Evidence from Asia, *Honors Projects.* Economics Department, Illinois Wesleyan University, Paper 121, 2013 http://digitalcommons.iwu.edu/econ_honproj/121.

Thakur, V. The Demographic Dividend in India: Gift or Curse? A State Level Analysis on Differing Age Structure and Its Implications for India's Economic Growth Prospects, *Working Paper Series No. 12-128*, Development Studies Institute, London School of Economics and Political Science, 2012.

Thornton, J. Population Growth and Economic Growth: Long-run Evidence from Latin America. *South. Econ. J.* **2001,** *68* (2), 464–468.

Toda, H. Y.; Yamamoto, T. Statistical Inferences in Vector Auto-regressions with Possibly Integrated Processes. *J. Econ.* **1995,** *66* (1), 225–250.

Tsen, W. H.; Furuoka, F. Population and Economic Growth in Asian Economies. *ASEAN Econ. Bull.* **2005,** *22* (3), 314–330

Yamaguchi, M. Technical Change and Population Growth in the Economic Development of Japan. *Economic Development Center Annual Report*; University of Minnesota: Minneapolis, MN, 1973.

Yamaguchi, M. *Population Growth and Economic Development*; Youhikaku: Tokyo, 2001.

Yao, W.; Kinugasa, T.; Hamori, S. An Empirical Analysis of the Relationship Between Economic Development and Population Growth in China. *Appl. Econ.* **2013,** *45* (33), 4651–4661.

Zhang, H.; Zhang, H.; Zhang, J. Demographic Age Structure and Economic Development: Evidence from Chinese Provinces. *J. Comp. Econ.* **2015,** *43* (1), 170–185.

CHAPTER 17

Challenges to Economic Growth and Development

MUSTAFA ÖZMUSUL*

Educational Administration, Supervision, Planning and Economics Department, Faculty of Education, Harran University, Şanlıurfa, Turkey

**E-mail: mustafaozmusul@gmail.com*

ABSTRACT

This study discovered some knitted dimensions to enlighten the challenges to economic growth and development. These are development policy based on human and social prosperity, institutional flexibility, developing human resource, transparent management and leadership, and sensitiveness to environment. More potential dimensions can be discovered particularly focusing on the national contexts. To cope with these challenges, writing prescription or developing a discourse for the solutions is not easy from chaos and complexity perspective. At least, recognizing and caring about them for policymakers and practitioners would be important. Reform initiatives are needed for social and economic development. Perhaps, one of the big question is "Are we really willing to accomplish this?" for all from leaders, politicians, business environments, public officials, the capital owners… to the others.

17.1 INTRODUCTION

Given the fact that the complexity and chaos are needed to understand well the phenomena in the social world, what kind of challenges might be possible to economic growth and development? This study is an attempt to enlighten this question. Initially, to establish a framework is important to discover the dimensions of economic growth and development, as a complex structure.

When considered that the dimensions of the elements that enable economic growth are multidimensional, this attempt at least would not seem slice of cake.

In order to explain how the economic activity happened, economists built many theories and concepts, such as invisible hand, technological innovation and investment in human capital, exogenous growth, endogenous growth, public spending, diminishing returns, free trade, open economy, new economy, etc. (The Economist, n.d.). Economic growth from one aspect is increasing the production and consumption of goods and services. It can be linked with national income such as gross domestic product (GDP) and gross national product (Czech and Pister, 2005). In short, explained economic growth as a quantitative improvement in country economy (Martins and Veiga, 2014).

On the basis of economic growth, the development can be explained as improving the social well-being (Shearman, 1990). Beyond improving the economic indicators, the term "development" is linked with improving the human factor. It is not only enlarging people's choices such as human rights, freedoms, capabilities, opportunities, long-healthy-creative lives but also needs quality work through ending child labor, forced labor, and trafficked workers (UNDP, 2015).

Moreover, the economic growth can become a hormone-injected growth through the shallow approaches that bases only the economic indicators, the policies focusing structures rather than processes, short term even daily and dispersed, and lack of big discourse and way of thinking. It is more likely to get favorable and organic results in terms of economic growth through the development policies that are healthy, consistent, integrated within its dimensions, particularly accepting human as a social value rather than economic asset, and connected a chaotic complex context.

To explore the dimensions of economic growth and development can shine light on the challenges to them. For that, this study focused on the following framework indicated in Figure 17.1. Determining the potential challenges, it should be said that these are not independent from each other but are knitted. Coping with one challenge may improve the other dimension/s. Similarly, ruining one dimension or bad effects on its structure may result in negative effect.

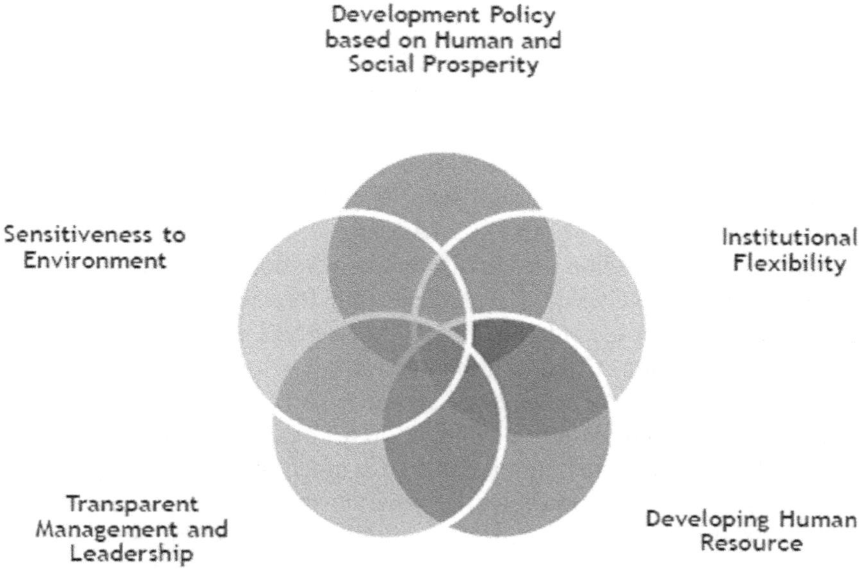

FIGURE 17.1 (See color insert.) The possible dimensions of the challenges to economic growth and development.

17.2 THE CHALLENGES TO DEVELOPMENT POLICY BASED ON HUMAN AND SOCIAL PROSPERITY

Development is a multidimensional term and complex structure since it entails establishing and continually improving a society of wealthy, prosperity, educative, healthy, democratic, solidarist, free, libertarian, respectable, innovative, lawful, ecologist, peaceful, and protective to cultural variety. For that, effective development policy needs wide and deep thinking. To cope with this challenge based on human and social prosperity, it enables to build a social development policy accepting the demands of all the social institutions, and distributing the resources to supporting all the groups, particularly the disadvantaged groups, so as to product social and economic value.

Establishing a development policy focusing only per capita income and GDP can be inadequate to support the disadvantageous groups; since these indicators do not show income inequality or provision of universal public services. Similarly, a development policy constrictedly increasing employment can be meaningless for a poor farmer who expects availability of seeds, protection from the effect of climate change, and services for children (Doane, 2014).

The countries are not only affected by the national decisions but are affected by international world. Coherence between national and international level, and among the various policy sectors, is critical to set up development policy. For instance, Finland has adopted a development policy as an integral part of foreign and security policy, and also it has declared that it cared about the UNESCO 2030 Agenda for Sustainable Development. Also, to reach the development goals, the cooperation between international organizations and NGOs is essential element for the countries during the developing and implementation process of effective development policy (Ministry for Foreign Affairs of Finland, n.d.).

Another point is to determine specific priorities for development. Finland (Ministry for Foreign Affairs of Finland, n.d.) has focused on four priority areas as follows: the rights and status of women and girls; the growth of developing countries' economies to generate more jobs, livelihoods, and well-being; democratic and better functioning societies; and food security, access to water and energy, and sustainable use of natural resources. Canada (Government of Canada, 2016) has adopted several priorities for development such as building effective partnerships, focusing efforts geographically and thematically, delivering and sustaining results, and increasing transparency and accountability.

Stated in the 13th Five Year Plan for 2016–2020, China determined major objectives such as maintain a medium-high rate of growth, achieve significant results in innovation-driven development, further coordinate development, improve standards of living and quality of life, improve the overall caliber of the population and the level of civility in society, achieve an overall improvement in the quality of the environment and ecosystems, and ensure all institutions become more mature and better established (NDRC, 2016).

Lack of law enforcement is another considerable challenge to be overcome for the third world countries (Institute of Ecolonomics, 2014). More important is maintaining the resistance to the rule of law and lacking perpetuity. Particularly the natural states leaders (i.e., Hitler) tend to use their power commonly compromise existing institutions. One of the major challenges for long-term economic growth is transition from the natural state to an open access order. Coping with this challenge needs to establish credible commitments for protecting the institutions and increasing the endeavors that they alive (Weingast, 2008).

In the systems that avoid the needed open-access transition the followings can be seen: the personal relationships in terms of politics and economics create a privilege coalition having more powerful members. The systems fail to provide public goods impersonally. Also they limit the accession to the organizations and the markets are controlled through cumbersome efforts. In the elections, the competition ability of opposition sides is comprised by the incumbents through various ways such as restricting the express of view ability. In addition, the absence of judicial system makes legislature negative in terms of passing the laws in the systems. Eventually they lack a perpetual state where the state officials dishonor the rules, laws, and the institutions that are vital in economic growth and social development (Weingast, 2008). As a result, combating with the lacking rule of law matter and transition to and protection of the open access order should be a specific priority not only in the development policy documents but also in practice.

Health of the people is also important element for economic development of a nation. A big challenge to improved public health outcomes is noncommunicable disease (NCD) epidemic (Hofman, 2014). NCDs, known as serious chronic diseases, kill 40 million people each year. A considerable part of these is premature deaths in low and middle income countries. In addition, cardiovascular diseases, cancers, respiratory diseases, diabetes, tobacco use, physical inactivity, the harmful use of alcohol and unhealthy diets, etc., are other serious NCDs. Especially unplanned urbanization, globalization of unhealthy lifestyles, and population ageing are increasing the risk factors to NCDs. On the other hand, there is a strong association between poverty and NCDs. For that, the initiatives of poverty reduction in low-income countries are hindered by increasing the household costs, preferring harmful products, or accessing limited health services (WHO, 2017).

According to one estimation, improving the health in developing countries in the period of 2000–2011 provided one quarter of economic growth. However, shortage of healthy workers has been a big challenge in the global level. Also the challenges to ensure healthy service delivery and its organization, to building capacity of health workforce, and health systems especially for public health risks and emergencies have become prominent in developing countries (WHO, 2016).

In addition, maternity protection is specifically important in terms of work health and safety. In this matter, some priorities that should be ensured as follows: well-being of mothers and their newborns, enabling women to successfully combine their reproductive and productive roles, preventing unequal treatment at work due to their reproductive role, and promoting the principle of equality of opportunity and treatment between women and men.

Moreover, sexual harassment, as a serious form of discrimination, decrease the quality of working life, jeopardize well-being, and have serious cost implications for workplaces. Coping with this serious situation, the legislations and needed revisions related to work safety and health for sexual harassment have become more of an issue (Gallup and ILO, 2017).

Millions of workers every year are confronted with an accident at work, and they have work-related health problems. Also each year, 4% of the global GDP is spent for work-related accidents and diseases. It seemed that over 2.3 million fatalities and 300 million accidents causing injuries in the workplace happen every year. These mean a huge cost for society, companies, and individuals. Moreover, there is a strong correlation between national competitiveness and occupational accidents in national levels. Today, poor working conditions have been considerable challenges and undermine the economic performance of the countries seriously (European Commission, 2011; ILO, 2009, 2017).

In developing countries, there is a little tendency to report the accidents and diseases. Many deaths and injuries due to occupational factors go unreported (ILO, 2013). On April 28, 2017, the ILO (2017) made a call entitled "accurate data will help to save lives." In this call, it is explained that the data on the number of the fatalities or accidents is insufficient, but it needs well-prepared national data that reflect the magnitude of the problem, the real impact of occupational accidents and diseases on workers, families, and economies (ILO, 2017). In turn, the challenge to perfection of the work health and safety through gaining qualitative data should be an important issue as priority in the process of preparing development policy.

The situation of natural resources of a country is considerable factor in determining specific priorities for development. Infertile land and insufficient natural resources will restrict income growth. For instance, dividing farmlands into small parcels can limit the usage of modern agriculture techniques. However, having much more natural resources is not warranty for growth. Due to the economic structure, some countries having rich natural resources can have poor growth performance. Nevertheless, through entrepreneurial ability some countries with low natural resources can have rapid rates (Lipsey and Ragan, 2005).

More than half of the vital ecosystems, which refer to complex biological mechanisms, are considerably damaged. In this matter, a big risk seems for development when considering the strong relations between health and environment. Coping with the matter entails development of integrated policies based on health, environment, and development goals. In turn, the policies on environment, health, and economic development should be designed together (WHO, 2008).

On the other hand, the mobility of people across the world has become an important challenge for policy makers. Especially 2030 Agenda for Sustainable Development draws attention to the link between migration and development. The labor market, agriculture, education, investment and financial services, and social protection and health issues are important elements of an effective development policy in the dimension of migration. Ignoring the migration reality and its needs can bring negative results (OECD, 2017).

One of the challenges to designing a good development policy is high military expenditures that cause bad governance and domestic conflicts. Especially, military spending can threaten supporting the socioeconomic-based development policies. For that, reducing the excessive military expenditures is a vital issue to take off the priorities of development policy particularly in less developed countries (Berthelemy et al., 1994). For that, to minimize the military expenditures, the effective development policy process entails both national and international dialogue and endeavors that will build up confidence to others. In turn, economic growth development policy cannot be considered without defense policies.

The landlocked countries, which have mostly insufficient human development and external trade when compared with the maritime ones, face considerable challenges due to their infrastructure, sound political relations, neighbors' peace and stability, and administrative practices. Especially they face high trade costs and low export rates (Faye et al., 2004). In turn, as a challenge, most of the third world countries are fragile to unexpected situations deriving from their both near and far neighbors. For that, it is important for the policy makers to search improving or changing the transportation infrastructure, and to develop the relations with their neighbors.

Having population growth rates in excess of the GDP growth rates can be serious challenge to economic development because of negative growth rates of per capita GDP. For that, the countries should monitor their population growth and take precautions to this challenge (Lipsey and Ragan, 2005).

As a result, to establish effective development policy, each country's strategic approach can differs from one to another because each country has unique problematic areas and needs in both local and national level. But no country lives in a vacuum; it should be open to the world and developments in the international development area. For that, the big challenge is in this context, to set up coherent development policy based on both in national and international direction.

17.3 THE CHALLENGES TO INSTITUTIONAL FLEXIBILITY

A good development policy document does not warrant its applicability. During the policy implementation process, one of the resulting challenges is strict institutional structure and approaches. Plan or policy is linked with the future. And the future is not deterministic, but chaotic. Therefore, none development policy or plan is perfectly designed; for that, its implementation process can go wrong, the unexpected challenges can be seen, and more cooperation and coherency among different and similar institutions can be needed to achieve the development goals. As a result, strict organizational structures, especially fusty bureaucratic and hierarchical approaches threaten implementing a good policy on developing process.

Coping with such restrictiveness needs establishing flexible institutional structure and processes such as giving up the structure image where some people make the said things from up, and the others sitting on the desk expect the outputs from down, setting up an interactive communication network, empowering people based on an ethical context, making innovative organization culture, increasing teamwork instead of individual works, supportive leadership styles, conducting common development works among different sectors for instance between education and health, and so on.

Institutional flexibility means adapting to the economic environment. Analyzed the role of flexibility in a model, Davis (2010) concluded that the institutional flexibility had a considerable role in economic growth. Institutional flexibility in the public sector also means an administrative reform. It requires focusing a transformation from bureaucratic organizational hardware to organizational software, which means management reform. Such flexibility brings institutional separation of the policy and operative functions. In this respect, agencification and decentralization can empower line managers and front-line staff on a basis of a new management philosophy (Pierre, 2012).

One of the considerable factor in enhancing innovation and entrepreneurship is flexibility of institutions. Organizational flexibility can create more individual choices and new behaviors that rush up an innovative culture (Eesley et al., 2014). When we consider that development is a dynamic process, we expect that especially today's public institutions should be flexible and avoid hierarchical structure and linear processes.

Traditional national and regional territorial institutions indicate some weaknesses because they are connected to other functionally based cross-border institutions and networks and are vulnerable to national and regional changes in function and territory. To cope with these challenges, flexibility for institutional structures and cooperation networks can be important solution (Hansen and Serin, 2010).

Maintaining good economy requires a flexible institutional matrix, which will enable coherence with technological and demographic changes. In particular, if we need long-run economic growth instead of short run with autocratic regimes, rule of law and protection of civil and political freedoms are vital to make the institutions flexible (North, 1986).

One way to enhance institutional flexibility especially public sector can be private–public sector partnerships. As well as making profit, private sector produces public outputs in terms of creating job, educating people, decreasing poverty, improving the services, etc. Considering that the private sector is more flexible than public sector, such partnership may convey the flexible structure and processes into public institutions.

Ensuring successful partnership between public and private sector requires serious analysis of the long-term development aims and risk allocation. This new model also means to flex the current structure through legal and institutional framework (World Bank, 2016). Private sector tends to produce vital contributions to economic growth and social development through various development benefits, such as employment creation, human resource development and improvement of technical capabilities, private sector activities, etc. Also it has a key role in solving poverty reduction, the environment, climate change, food supply, and resources. In this respect, The Fifth Tokyo International Conference on African Development (TICAD V) in 2013 was a good sample with a call attention to the importance of promotion of private sector-led growth in Africa (Japan International Cooperation Agency, n.d.)

Democratization of the institutions and more openness to stakeholders both in decision-making and developing reform choices are vital issues for social and economic development. Nevertheless, this process is not limited within the central structures, but it should be disseminated into the rural area. In this matter, the intermittent repression, a serious gap between social organizations and political parties, and contradictory elements in the official "modernization" project are considerable challenges to rural democratization (Paré, 1990).

Some cultural obstacles that affect traditional and habitual ways of doing business seriously can bring inefficiency. For instance, personal considerations of family can be paid more attention than market incentives in motivating behavior. In addition, the perspective on children is important factor in the economic structure. So, preparing the development plans without consideration of the local cultural, social, religious factors, or legal patterns can wreck the development goals. It also can end up that local populations abstain from supporting the growth and development endeavors (Lipsey and Ragan, 2005).

17.4 THE CHALLENGES TO DEVELOPING HUMAN RESOURCE

A good development policy cannot be successful merely through institutions. Since the institutions live in a human made society and they are meaningful with their own human resource. Ignoring human resource and the main ambient factors influencing people both in the institutions and society fails the objectives of a good development policy. The humans in the institutions, neither public nor private ones, come from a world covering society, family, schools, and universities, and they are also a reflection of cultural context. Among the factors influencing a human, having the educational quality is a considerable one. Taken together, it can be said that human quality in a society potentially equals institutional quality. The big question is that how can we improve the human quality?

We know that first formal place to make people more qualified is the school. After or simultaneously people can continue to more education choices such as vocational and technical courses and seminars through formal or informal institutions. Then, educated people through a school system work in the various public and private organizations. The institutions or working places also maintain to educate their people to make them ready to adopt the changes requiring new information and skills not merely technically but socially. In this respect, today and next organizations normally expect that people are qualified, educated, and open to the new information and skills particularly team working and innovative capacity into the future. Because, the life is chaotic and changing continuously, today's information and skills are can be invalid in the future.

Also having diploma, certificate, or another formal documents, which indicate individuals' education level, do not show their all information and skills. In addition, the graduation rates in a school system can be insufficient indicator for determining the quality of people. In this respect, we need available data indicating the qualification level. As an international indicator, the OECD's Program for International Student Assessment (PISA), started in 2000, measures the qualification level of 15-year-old students, near the end of their compulsory education. According to the report published by OECD (2016a), in three domain of reading, science, and math scores, at least 50% of students in Brazil, Algeria, Dominican Republic, FYROM, Georgia, Indonesia,

Kosovo, Lebanon, Peru, Qatar, and Tunisia among the participated countries, perform below level 2, which is expected from all the students to achieve because of baseline for full participation in modern societies. From one aspect, as a result of insufficient primary education, the quality of secondary education can be insufficient. And its effect can be seen in university education or in a working life. For that, human quality in the institutions is linked with each previous education level, particularly with compulsory education.

The qualification of human resource is not limited to the endeavors in the formal school system. When a person starts to work in the organizations, her or his teaching and training needs should be continued. More importantly, human is a valuable asset of an organization and should be motivated, empowered, even supported socially and psychologically. In this regard, today's organization is a school which consists of administrators, teachers, other staff, learners and their families, and environment. Taken together, investment in human resource is a critical issue for both public and private sector through the school systems and the organizations.

Improving vocational education and training systems, and also lifelong learning programs is crucial to higher skills through drawing considerable attention into the share of workers who are either over- or underskilled for their job (OECD, 2017). Poor adult skill is a big challenge for social, cultural, politic, and economic development in both individual and institutional level. Especially, boosting the economic growth and development cannot be realized without good adult skills, which is vital capital and value for today's and tomorrow's organizations. As a considerable international indicator, Programme for the International Assessment of Adult Competencies (PIACC) conducted by OECD (2016b) measures the adult skills in literacy, numeracy, and problem solving for the job. It revealed that 18.5% of adults have poor reading skills and 22.7% of adults have poor numeracy skills in almost all countries/economies in the PIACC study.

The role of education and training are critical for improving the human capital. Also it is well known that the relation between investments in human capital and economic progress (Becker, 1994). Focusing creating sustainable futures all, UNESCO (2016) in the agenda of 2030 called three starkly clear messages: (1) it seems that only 70% of children in poor countries will complete primary school in 2030. (2) Failing to ensuring the "inclusive and equitable quality education and promote lifelong learning opportunities for all" will poor education results and will posses unachieved development goals. (3) It is compulsory to change the educational paradigm in human well-being and global development so as to foster the skills, attitudes, and behavior for sustainable and inclusive growth.

The cost of investing in skills is inexpensive when considering the high cost of poorer health, lower incomes, unemployment, and social exclusion. As a result, coping with the challenge to develop the skills entails a robust connection between the world of learning and the world of work (Taylor, 2012).

17.5 THE CHALLENGES TO TRANSPARENT MANAGEMENT AND LEADERSHIP

A well-designed and produced automobile or music instrument is insufficient, useless, and meaningless unless it is well managed. Similarly, a well-prepared development policy based on human and social prosperity is deficient without a good management that focuses on human factor, systemic view, ethical values, as well as technical and intellectual theoretical framework. What can be the challenges to a well management for the organizations that are expected to achieve the big development policy goals? The answer is both easy and difficult. The easy answer is changing the management style. But, if we do not satisfy the current management style or paradigm or they are not available for today or future, the situation allows us to search new one. Perhaps the big challenge is how to change the management style. If there are big development goals waiting seriously to be achieved, the management endeavors like the other factors (policy, institutional structure, human source, resources, environmental factors, etc.) should be available for them. Because the big goals are linked with the future vision and they mean unachieved ones today. And, today's or past management approaches might be considerably responsible for today's situation. In short today's approaches cannot be adequate for tomorrow.

After a chaotic starting to this title, let us continue to the challenges through the framework of the management theory. We know that Taylorist approaches focus only on work, productivity, and profit, but do not draw attention in human factor as the social and psychological factors. We know well from total quality theory and its implementations that focusing only on short-term goals while ignoring long-term ones and institutional vision will disallow the goals of development policy. Also from the system theory or chaotic paradigm, the management of the organizations entails to take into consideration multiple factors. In this respect, using the narrow perspectives focusing only one factor, such the universal classic management theories did advocate, can bring serious challenges. If we dream a big photograph for the future such as becoming innovative and inclusive society, the management should be different. In turn, we are humans, the staffs in the various positions in the organizations are human, the organizations are made of human, and finally we need the humans in the management. This means some codes lasting forever such as working together, achieving together, learning together, behaving together, and managing together….

In terms of the management, one of the big challenges to the development is nontransparent management. In a nontransparent management context, there is a strict management style, limited communication is seen, one or very few people make decision, many people are disengaged and unmotivated, the cooperation and solidarity among the people are weak, task-oriented leadership is given importance rather than people-oriented, corruption tendency occurs, people tend to work only individual goals, ambitions, and mentality rather than the future of organization, even the organizational anomalies can be increased.

The corruption in the public sector develops through abusing authority or position for gaining individual benefits. It damages the public sensitiveness through low quality of administrative service, raising the service costs, and decreasing the credibility. Also the wrongful management, such as collusive ties with the politics practices, causes decrement of the savings, poor national competiveness, and growth potential of the national economy. More importantly, the corruption can spread from one area to another, even the whole society. To illustrate, the corruption in the public administration can be seen conveniently in the education, military, legal, and medical sector (Lee, 2004). In particular, the corruption tendency in the nontransparent management damages social development, as well as economic growth.

The report as a product of the project funded by the European Union indicated the followings (Pring, 2016): one in three respondents considers that corruption is one of the biggest problems faced by their country. More than half of the citizens see that their government is insufficient to fight against corruption. Especially, it is seen that the most corrupt is derived from the politicians and public officials. The wealth influence on government decisions is too much. Another important finding is that bribery is still common, particularly in the commonwealth of independent states. Coping with these challenges "standing up" and "speaking out" are the best potential solutions. Moreover, another remarkable point in this report is a need of adoption and enforcement of legislation to protect the whistleblowers within international standard. As a result, the corruption pest that damages seriously the social and economic development should be an issue to handle both in government level and the organizations level.

From an economic view, it can be said that the ethical behavior decreases the transaction costs, internalizes negative externalities, but improve efficiency (Argandona, 2004). Such behavior entails to respect for each employee regardless of their religious and cultural beliefs, to establish mutual respect through each person's sensitiveness to others, procedural fairness implementing procedures fairly, and decision-making transparency avoiding arbitrary attitudes and processes (McPheat, 2009).

The ethical perspective in the organizations for the management entails to see the people as having a family, social and cultural background, emotions, missing, dreams, beliefs, attitudes, tendencies showing group behavior, pains, health problems, strong and weaknesses sides, etc. Such perspective is not only valid for the managers but the staff, because all people in the organization interact with the others. And the critical role of the managers is the enabling and disseminating this perspective to all organizations for producing a cultural context of respect to and understanding each other.

In a transparent management context, democratic processes that avoid an authoritarian approach are seen. In this matter, enabling consensus instead of searching to have plurality in the decision-making process is needed. Different ideas and views should be given importance and the arguments behind them should be discussed with the stakeholders. Since the thinkers differently are wealth of the organization, the management should search to understand and encourage them. The cost of neglecting the others who think differently can be too much. For instance, the people whose ideas are ignored can start to feel themselves worthless and alienation from their organization, abstain their ideas particularly in the group meetings, and even become pessimist to the organization, the other people, and environment. Shortly, losing people can be losing the organization.

Since the life consists of complex relationships, everything is connected with the other things. Management also is not limited with its own structure and the immediate environment, but it can need to close relationship with the similar or different organizations. For instance, managing the education institutions entails to work together with the business organizations. The demanding knowledge and skills by the business world in a country bears meaning for the education and training world. Moreover, managing the agriculture sector can need to cooperate with education, security, economic, politic, social, and cultural organizations. A management of municipality that desires to develop the human source can work with a university. Similarly, a school aiming to protect its students from substance abuse can work together with provincial security directorate, social services, specialists from a hospital, or another school on common action plan.

One way to cope with the challenges of the management in the organization can be an establishment of management style, which can be entitled as managing together rather than managing isolated. We well know that no one knows all, no one does all. Consulting to people not only in the horizontal or vertical positions but in the low in the management processes is particularly vital for the managers in the organizations. Especially the managers can held continual meetings with the staff and take their suggestions in the important decisions without dictation. The photograph indicating that the managers sit at the table in most of the work time, take the decisions alone, call the staff into her or his room and give commands to them, mitigate the complaints or demands temporarily, implementing the directions given by up positions, or any residuals as a result of passive management has been fusty. The new photograph at least indicates that the managers accept every person in the organization as a considerable part of the organizational team based on an ethical, professional, and sensitiveness to environment context.

In the new photograph also it is seen that the management is nested within the leadership. As a fact or need which gets stronger the management, the leadership would be a source of inspiration, a sparkle for disclosing the potential of people, unblocking the obstacles for the work and people, help to bring the needed parts to solve the puzzle, mediator for people who are in conflict, and more importantly creating a culture stimulating all of these and similar leadership elements. On the other hand, it is difficult to say that a universal management or leadership theory is valid in each situation and every organization. For that the management and leadership are human activities requiring to care about all the factors that affect people.

To cope with the challenges to management, the monitoring and evaluation systems in the organizations can be helpful. Monitoring refers to an ongoing process reflecting the feedback on the progress for achieving the aims. Asking to what extent do we make progress on achieving the results rather than asking do we take the actions we said we would take? Evaluation can be seen as an assessment of the finished or ongoing activities. Taken together, monitoring and evaluation contribute to the decision-making process and improve performance while implementing the prepared plans (Menon et al., 2009).

It seems insufficient that to start the works for achieving the concrete goals through distributing the task to each team or person, and then waiting the tasks to be completed for effective management. Particularly, through losing the big photograph, dividing the work into small parts and making people responsible for the related part, and ignoring the interactions between them can damage the integrity of organization. The organizations or institutions consisting of the staff, tasks, needs, teams, relations, goals, emotions, values, attitudes, etc., are not structures that can be abandoned to their fate. In this regard, the role of the monitoring and evaluation is vital to protect the organization against the entropy. Establishing an ethical and professional monitoring and evaluation, it can be possible to determine the inconveniences and the potential solutions for them, the demands from the departments and the staff for their task and relations or the other departments, deviation from the goals, and the measures to be taken for them. It also should not damage to people and their respectability, but should be tools rather than aims to develop the organizational values.

In the monitoring and evaluation process, a considerable scientific data is needed to enlighten well the situation at least. The data requires various scientific method and techniques. In this point, obtaining the data cannot be limited with one method. Since, each method has both strong and weak aspects. It should be empowered with the others. In turn, all the techniques of obtaining scientific data, such as questionnaires, scales, quantitative indicators, descriptive statistics, the interviews, observations, field notes, etc., are valuable in the monitoring and evaluation process. Moreover, to evaluate various data sources and turn into knowledge are not easy and cannot be given only a person to evaluate all. Collecting and evaluating data needs teamwork. Also it is not a flash in the pan, but is a process. Monitoring and evaluation process and the scientific data collection as a considerable part of this process should be done in a systematic way. In particular, managing the organization without monitoring and evaluation resembles driving a car without looking the fuel indicator, heat indicator, rear view mirror, or missing the abnormal voices coming from the motor.

17.6 THE CHALLENGES TO SENSITIVENESS TO ENVIRONMENT

The institutions or organizations that hold the human and other sources and are expected social and economic development do not live in an isolated environment, but in a complex environment that is affected from near and far environment, namely, from local, regional, national, and even global environment. Thus, it is not enough that they implement their own plans, mostly given by the top authorities. Coping with the challenges that will be source of the global or local developments mainly needs sensitiveness to the environment. Ignoring the environment today can lead to a crisis tomorrow. In particular, the top and local authorities that have a considerable role in preparing the national development policy should be sensitive to the developments in the environment. Even, one of the conditions to survive for the public and private organizations is to take the environment seriously. This may entail to protect the organization and take measure against the hazardous effect of the environment or to adapt to the environment.

On the other hand, the exogenous shocks and the practices derived from international development finance can be serious challenges for economic growth. The exogenous shocks can be divided in two groups: (1) natural shocks such as earthquakes or volcanic eruptions, typhoons and hurricanes, droughts, and floods. (2) External shocks such as slumps in external demand, world commodity prices instability, and international fluctuations of interest rates. Additionally, some domestic shocks may be derived from political instability or unforeseen political changes. To understand this phenomenon and produce concrete solutions, the Economic Vulnerability Index developed by United Nations (UN) can be utilized (Guillaumont, 2007, 2011).

From the big picture, the Global Risk Report 2017 published by World Economic Forum (2017), which indicates five key challenges for the world which can be described as follows: (1) rising income and wealth disparity, which is seen as the most considerable risk over the next 10 years in terms of global development. (2) Changing climate that affects the composition of the global atmosphere, natural climate variability as well. (3) Increasing polarization of societies that refers to conflict due to diverging or extreme values as well as political or religious views. (4) Rising cyber dependency through increasing digital interconnection of people, things, and organizations. (5) Ageing population through declining fertility and decrease of middle and old age mortality. In brief, the global report reveals the potential hazards in the next years for the world. For that, these are binder for both the national organizations and local organizations. Understanding and taking measures for the global issues rather than ignoring them will make more contributions to the social and economic development.

The global development report by OECD (2016c) draws attention to the migration matter. The public policies can lead the migration flows and patterns in various ways. Whereas most of its source is voluntary oriented, some are armed conflict or violence oriented. By the end of 2015, about 16.1 million refugees were reported, and most of them live in developing countries. It is also reported that the international migration could be a driver of development. Consequently, coping with the challenges of migration facing the countries, the following dimensions were suggested by the report: designing policies that incorporate the development dimension so as to maximize its benefits; fostering policy and institutional coherence, particularly in facilitating intragovernmental coordination, encouraging local initiatives due to their frontal position in managing the migration fact; and empowering international cooperation for present and future migration challenges to cope through positive ways.

17.7 CONCLUDING REMARKS

In conclusion, this study discovered some knitted dimensions to enlighten the challenges to economic growth and development. These are development policy based on human and social prosperity, institutional flexibility, developing human resource, transparent management and leadership, and sensitiveness to environment. More potential dimensions can be discovered particularly focusing on the national contexts. To cope with these challenges, writing prescription or developing a discourse for the solutions is not easy from chaos and complexity perspective. At least, recognizing and caring about them for policymakers and practitioners would be important. Reform initiatives are needed for social and economic development. Perhaps, one of the big question is "are we really willing to accomplish this?" for all from leaders, politicians, business environments, public officials, the capital owners… to the others.

KEYWORDS

- **economic growth**
- **economic development**
- **human prosperity**
- **social prosperity**
- **institutional flexibility**
- **transparent management and leadership**
- **sensitiveness to environment**

REFERENCES

Argandona, A. On Ethical, Social and Environmental Management Systems. *J. Bus. Ethics* **2004,** *51* (1), 41–52.

Becker, G. *Human Capital Revisited*; 1994 [Online]. http://www.nber.org/chapters /c11229.pdf. (accessed Apr 22, 2017).

Berthelemy, J. C.; Mcnamara, R. S.; Sen, S. *The Disarmament Dividend: Challenges for Development Policy*. OECD Development Centre, Policy Brief No. 8, 1994.

Czech, B.; Pister, P. *Economic Growth, Fish Conservation, and the American Fisheries Society: Introduction to a Special Series*; 2005 [Online]. https://supplyshock.files. wordpress. com/2014/09/economic-growth-fish-conservation-and-the-american-fisheries-society.pdf (accessed Mar 10, 2017).

Davis, L. S. Institutional Flexibility and Economic Growth. *J. Comparative Econ.* **2010,** *38* (3), 306–320.

Doane, D. *What's so Bad About Development*? 2014 [Online]. https://www.theguardian.com/ global-development/poverty-matters/2014/sep/01/development-ngos-third world-global-south. (accessed Mar 22, 2017).

Eesley, C.; Roberts, E. B.; Tian, X.; Yang, D. *Institutional Flexibility and Entrepreneurship*. Stanford Center for International Development Working Paper 543; 2014 [Online]. http:// scid.stanford.edu/sites/default/files/publications/543wp.pdf. (accessed Mar 22, 2017).

European Commission. *Socio-economic Costs of Accidents at Work and Work-related Ill Health: Key Messages and Case Studies;* 2011 [Online]. http:*//ec.europa.eu/social/BlobServlet?docId=7417&langId=en. (accessed Mar 22, 2017).*

Faye, M. L.; McArthur, J. W.; Sachs, J. D.; Snow, T. The Challenges Facing Landlocked Developing Countries. *J. Human Develop.* **2004,** *5* (1), 31–68.

Gallup; ILO. *Towards a Better Future for Women and Work: Voices of Women and Men*; 2017 [Online]. http://www.ilo.org/wcmsp5/groups/public/---dgreports/---dcomm/---publ/ documents/publication/wcms_546256.pdf. (accessed Mar 22, 2017).

Government of Canada. *Development Priorities;* 2016 [Online]. http://www.international. gc.ca/development-developpement/priorities-priorites/aidagenda-planaide.aspx?lang=eng. (accessed Mar 22, 2017).

Guillaumont, P. Economic Vulnerability, Persistent Challenge to African Development. *African Develop. Rev.* **2007,** *19* (1), 123–162. DOI: 10.1111/j.1467-8268.2007.00157.x. (accessed Mar 22, 2017).

Guillaumont, P. An Economic Vulnerability Index: Its Design and Use for International Development Policy. CERDI, Etudes et Documents; 2011 [Online]. http://unohrlls.org/ UserFiles/File/LDC%20Documents/AHWG%20on%20smooth%20transition/EVI%20 by%20Guillaumont%202009.pdf. (accessed Mar 22, 2017).

Hansen, P. A.; Serin, G. Rescaling or Institutional Flexibility? The Experience of the Cross-Border Øresund Region. *Regional Federal Stud.* **2010,** *20* (2), 201–227.

Hofman, K. Non-communicable Diseases in South Africa: A Challenge to Economic Development. *S. Afr. Med. J.* **2014,** *104* (10), 647. DOI: 10.7196/Samj.8727.

Institute of Ecolonomics. *Challenges of Third World Economic Development*; 2014 [Online]. http://ecolonomics.org/challenges-of-third world-economic-development/. (accessed Mar 22, 2017).

International Labour Organization (ILO). *World Day for Safety and Health at Work 2009 Facts on Safety and Health at Wor;* 2009 [Online]. http://www.ilo.org/wcmsp5 /groups/

public/@dgreports/@dcomm/documents/publication/wcms_105146.pdf. (accessed Mar 22, 2017).

International Labour Organization (ILO). *Safety and Health at Work: Hopes and Challenges in Development Cooperation. The Example of an EU-ILO Joint Project "Improving Safety and Health at Work Through a Decent Work Agenda*"; ILO Publications: Geneva, Switzerland, 2013.

International Labour Organization (ILO). *World Day for Safety and Health at Work 2017: Accurate Data Will Help to Save Lives*; 2017 [Online]. http://www.ilo.org/global/about-the-ilo/how-the-ilo-works/ilo-director-general/statements-and-speeches/WCMS_551573/lang--en/index.htm. (accessed Mar 22, 2017).

Japan International Cooperation Agency. *Public-Private Partnerships*; n.d. [Online] https://www.jica.go.jp/english/our_work/types_of_assistance/partnership/. (accessed Mar 22, 2017).

Lee, N.J. *Ethics Management and National Development. Signing Ceremony of Business Ethics Pact with Public Corporations;* 2004 [Online]. http://unpan1.un.org/intradoc/groups/public/documents/apcity/unpan019147.pdf. (accessed Mar 22, 2017).

Lipsey, R. G.; Ragan, C. *Economics*; Pearson Education Canada Inc: Toronto, Canada, 2005.

Martins, S.; Veiga, F. J. Government Size, Composition of Public Expenditure, and Economic Development. *Int. Tax Public Finance*. **2014,** *21* (4), 578–597.

McPheat, S. *The Four Principles of Ethical Management*; 2009 [Online]. http://www.mtdtraining.com/blog/the-four-principles-of-ethical-management.htm. (accessed Mar 22, 2017).

Menon, S.; Karl, J.; Wignaraja, K. *Handbook on Planning, Monitoring and Evaluating For Development Results*; UNDP Evaluation Office: New York, NY, 2009.

Ministry for Foreign Affairs of Finland. *Goals and Principles of Finland's Development Polic*; n.d. [Online]. http://formin.finland.fi/public/default.aspx?nodeid=49312& contentlan=2&culture=en-US. (accessed Mar 22, 2017).

National Development and Reform Commission (NDRC), People's Republic of China. *The 13th Five-Year Plan for Economic and Social Development of the People's Republic of China*; 2016 [Online]. http://en.ndrc.gov.cn/policyrelease/201612/P02016 1207645766966662.pdf. (accessed Mar 22, 2017).

North, D. C. The New Institutional Economics. *J. Inst. Theor. Econ./Z. Gesamte Staatswiss.* **1986,** *142* (1), 230–237.

OECD. *PISA 2015 Results (Volume I): Excellence and Equity in Education, PISA*; 2016a [Online]. http://www.oecd.org/education/pisa-2015-results-volume-i-9789264266490-en.htm. (accessed Mar 22, 2017).

OECD. *Skills Matter: Further Results from the Survey of Adult Skills, OECD Skills Studies*; OECD Publishing: Paris, 2016b [Online]. http://dx.doi.org/10.1787/9789264258051-en. (accessed Mar 22, 2017).

OECD. *Perspectives on Global Development 2017: International Migration in a Shifting World;* OECD Publishing: Paris, 2016c.

OECD. *Economic Policy Reforms 2017: Going for Growth*; OECD Publishing: Paris, 2017a.

OECD. *Interrelations Between Public Policies, Migration and Development;* OECD Publishing: Paris, 2017b [Online]. http://dx.doi.org/10.1787/9789264265615-en. (accessed Mar 22, 2017).

Paré, L. The Challenges of Rural Democratisation in Mexico. *J. Develop. Stud.* **1990,** *26* (4), 79–96. DOI: 10.1080/00220389008422174.

Pierre, J. Governance and Institutional Flexibility. In *The Oxford Handbook of Governance*; Levi-Faur, D., Ed., Oxford University Press: New York, NY, 2012; pp 187–200.

Pring, C. *People and Corruption: Europe and Central Asia.* Transparency International; 2016 [Online]. http://www.transparency.org. (accessed Mar 22, 2017).

Shearman, R. The Meaning and Ethics of Sustainability. *Environ. Manag.* **1990,** *14* (1), 1–8.

Taylor, K. *Why is Human Capital Important for Development?* 2012 [Online]. https://community.oecd.org/docs/DOC-41754. (accessed Mar 22, 2017).

The Economist. *Economics A-Z;* n.d. [Online]. http://www.economist.fcom/economics -a-to-z, (accessed Mar 10, 2017).

UNDP. *Human Development Report 2015: Work for Human Development*; 2015 [Online]. http://*hdr.undp.org/sites/default/files/2015_human_development_report.pdf.* (accessed Mar 22, 2017).

UNESCO. *Global Education Monitoring Report: Education for People and Planet: Creating Sustainable;* 2016 [Online]. http://unesdoc.unesco.org/images/0024/002457/245752e.pdf. accessed date 22.04.2017

Weingast, B. R. Why Developing Countries Prove so Resistant to the Rule of Law. In *Global Perspectives on the Rule of Law*; 2008. http://jenni.uchicago.edu/WJP/Vienna_2008/Weingast_ROL_MS_2%2000_08-0519.pdf (accessed Mar 22, 2017).

WHO. *Health and Environment: Managing the Linkages for Sustainable Development*; The WHO/UNEP Health and Environment Linkages Initiative: Switzerland, 2008.

WHO. Noncommunicable Diseases. Fact Sheet Updated April 2017, 2017 [Online]. http://www.who.int/mediacentre/factsheets/fs355/en/. accessed date 22.04.2017

World Bank. *About Public-Private Partnerships*; 2016 [Online]. https://ppp. worldbank.org/public-private-partnership/node/335/. (accessed Mar 22, 2017).

World Economic Forum. *The Global Risks Report 2017*, 12th ed.; 2017 [Online]. http://www3.weforum.org/docs/GRR17_Report_web.pdf. (accessed Mar 22, 2017).

World Health Organization (WHO). *Working for Health and Growth: Investing in the Health Workforce.* Report of the High-Level Commission on Health Employment and Economic Growth; 2016 [Online]. http://apps.who.int/iris/bitstream/10665/250047/1/9789241511308-eng.pdf?ua=1. (accessed Mar 22, 2017).

CHAPTER 18

Reflections of Lifelong Learning on Economic Growth of Countries

EMEL TERZİOĞLU BARIŞ*

Educational Sciences Department, Karabük University, Karabük, Turkey

**E-mail: emelterzioglu@gmail.com*

ABSTRACT

Education is a phenomenon that is affected in life all over the world with all dimensions and is a necessity to all individuals during lifespan. Today's world, after the rapid changes and developments in science and technology, the world looks like a little village. An event that realize in anywhere in the world is known after a few minutes because of the internet. So, any event that occurs in somewhere in the world, can be the cause of the upside of the economy by causing fluctuations in the stock market on the other side of the world. For example any terrorist event or an election in anywhere in the world can be rise and falls in currency and inflation rises in a day. We can be named this situation as butterfly effect from chaos terminology. In this context, it is imperative that lifelong learning at the forefront of lifelong and supporting methods be taken into the programs of the countries and the development and implementation of educational policies. Strengthening of economies is possible with the training of qualified personnel. So, how will the new century handle lifelong learning in economic growth?

18.1 INTRODUCTION

When we consider that education and learning are cultural and socially structured phenomena, also have political nature and are based on world-views (Lima et al., 2009). Education is a phenomenon that is affected in life all over the world with all dimensions and is a necessity to all individuals during the lifespan. It is not surprising; we can say that this is lifelong learning. Thus, lifelong learning has become a necessity in many countries, so it is more than just training and education beyond formal schooling (World Bank, 2003).

Education, especially lifelong education is an interdisciplinary concept, so education is directly related to the economy. The main factor that keeps the economy running is people, and education and training of people who are in working life is extremely important. Lifelong learning is important for preparing employees to compete with the global economy (World Bank, 2003). Furthermore, today, *knowledge* is one of the main forces of the economy and economy of global knowledge has transformed labor demand all over the world (Singai et al., 2016). Furthermore, knowledge influences global knowledge development, learning, abilities, social life, and social status of individuals (Mandl and Krause, 2001). Preparing workers competing with knowledge economy requires education and training throughout the life cycle (Singai et al., 2016).

A lifelong learning framework encompasses learning during lifespan in different learning environments (World Bank, 2003). As Torres (2002) stated, "lifelong learning has been adopted in the global North as learning strategy in active citizenship and getting necessary knowledge, skills, values, attitudes in employment and work, but for the Global South lifelong learning is basic education that matters most." Understanding the lifelong learning on, in other terms, lifelong learning perspective can change place to place in the world. Sharing points in using lifelong learning in the development policy of countries are supporting active citizenship, creating qualified human capital, and building knowledge society.

In international era, lifespan development opportunities which are given to people of all ages by governments are important for any societies' development; this is a criterion in the growth of countries. Lifelong learning opportunity and development level of the countries increase in direct proportion. Furthermore, this is critical for the global knowledge economy of countries (World Bank, 2003). Thus, European Employment Guidelines accept the importance of lifelong learning and human capital policies to maintain older people in the labor market (Hutsebaut, 2005).

All in all, knowledge economy can be defined as: "production and services based on knowledge-intensive activities that contribute technical and scientific development" (Powell and Snellman, 2004: 199). Lifelong learning concept can be thought as preventive and lifespan process that national experiences show the general will to tackle issues at the end of someone's career (Hutsebaut, 2005). Lifelong learning activities are useful for people of all ages with different necessities. It can do awareness and training work on many topics including reducing lifetime labor market participation policies and reducing exclusion or addiction risk in the labor market throughout the lifespan (Luchsinger, 2016). How will the new century handle lifelong learning in economic growth?

18.2 LIFELONG LEARNING AND ECONOMIC GROWTH

Continuing professional development, leadership development models in lifelong learning and career of the workers are necessary for all people (WHO, 2016). Adult learning and education policies always result from discourses, practices, guidance, actions, priorities, and rules located in the international, national, and organizational contexts (Lima et al., 2009).

All countries leave their limited expertise and geographical inequality and giving priority to investment in lifelong learning by providing demographic transition, providing the opportunity to strengthen vocational training in the education of youth can achieve a better educational and economic performance (WHO, 2016).

The demographic transition has become a dramatic global phenomenon; because of mortality and consequently a continual decline in fertility so that high and/or equal death and birth rates ultimately lead to low and/or equal rates; as a result, death and birth rates in developed countries are low (McNay, 2003). Declining mortality following declining fertility during demographic transition leads to decades of child dependency at first, followed by decades of child dependency, and finally population aging (Hock and Weil, 2012; Mason et al., 2016).

Demographic transition requires lifelong learning; because it is overlaid in a rapidly changing world, it has to be structured to allow lifelong learning, more frequent changes in professions, and may allow for inclusion in the labor market much sooner than it is now (von Nordheim Nielsen, 2005). If it is exemplified, women's education has been repeatedly shown to have a powerful, depressing effect on both fertility and child mortality (McNay, 2003).

In this context, to create a structure based on the knowledge society searching for the perfection, to imply politics with all the ambiguities and the characteristic search of a truth, interest, and uncertainties for everybody's general life-long education (von Nordheim Nielsen, 2005). Most women undoubtedly win from demographic transition with the decline of fertility, the most important demographic change associated with increased freedom of education and employment (Bongaarts, 2009; McNay, 2003).

The importance of lifelong learning is increasing day by day in today's world. Two decades after lifelong learning theory, partly due to the economic crisis European countries show integration in economics and politics worse than expectations (Singai et al., 2016). Lifelong learning has gained importance throughout globalization. Nonetheless, discourse and practice changed between countries according to the characteristics of global regions (Singai et al., 2016).

The provisions of an aging population are important for economic growth, demographic transitions, and lifelong learning. The aging population is not the same between countries and varies between developed and developing countries (Fig. 18.1).

Most developing countries now have large elderly cohorts: 20% of developed countries will be over 60 years old and 30% over the next 40 years. According to current estimates, less than 10% of the population in developing countries will change soon, over the age of 60; it is expected that the rate will more than double when it reaches 2050 (Bloom et al., 2010). So, lifelong learning of immigrants and native-born is expected to improve their potential of participation in the knowledge economy (Singai et al., 2016).

Lifelong learning is important for the knowledge economy in education. Lifelong learning programs can help this effort by helping people adapt their knowledge and skills to the demands of changing economy (Bloom et al., 2010). In recent years, it is difficult to discuss and compare the issues of lifelong learning politics due to the growing prevalence of lifelong learning in adult education (Singai et al., 2016).

Developing countries and countries with transition economies bear the risk of marginalization in a competitive global information economy; for this reason education systems are unable to gain the skills they need, and need to take the place of a knowledge-based, teacher-directed learner in the framework of an official program and the education system is governed by directives with a new type of learning that emphasizes the creation, application, analysis, and synthesis of knowledge, and learning based on cooperative learning throughout life (World Bank, 2003). Today, after the rapid changes and developments in science and technology, the world looks like a little village. An event that occurs anywhere in the world is known after a few minutes because of the internet. So, any event that occurs in somewhere in the world can be the cause of the upside of the economy by causing fluctuations in the stock market on the other side of the world. For

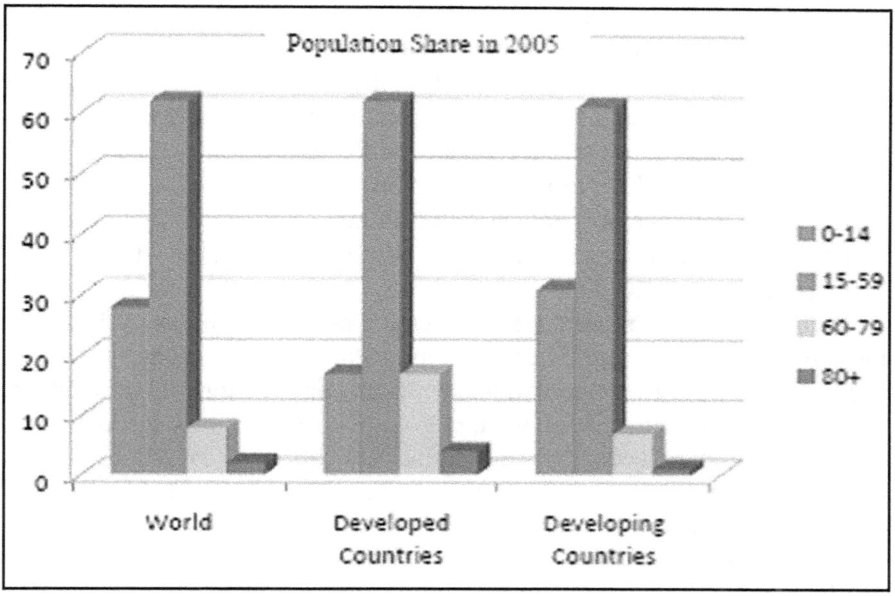

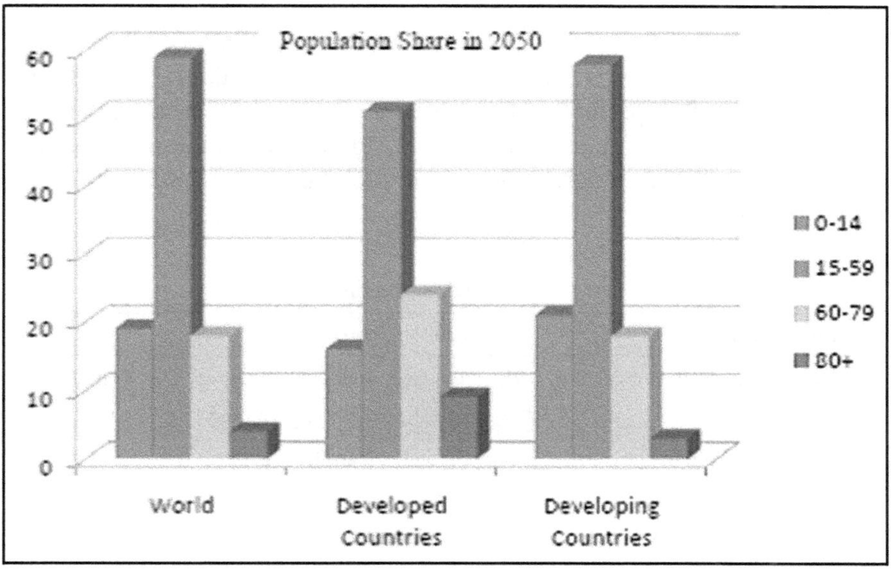

FIGURE 18.1 (See color insert.) World and development-group age-structure changes.

Source: Adapted from UN (2009); Bloom et al. (2010).

example, any terrorist event or an election anywhere in the world can lead to the rise and falls in currency and is reason for the inflation rises in a day. We can name this situation as a butterfly effect from chaos terminology. As Erçetin et al. (2014) said "chaos is a source of life and is a reality as living and self-organizing systems for organizations and people." The sensitive dependence on initial conditions shows how a small change at one place or moment in a nonlinear system can result in large differences to a later state in the system (Açıkalın and Erçetin, 2014; Erçetinet al., 2016)—the butterfly effect (Erçetinet al., 2013).

At this point, the importance of education as a factor that directly affects the economy is increasing. The final component—adult learning—takes into account a wide range of populations, while the main sectors of initial education before entry into the labor market with linear progress through early childhood, primary, secondary, and higher education are clearly defined. It consists of a large number of fragmented items that are divided into formal education, work, home, or elsewhere (Ó Riagáin, 2002).

Education is inadequate in many developing countries. The scope is insufficient, access is unequal (especially in higher education and employment and adult education), and educational quality is insufficient (World Bank, 2003).

Older workers often struggle with discrimination in the workplace, which is why discrimination is particularly prevalent regarding education and recruitment, so it is preferable to employ young workers who will assume that they will work longer (Luchsinger, 2016). The traditional learning model (Table 18.1) differs significantly in lifelong learning methods (World Bank, 2003):

TABLE 18.1 Traditional Learning and Lifelong Learning.

Traditional learning	**Lifelong learning**
Teachers: information source,	Teachers: guiding the sources of information,
get in-service training	initial and ongoing training
Students: learn from the teacher,	Students: learn by practice,
work on their own,	learn in groups from each other,
all students do the same	people have got lifelong learning opportunities
Tests: it is provided to prevent students from progressing until they fully learn the skill set	Assessment: used to lead learning styles and identify ways of learning in the future

Source: Adapted from World Bank (2003).

Although the concept of "lifelong learning" has proved to be very attractive to policy-makers faced with the shifting demands of contemporary labor markets, operationally and structurally it remains somewhat elusive (Ó Riagáin, 2002). The legal age requirement for retirement, lack of familiarity with modern technology, limited options for lifelong learning, and investment in training or older employees are among the institutional, legal, social, and technical barriers which limit workforce participation for the elderly in the Asia Pacific (Luchsinger, 2016).

18.3 CONCLUSION

One of the most important and effective coping, supportive, and developing methods of economies of countries is lifelong learning. In this context, it is imperative that lifelong learning at the forefront of lifelong and supporting methods be taken into the programs of the countries and the development and implementation of educational policies. So, strengthening of economies is possible with the training of qualified personnel.

KEYWORDS

- **life-long learning**
- **education**
- **economic growth**
- **teacher**
- **traditional learning**

REFERENCES

Açıkalın, Ş. N.; Erçetin, Ş. Ş. Great Illusion in Twenty-first Century-Chaos Knocking Door. In *Chaos, Complexity and Leadership 2012*; Banerjee, S., Erçetin, Ş. Ş., Eds.; Springer Proceedings in Complexity Springer: Germany, 2014. DOI 10.1007/978-94-007-7362-2_1

Bloom, D. E.; Canning, D.; Fink, G. Implications of Population Ageing for Economic Growth. *Oxf. Rev. Econ. Policy* **2010,** *26* (4), 583–612.

Erçetin, Ş. Ş.; Hamedoğlu, M. A.; Açıkalın, Ş. N. Can Managerial Narcissism which Likely Leads to a Chaotic Process be Measured? Managerial Narcissism Scale (MNS). In *Chaos and Complexity in World Politics*: *Economic Growth and Demographic Transition in Third World Nations;* Erçetin, Ş. Ş., Banerjee, S, Eds.; IGI Global: USA, 2014.

Erçetin, Ş. Ş.; Açıkalın, Ş. N.; Bülbül, M. Ş. A Multi-Dimensional Approach to Leadership in Chaotic Environments. In *Chaos and Complexity Theory for Management: Nonlinear Dynamics*; Banerjee, S., Ed.; IGI Global: USA, 2013; pp 89–104.

European Commission. *Commission Staff Working Paper: A Memorandum on Lifelong Learning*. European Commission: Brussels, 2000.

Hutsebaut, M. How to Reconcile Employees Interest with the Increasing Older Workers Employment Policies. In *Health, Ageing and Work, The Counter-ageing Society*, European Papers on the New Welfare 1, The Risk Institute, Trieste: Geneva, 2005; pp 116–122,

Lima, L. C.; Guimarães, P.; Touma, N. Comparing Policies in Lifelong Learning. Adult Learning and Education Policies in Germany, Portugal and Sweden: An Analysis of National Reports to CONFINTEA VI, 2009; pp 29–65.

Luchsinger, G. Shaping the Future: How Changing Demographics Can Power Human Development. Asia-Pacific Human Development Report. UNDP United Nations Development Programme, New York, 2016.

Malitza, M. The Double Helix of Learning and Work. In *Health, Ageing and Work, the Counter-ageing Society*; European Papers on the New Welfare 1, The Risk, Institute, Trieste: Geneva, 2005; pp 123–127.

Mandl, H.; Krause, U. M. (2001). Lernkompetenz für die Wissensgesellschaft. http://core.ac.uk/download/pdf/12161788.pdf (accessed Dec 31, 2016).

Mason, A.; Lee, R.; Lee, S. H. (2016, August 16). The Demographic Transition and Economic Growth in the Pacific Rim. The East Asian Seminar on Economics (EASE), Seoul, Korea, June 19–21, 2008.

McNay, K. Women's Changing Roles in the Context of the Demographic Transition. Gender and Education for All: The Leap to Equality. Background Paper Prepared for the Education for All Global Monitoring Report 2003/4, The Leap to Equality, 2003.

Ó Riagáin, P. The Consequences of Demographic Trends for Language Learning and Diversity. *Guide for the Development of Language Education Policies in Europe from Linguistic Diversity to Plurilingual Education*; Linguistics Institute of Ireland, Council of Europe: Strasbourg, 2002.

Powell, W.; Snellman, K. The Knowledge Economy. *Ann. Rev. Soc.* **2004,** *30*, 199–220.

Rubenson, K. Lifelong Learning: Between Humanism and Global Capitalism. In *The Routledge International Handbook of Lifelong Learning*; Jarvis, P., Ed.; Routledge: 2011; pp 411–422.

Singai, C.; Gioli, G.; Riemer, E.; Regmi, K. D.; Mastrokoukou, S.; Singh, S. Knowledge Economy and Demographic Change: Comparative Case Study of Europe and South Asia. https://www.peterlang.com/search?f_0=author&q_0=Regina+Egetenmeyer Egetenmeyer, R., Ed.; Adult Education and Lifelong Learning in Europe and Beyond. https://www.peterlang.com/view/serial/SPAG Studies in Pedagogy, Andragogy, and Gerontagogy 2016; pp 66–84. http://dx.doi.org/10.3726/978-3-653-05973-1

Torres, R. M. Lifelong Learning in the North, Education for all in the South. In *Integrating Lifelong Learning Perspectives;* Medel-Añonuevo, C., Ed.; UNESCO Institute for Education: Hamburg, 2002.

United Nations (2009). World Population Prospects: The 2008 Revision. CD-ROM Edition: Extended Dataset, United Nations.

von Nordheim Nielsen, F. (). Active Ageing: a Core Policy Priority for the European Union. In *Health, Ageing and Work*; The Counter-ageing Society:European Papers on the New Welfare 1, The Risk Institute, Trieste, Geneva, 2005; pp 66–78.

WHO (2016). High Level Commission on Health Employment and Economic Growth Working for Health and Growth: Investing in the Health Workforce. Geneva: WHO Library Cataloguing-in-Publication Data.

World Bank (2003). Lifelong Learning in the Global Knowledge Economy: Challenges for Developing Countries. *A World Bank Report*. Washington, D.C.

CHAPTER 19

Effects of Syrian Refugees on Demography and Economics and Education

ŞEFİKA ŞULE ERÇETİN* and DURAN MAVİ

Department of Educational Sciences, Faculty of Education, Division of Educational Administration, Hacettepe University, Ankara, Turkey

**Corresponding author. E-mail: sefikasule@gmail.com*

ABSTRACT

The purpose of this research in this chapter is that discuss in term of education the effects of Syrian aslyum seekers who who migrated to Turkey because of civil war and political turmoil on demography and economy. Within this scope, it is aimed to be able to reveal the problems, to be able to clarify the solution proposals and to help them. Participants of this research consist of 12 people in Elbistan county of Kahramanmaraş, 6 of whom are vegetable and fruit storage, 6 of which are artisans in small industrial area. Participants were voluntarily determined by snowball sampling method. The data were collected through interviews. The data obtained in the research were analyzed by content analysis method. As a result of the research, the following information is presented, taking advantage of the information provided by the participants: (1) establishing and developing a data bank showing the demographic status of the Syrians, continuing to record and organizing the domestic migration/distribution through needs, (2) compulsory language and job security training for all working Syrian individuals, (3) inclusion of asylum seekers in vocational training, (4) adoption of the necessity of providing Syrian individuals with their own livelihoods and the employment of Syrian individuals in local businesses is a work done in this context, (5) reduction of social security burden on SMEs, which constitutes a significant part of tax and workload.

19.1 INTRODUCTION

Migration phenomena arising from different causes affect migrants, migrated country, and emigration country from various dimensions. Warfare is also the cause of mass migration movements. People who migrated because of wars think of neighboring countries for migration (Tunç, 2015).

It is possible that large international migrations are likely to lead to social, economic, and educational problems in the host countries and may lead to transformations that could cause indigenous peoples to react to migrants (Deniz, 2014: 177). The situation of Syrian refugees who came to Turkey because of the war has been increasingly being permanent with their problems. Therefore, it is necessary to analyze the subject thoroughly and to produce various solutions to the problems.

19.2 CONCEPT OF MIGRATION

Migration is the process of individuals moving from one place to another temporarily or permanently due to social, political, economic, etc. reasons (Deniz, 2014; Çopur and Demirel, 2017; Sayın et al., 2016:2). It can be examined under two headings: *internal migration* based on population mobility among the various centers within the country boundaries, and *external migration* carried out by people or persons across country borders. Seasonal migrations for activities such as holidays and excursions; permanent migrations from compelling reasons such as violence, natural disaster, and education; labor migrations arising from appointments and compulsory/voluntary migrations made to have better opportunities are considered within the context of internal migration (Sayın et al., 2016). International migration is a permanent type of immigration that increases the population of the country of origin, which reduces the population of the person's or persons' own countries. The *brain drain* of talented individuals toward developed countries, from developing or undeveloped countries, or from developed countries to other developed countries, *worker migration* that include migration of employee, *exchange migration* that involve mutual exchange of citizens between two countries due to various reasons are types of external migration. Internal migration does not change the population of the countries, but it can affect the population density of regions and cities. It is often encountered from the sea to the city. This is not the case for external migration. In countries that receive external migration, population growth will occur (Canbay and Gülden, 2016).

19.3 TURKEY AND MIGRATION

Arab Spring, which started in Tunisia and influenced many Arab countries, caused internal disturbances in Syria and triggered the migration of millions of people (HKU, 2013; Seydi, 2014; ORSAM, 2015b; KTO, 2016; Sayın et al., 2016). This wave has seriously affected the countries of the region in every field. Naturally, Turkey has taken its share of this human tragedy

Because of its geographical location, Anatolia is known to be affected by international migration throughout the history (Deniz, 2014: 175). Today, for refugees who do not know how to get admission into European countries, Turkey has become a secondary target. A number of Syrian refugees who are an important part of the millions have found themselves in Turkey, in this context (Çopur and Demirel, 2017). So much so that even the Syrians are not official documents, were taken to camps in Turkey as refugees, settled in the city is determined by the Immigration Administration General Directorate (Uzun, 2015).

Load on Turkey as a country that receives a lot of migrants, on the migration routes, trying to solve a significant part of financial problems with their own resources is heavier than other countries neighboring Syria (Güven, 2013; ORSAM, 2015a; Paksoy et al., 2015; Uzun, 2015; Sayın et al., 2016; Khalaf and Ilgar, 2017). Turkey's attitude toward the situation of Syrian refugees is to be commended (Seydi, 2014). However, it should be noted that this situation also imposes responsibilities in a variety of meanings to Turkey (Uzun, 2015). This is the inevitable result of migration (Tunç, 2015).

19.4 SYRIAN REFUGEES

Turkey's position makes it the *migration corridor* (Deniz, 2014, 177; Koç et al., 2015; Khalaf and Ilgar, 2017). Syrian individuals in Turkey will be more accurate to define the asylum seekers rather than refugees according to the provisions of the *Geneva Convention (1951), the Protocol on the Status of Refugees (1967) and the International Law on International Protection (2014)* (Koç et al., 2015; Tunç, 2015; Çopur and Demirel, 2017; Khalaf and Ilgar, 2017).

Despite the fact that there is social mobility at its core, migrations may have financial, demographic, and cultural influences. The effects of prolonged civil war in Syria have been increasing over the population and the economy and working life in Turkey (Çopur and Demirel, 2017). Tendencies to go to

Europe of Syrian refugees that arrive in Turkey is seen to be low (Khalaf and Ilgar, 2017). At the same time, the desire to return to their countries seems to be gradually decreasing (Tunç, 2015).

Undoubtedly, the economic, medical, educational, and social needs of asylum seekers should be met like every individual (Uzun, 2015). The temporary protection they receive helps to keep the wounds of the Syrian asylum seekers for a while (KTO, 2016). But it should not be forgotten that the privileges granted by Syrian asylum seekers through temporary protection can be considered as a facility rather than a legal right. Therefore, it may be an appropriate approach to consider the expenditure of public resources being used and the effort spent on asylum seekers from the perspective of population, economy, and education.

19.4.1 REFUGEES AND DEMOGRAPHIC SITUATION

Turkey is the most asylum seekers receiving country between neighboring countries to Syria (ORSAM, 2015a). The total number of asylum seekers arriving in Turkey in different sources is reported that up to five million (TEDMEM, 2016). However, the number of Syrian asylum seekers cannot be precisely stated; because a significant number and intensity of asylum seekers soon have entered to Turkey in regular and irregular manner of legal and illegal ways by the Open Door policy (Güven, 2013; Tunç, 2015; Çopur and Demirel, 2017; Kaygısız, 2017).

Population growth is felt in all areas of public and private (Karakütük and Kavak, 2017). The available official data reveal that Syrian asylum seekers in Turkey increase the population by 4% (Sayın et al., 2016). This rate has risen above 10% in the cases of illicit asylum seekers in Syria (ORSAM, 2015a; Tunç, 2015). Turkey's population increased by 1.35% already (TÜİK, 2017). Naturally, the rate of population growth has also changed (Canbay and Gülden, 2016). Therefore, a serious impact is inevitable on all areas of emigrants in Turkey in all sectors across the country.

It seems that there is no systematic situation in the distribution of Syrian asylum seekers in the country. A very large part lives outside of the camps (ORSAM, 2015a; Kaygısız, 2017; TEDMEM, 2016; Khalaf and Ilgar, 2017). According to data of The Immigration Administration General Directorate (2016), 89% of Syrian asylum seekers in Turkey (about 2,350,000 people) are scattered into the country. Transportation facilities provided by Istanbul also cause immigration (Deniz, 2014: 193). Following Istanbul, Bursa, Mersin, Adana, and Gaziantep are the cities preferred by Syrian asylum seekers due to their employment advantages. The main mass of Syrian asylum seekers who are experiencing adaptation problems are not those living in the camps but those scattered in the city. The settlement centers where Syrian asylum seekers are located are in Gaziantep, Şanlıurfa, Kilis, Mardin, Kahramanmaraş, Osmaniye, Adıyaman, Malatya, and Adana (TEDMEM, 2016).

It is stated that 63% of Syrian asylum seekers are at working age (Kaygısız, 2017). In addition, the number of women and children is remarkable. It is known that half of the Syrian asylum seekers living outside the camps are not profession seekers and seek jobs (AFAD, 2014a). This affects economic problems as well as the unwanted rates of violence and crime (Canbay and Gülden, 2016).

19.4.2 REFUGEES AND ECONOMIC SITUATION

Asylum seekers/refugees do not want to be admitted in places that the migrate by local people because they are in need of economic support, the pressures on the budget, they are employed in heavy jobs with low wages without the social security costs (Tunç, 2015; Canbay and Gülden, 2016). Because being a stranger means to be excluded and to be otherized; in many places and at any time (Koç et al., 2015). But it can be said that the situation is not so worse in Turkey. Syrian asylum seekers have free examination, treatment, and medicine rights in public hospitals; education rights from primary school to higher education, including hobby courses, scholarships, and dormitories for those who meet certain conditions; free translation rights and work permits in sectors and locations determined by the council of ministers (Uzun, 2015).

It is seen that the greatest impact on the economy of Syrian asylum seekers is the expenditures. Only 5% of the general expenditures of the aids have been coming from abroad for Syrian asylum seekers (Khalaf and Ilgar, 2017). The rest of the expenditure is subsidized by Turkey.

The economic burden on the health sector has become more aggravated, health services have become difficult to obtain, the number of students per teacher has increased, and there has been a decrease in wages and salaries with the arrival of Syrian asylum seekers to the country, (ORSAM, 2015a; ORSAM, 2015b; Tunç, 2015; Canbay and Gülden, 2016; Kaygısız, 2017). The presence of Syrian refugees in Turkey has affected the labor market situation in our country and has an adverse impact on the unemployment figures in Turkey (HKU, 2013; ORSAM, 2015a; Paksoy et al., 2015; Sayın et al., 2016; Kaygısız, 2017). In addition, the rate of inflation in the border regions where the Syrian asylum seeker population is exceeded rises above the country average (ORSAM 2015; KTO, 2016; Sayın et al., 2016). Similar effects are expected to increase in the coming years (HKU, 2013).

An important issue is that they are used as cheap labor in places where Syrian asylum seekers are (Güven, 2013; HKU, 2013; ORSAM, 2015a; Paksoy et al., 2015; Kaygısız, 2017). Syrian asylum seekers are generally employed in sectors where unskilled labor is used (Deniz, 2014; Tunç, 2015; Sayın et al., 2016; Kaygısız, 2017). This situation leads to the closure of employee vacancies in the relevant sectors through Syrian asylum seekers (Tunç, 2015). So that they create small-scale *job markets* in places where they live (Kaygısız, 2017).

Syrian refugees in Turkey are being seen as a poor image they draw in the economic sense. Syrian asylum seekers find themselves in areas that require more unqualified labor (ORSAM, 2015a; Kaygısız, 2017). They are trying not to live a special life in high-quality places but they try to meet basic needs like nutrition and security (Koç et al., 2015; ORSAM, 2015a). This is pushing them to look for jobs, even at low prices and off the record. Asylum seekers work in sectors like construction, textile, and agriculture without social security (Deniz, 2014; Kirişçi, 2014; Koç, Görücü and Akbıyık, 2015; Paksoy et al., 2015; ORSAM, 2015b; Uzun, 2015; Sayın et al., 2016; Kaygısız, 2017).

In the current situation in Turkey (except for the Syrian refugees), a significant portion of total employment is unregistered (Kaygısız, 2017). It would be a good idea to say that businesses that fail to fulfill their obligation for unregistered Syrian asylum seekers lead to unfair competition (ORSAM, 2015a; Tunç, 2015; KTO, 2016; Sayın et al., 2016; Kaygısız, 2017). It can be said that this subject brings a significant burden to the economic structure in Turkey. Informal economic initiatives are also causing the exploitation of Syrian asylum seekers (Deniz, 2014). It should not be forgotten that such an understanding will mean that the local people will pay more taxes (Canbay and Gülden, 2016).

Problems such as the lack of expected increase in salaries and the decline in salaries have become noticeable with the advent of Syrian asylum seekers (ORSAM, 2015a; Tunç, 2015). This restricts local people's job opportunities and causes reactions (Erdoğan, 2014; Sayın et al., 2016). This opens the door to possible tensions between Syrian asylum seekers and the local population (HKU, 2013; Reçber and Ayhan, 2013; Uzun, 2015; Canbay and Gülden, 2016; KTO, 2016; Sayın et al., 2016; Kaygısız, 2017). The rate of job loss is positively correlated with this perception (ORSAM, 2015b; KTO, 2016). It is observed that in cities where Syrian asylum seekers are located, perceptions of individuals losing their jobs have been rising (ORSAM, 2015b; Tunç, 2015).

Syrian asylum seekers to come to Turkey has led to a position that the advantages and disadvantages closing to each other well (ORSAM, 2015a). Syrian population has shifted many industrial plants or capital to Turkey (Reçber and Ayhan, 2013; ORSAM, 2015a). Syrian asylum seekers find jobs in smaller businesses and establish similar businesses to meet their needs. It can also be said that Syrian asylum seekers have brought mobility to the real estate/construction market. It should be noted that Syrian asylum seekers did not only have the cheapest labor force, but also had the effects of reducing costs and expanding markets (Güven, 2013; ORSAM, 2015a; ORSAM, 2015b; Paksoy et al., 2015). In this respect, it can be assumed that Syrian asylum seekers meet the labor requirement. The economic impact caused by the migration of Syrian refugees to Turkey that experienced price increases in some areas also have a negative impact on local people's budgets would be accurate (ORSAM, 2015b). Local businesses indicate that there are not enough customers of Syrian asylum seekers and that their profitability has not increased sufficiently (HKU, 2013). But it is a fact that the local economy grows numerically in migrating regions (Reçber and Ayhan, 2013). The impact of Syrian asylum seekers in the economy of Kahramanmaraş has reached the level that will affect the export and import figures of the city (ORSAM, 2015b). However, it should not be forgotten that the economic incomes of the first years of migration have decreased with the exhaustion of the resources that Syrian asylum seekers have brought with them (Güven, 2013).

The proportion of Syrian asylum seekers who are healthy and able to work to children within the population is important. Almost half of the registered Syrian asylum seekers are individuals under the age of 15 (Kaygısız, 2017). This situation coupled with livelihoods and adjustment problems and has led to an increase in the number of child workers (Güven, 2013; ORSAM, 2015a; Tunç, 2015). Syrian asylum seekers encourage their children to work under severe conditions rather than sending them to the school (Türüt, 2015). This affects the sociocultural field.

19.4.3 REFUGEES AND EDUCATIONAL SITUATION

Extraordinary changes in influence can lead to unpredictable outcomes in education policies and budgets, because population changes are directly related to future student numbers (Karakütük and Kavak, 2017). It seems that

Syrian asylum seekers are also enforcing the Ministry of National Education (MONE) in financial and financial sense (ERG, 2016; TEDMEM, 2016).

The number of Syrian refugees in Turkey in the field of educational services is over 300,000 (including 80,000 in the camps) (ERG, 2016; UNICEF, 2016). It is seen that the smallest part of this population educated in Syrian asylum seekers is the individuals in the age of secondary school (TEDMEM, 2016). When the numbers are added to other Afghan, Somali, and Middle Eastern immigrants, the complexity becomes even clearer.

Educational participation of migrant individuals and their children is low compared to local children (Karakütük and Kavak, 2017). It is known that 60% of Syrian asylum seekers' children do not go to school in Turkey (ERG, 2016). The fact that this large population is not included in the scope of education may lead to various risks in the future (Tunç, 2015).

The process of integration of the Syrian asylum seekers into Turkish schools will soon be completed and the settlement process for asylum seekers in public schools will be completed. As of the end of 2017, MONE's ongoing inclusive education activities for teachers are ongoing. Some countries receive well-trained migrants with a quality socio-cultural heritage (Karakütük and Kavak, 2017). The situation in Turkey seems not to be in this case. It is known that the education level of a significant part of asylum seekers is low (ORSAM, 2015b; Tunç, 2015; Kaygısız, 2017). So much so that 21% of Syrian asylum seekers do not even know how to read and write (AFAD, 2014b). While the proportion of Syrian asylum seekers who graduated from universities is less than 10%, the proportion of those who have primary and lower school graduates is more than half of the total population of asylum seekers (AFAD, 2013). While Syrian asylum seekers with high education level are traveling to cities such as Istanbul and Izmir, those with low educational level are found to be located in cities near the border of Syria (KTO, 2016).

Separation tendencies of qualified Syrian asylum seekers from Turkey are higher than that of unskilled Syrian refugees (Kaygısız, 2017). This situation arises from thoughts of them that they are not worthy of the position they are in and the lack of citizenship rights. Equality problems of some Syrian asylum seekers can push them to informal employment.

19.4.4 EXPECTATIONS AND PROBABILITIES

The demographic, economic, and cultural characterization of the period of population growth is a natural process (Karakütük and Kavak, 2017). It is evident that this serious migration will lead to a number of human and economic problems that will push the Turkish economy further in the long run (Uzun, 2015; Paksoy et al., 2015; Karakütük and Kavak, 2017). The gravity of the situation obliges the situation of Syrian asylum seekers to be taken into account in relation to demography, economy, and education. Requirements of international agreements and neighborly relations are directed Turkey to be precise against Syrian refugees. It is known that there will be various prices.

For now, Syrian refugees in Turkey do not appear to make investments that will provide employment and such initiatives are not expected in the short term (Güven, 2013; HKU, 2013; Kaygısız, 2017). Moreover, local people are not willing to sign economic projects and partnerships with the Syrian population (HKU, 2013). The low level of education can reduce the wishes of the Syrian asylum seekers to contribute to the economy. It can even be said that this will create a cost in itself. Syrian asylum seekers' expectations from Turkey are likely to increase in the future (Tunç, 2015).

All these developments necessitate the social, economic, and cultural adaptation process of Syrian asylum seekers, because asylum seekers who are able to survive on their own feet in economic terms are less likely to suffer negative impacts (HKU, 2013; KTO, 2016). But it is also a difficult process for harmony since it is not easy for individuals of two communities with different backgrounds.

19.5 AIM OF THE RESEARCH

The purpose of this research is that discuss in term of education the effects of Syrian asylum seekers who migrated to Turkey because of civil war and political turmoil on demography and economy. Within this scope, it is aimed to be able to reveal the problems, to be able to clarify the solution proposals and to help them.

19.6 METHODOLOGY

19.6.1 PARTICIPANTS

Participants of this research consist of 12 people in Elbistan county of Kahramanmaraş, 6 of whom are vegetable and fruit storage, 6 of which are artisans in the small industrial area. Participants were voluntarily determined by snowball sampling method. The youngest participant is 30 years old, the oldest one is 60 years old and the average age is 42.8. There are no participants in the related professions whose working time is less than 15 years. In addition, one of the participants has a bachelor's degree, 10 graduated from high school, and 1 graduated from secondary school.

19.6.2 COLLECTION OF DATA

The data were collected through interviews. Appointments were made for the interviews to be effective. Interview periods are 13, 10, 16, 11, 17, 14, 14, 11, 18, 10, 15, and 15 min, respectively. Participants were coded S1, S2, S3, S4, S5, and S6 in the small industrial sites while the participants in the vegetable and fruit store were coded H1, H2, H3, H4, H5, and H6. Since the participants were all male, there was no need to encode sex. The interviews were backed up with a voice recorder to be transferred later. The voice recorder was turned off at the moment the participants requested. A note is also kept when it is needed. Participants were given full permission and their names and business names were kept confidential.

The questions asked to the participants are as follows, in the context of the possibilities provided, the problems encountered and the solution proposed.

1. What can you say about the pre- and post-migration situation of the sector (population, demographic characteristics ...)?
2. (If any) Do you employ Syrian asylum seekers, local people? Would you talk a little of the reasons with the facilities you provided?
3. What can you say about the qualities of Syrian asylum seekers? Can you give information about the difficulties/inconveniences/aspirations of asylum seekers?
4. Are the local people unemployed because of Syrian asylum seekers? If so, what could be the reasons for this and which socio-cultural strata are affected by this situation?
5. Do you have any suggestions for solutions? If so, what are your suggestions?

19.6.3 ANALYSIS OF DATA

The data obtained in the research were analyzed by content analysis method. This process is done through the extraction of concepts and sub-concepts, themes, and sub-themes. Research questions have been references used for themes. Sub-themes, concepts, and sub-concepts are provided from the interviews. Once the necessary codes have been extracted, those who are in a relationship with each other are brought together and arranged.

19.6.4 VALIDITY AND RELIABILITY OF THE STUDY

The interviews made within the scope of the research were backed up by voice recorder. Two audio devices have been used to increase reliability. Intermediate questions were asked and interim answers were taken in order to reduce the lack of meaning and reach the right information. It was also benefited from the notes taken during registration during the conversion of the interview records into texts. In this way, work has been tried to be strengthened.

19.7 FINDINGS

19.7.1 SECTORAL FINDINGS BEFORE AND AFTER MIGRATION

Participants are in a consensus that there is no improvement in the situation in the sectors after the Syrian crisis and that the situation is better in the first place. Although Syrian asylum seekers have brought mobility to the market in the first place, they do not have it in time, and they have left their place unemployed. In this context, the sectoral codes for migration before and after migration are given in Table 19.1.

TABLE 19.1 Sectoral Codes Related to Pre-migration and Post-migration.

THEME-1: Sectoral codes related to pre-migration and post-migration	
Sub-theme 1: population and demographic characteristics	**Sub-theme 2: economic features**
Crowded	Increased borrowing
Unemployment	Downsizing
Unqualified	Falling business volume
Job dislike	Longing for the past
Incompatibility	Low expectations
Lack of education	Increase in expenses

Among the participants, H3, H4, H5, S3, S4, and S5 suggest that the confusion of migration is still continuing even if its initial confusion diminishes. Regarding the issue, H4 declares that it has observed that the young but unemployed group in its sector is growing even more than the old one. S4 refers to this population as an army, and its most prominent feature is the lack of craftsmanship as a subtle job absence from its hands. Here is the summary of S5, he is an industrial artisan since childhood, stating that:

> (…) I do not see anything other than the ones that are constantly growing, queuing in front of the hospital. 'With a sad and complaining expression' Salt pepper is the case of our country. We made it difficult for ourselves. When you shake your hands, Syria has hit the bazaar on the market. No arts, no business to build. They will pass you through me at this speed, saying 'Syrian refugees.' We are not able to find apprentice-talents from young people, and that's it. Syrians crossed ours. I'm not a racist man. Do not step on my name, my teacher. The work of the Arabs, give me food, I have covered my back. We still have what we have.

H5, who has been in business for more than 40 years, and who has made commercial success on an international scale, says something similar to what S5 says:

> (…) These are the poor. But they do not care a little bit … They do not have trainings but they do not go to school, they do not have money, but they do not work, they have livelihoods, but every day the numbers are increasing. In the red lights, in front of the mosque, in the parks … These things get worse. (…)

Participants mention that Syrian asylum seekers did not make a positive contribution to the situation in the sectors; and even directly and indirectly on themselves. So much so that H3 would say, "I wish they had never come. Everywhere is Syrian." This summarizes the increase in population density and proportion of Syrian asylum seekers.

19.7.2 FINDINGS RELATED TO THE EMPLOYMENT OF SYRIAN REFUGEES

Employers employing Syrian asylum seekers have shown themselves to be pushing higher costs for this situation. They are saying that Syrian asylum seekers are already opening up their limited opportunities. In this context, the codes of employment of Syrian asylum seekers are presented below.

TABLE 19.2 Findings Related to the Employment of Syrian Refugees.

THEME-2: Findings related to the employment of Syrian refugees	
Sub-theme 1: provided facilities	**Sub-theme 2: reasons**
Shelter	Insufficient income
Clothing assistance	Bad conditions
Food aid	Reluctance of indigenous individuals
Positive discrimination in wage	High social security costs
Job	Anxiety about lowering costs
	Sectoral fluctuations in demand for labor

H1, H3, H4, H5, and H6 stated that they give jobs to Syrian refugees at their workplace. H2, on the other hand, has an economic sense of difficulty and does not transcend this narrow strait, saying it will give a job to a Syrian asylum seeker. In S1, S2, S3, S4, S5, and S6, they stated that they did not employ Syrian asylum seekers during the research but they had already entered it. The first causes of asylum seekers' work related to the issue are low working fees. Individuals who do not work asylum seekers argue that high job risks and legislation prevent them from doing so. H2 states below:

> (…) There was Syria in my shop. But now things are bad. That's the case with our business. It works in the summer and spends the winter. As for the summer, I will recruit Münir, referring to the Syrian asylum seeker. He works a little money, Münir. That's my opportunity. If I have what I am putting together, we manage to live together. His wife, "Münir's wife" goes to my house daily cleaning. If possible, will I employ Turkish workers? But it is very difficult. Insurance, the record … It's not getting underway. I've done it before. At this time, not at all. The Syrians are good at doing the job. The Turkmen ones are more hard working. I know this is not legal I do. But it is not legal for me to fall into this situation.

The S6, who has been a craftsman since a young age, stated that:

> My work requires skill. That's why I bought one or two people. But they could not hold back. If everyone is paying for the weekend, I gave it to him "as wage," and he did not come on Monday. I took shoe, brought him from my own dresser at home, and I buy him meal where I ate the meal. No, he left the job. (…) The people of ours do not work on those wages anymore

in works like my work. Their minds were not in place anyway. The Syrians were also an excuse to escape from the work. I need the staff "smile" but cheap. (…)

S2 is a major operator. Previously, he says he was out of a Turkish worker from his job because he learned that he was staying in prison before. He said that he held the employee's credibility in front, saying, "No ID, no passport. But I made him get a job elsewhere."

If all participants were on equal terms, Syrian refugees would be replaced by local people instead. To the extent of their potential, they are holding the hand of asylum seekers. They emphasize that the majority of employee costs and expenses are important factors in determining wages. S3 and S1, who also manage in an institution, summarize the situation as "We have such conditions."

19.7.3 FINDINGS RELATED TO WORKING CONDITIONS AND COMPETENCIES OF SYRIAN REFUGEES

All of the employers interviewed during the study period complain about the low quality of Syrian asylum seekers. Participants pointed out that the low level of educationed local people is similar to Syrian asylum seekers, but that the low salaries of Syrian asylum seekers are at the forefront. In this context, the coding of the working conditions and competences of Syrian asylum seekers are given in Table 19.3.

TABLE 19.3 Codes Regarding the Working Conditions and Competencies of Syrian Asylum Seekers.

THEME-3: Codes regarding the working conditions and competencies of Syrian asylum seekers	
Sub-theme 1: employee qualifications	**Sub-theme 2: challenges encountered**
Low educational level	Fear of punishment
Bored and sloppy	Low productivity
Unqualified	
Resistance to learning Turkish	Low level of work
Limited communication skills	Legal difficulties
	Low performance
	Non-compliance with working rules
	High labor turnover

All participants indicate that the quality of the work put forward by Syrian asylum seekers is low. In addition, all participants describe Syrian asylum seekers as individuals with *low skilled* or *unskilled workforce* in their working life. While H4, H6, S1, S2, S4, and S5 suggest that the main reason for this is uneducation, H1 and S3 point out individual characteristics and language problem.

"They do not know … They can not do a job here. They can not weld, install, maintain and repair. So they can not find a job here, too." H5 and H6 parallel to S5, which is trying to summarize the situation with the census, are also presented with the statement "Only thing that comes from their hands is portering." The same participants indicate that Syrian asylum seekers did not participate in organized courses and did not comply with the rules and regulations applied at the workplace.

H5, who has taken over his profession from his father and told him that he is doing this job he knows well, states that he has employed four Syrian asylum seekers beside him for being a kindness. His expressions about the difficulties of employing Syrian asylum seekers are as follows:

> (…) My situation is better than my neighbors. But this Syrian confusion has also hit me (in the sense that it affects). I took Syria beside the Turkish workers and they were the ones who cut off the trade because I did not give job the Turkish workers (referring to the local individuals). (Shaking his head) It does not matter. (By pointing to the Syrian asylum-seeker with his hand) It would be a lie to say that they are not working as much as ours. But it's not like I expected. There is a permanent hiring entry-exit situation in the work places in the residence. Among the Syrians, however, I continue to work with the same people. (By referring to the social security record and transactions) There is no SSK. Yes it is not true. I am afraid of punishment, but my workers have no ID or passport. If I could hire someone instead of them, the costs would increase. I have literacy in my workers, one of them probably read secondary school or something. So it is not different from our other workers. Well-educated people do not stop here. At the beginning of the war, there was a doctor here. I saw from Facebook, doctor went to Germany. (…)

S3 and S4 indicate that the Syrian youth sometimes come to their side. S3 "When I took him as a worker, he was no literacy, he did a few wrong things, so I got angry and never came back." S4 "They will not work. If we give charity, they will not learn. The man who harmed the property of the customer who came to work at noon, what do I do?" It describes the state of compliance and performance shown to the working rules. H2 is regrettably saying, "Syrians will be big trouble for us. They do not know jobs, they steal, they do not want to go, they do not want to do what they need to stay... Even our ignorant is more educated than Syrians." S1 say that the employment of Syrian asylum seekers is not very attractive with these sentences: "I do not want to get into trouble by hiring Syria. He tires me, slows me down."H1 is trying to explain the educational and qualitative calibrations of Syrian asylum seekers with a refutation of "I do not think that there will be more than the working hours in construction, neither the diploma nor the reading ..."

19.7.4 FINDINGS RELATED TO WORKING CONDITIONS OF INDIGENOUS PEOPLE

Participants in the survey seem to have accepted the idea that local people can not find work because of Syrian asylum seekers. But the reasons are basically economical. It is stated that individuals who are not college graduates or professionals are not able to work because of asylum seekers. In this context, the coding of the working status of the local people is presented in Table 19.4.

All participants agree that Syrian asylum seekers are making it difficult for local people to find work. However, there has not been found any organization that recruits local individuals to hire Syrian asylum seekers. It is declared by all participants that the inclusion of asylum seekers in the labor market led to a decline in wages and reduced costs.

TABLE 19.4 Codes of Working Conditions of Local People.

THEME-4: Codes of working conditions of local people	
Sub-theme 1: situation of local people	**Sub-theme 2: those affected**
Job loss	Those with a degree of high school and below
Decline in wages and wages	Young people and adults
Discrimination	
Rejection (indifference)	
Risk perception	

Among those who participated in the survey, H1 states that "local people are employed together with Syrian asylum seekers, and that this brings some tensions between the workers, which may cause debate in some places." The situation is trying to make it more clear with the statement "(referring to the Syrian asylum-seekers) These came and our Turkish Turks are irritable." H2, H3, H4, S5, and S6 indicate that Syrian asylum seekers are only taken to unskilled jobs. It is explained by S1 that the greatest domestic rival of this group threatens those with a high school and secondary school graduation degrees:

> (…) The Syrians are not a threat here for me and like me, they can not be. I am graduate of university. I have a diploma in business administration. I'm using calculation software. Bank transactions, purchases of thousands of liras, Customer relations are running over me. They are (for Syrian asylum seekers) risk for tea seller, soup seller, construction worker. If I were an untrained person, they would be a rival for me in these market conditions. But in such places, individuals like me are kept on the hand. Suriyeililerin paper collecting, scrapping jobs remain (sad).

This reflects that Syrian asylum seekers are more effective on low socio-cultural strata, where unemployment is high. S4 is describing the situation of the industry with the statement "The tradesmen, the craftsmen have the dry salt." H4 is similar to H2, but the local unemployed think that Syrian asylum seekers have a similarity in their skills, but that they are not worried about this problem as much as they should.

TABLE 19.5 Codes on Solution Suggestions.

THEME-5: Codes on solution suggestions			
Sub-theme 1: sectoral solutions for pre- and post-migration recommendations	**Sub-theme 2: solutions for the employment of Syrian asylum seekers**	**Sub-theme 3: solutions for working conditions and competencies of Syrian refugees**	**Sub-theme 4: Yerli solutions for for public working conditions**
Import preventive, production increasing measures	Reducing the economic burden of local employment	Increase employee quality	Prevention of informal employment
Asylum seekers' return	Improving the conditions of asylum seekers	Encouragement/ obligation of participation in language and vocational courses	Increase incentives
Structural reforms			Explanation of labor supply–demand balance
			Encouragement/ obligation of participation in vocational courses

19.7.5 FINDINGS RELATED TO PROPOSED SOLUTIONS

In the last part of the research, participants' proposal of a solution to the topic was discussed. All participants addressed the importance of new approaches to education and political institutions in solving this problem. But they also acknowledge that they have a despair over the solution, and they admit that their belief in solution is not at a high level.

Participants point out that the bottleneck in the sectors is even more complicated by the advent of Syrian asylum seekers and that it is very difficult to solve these problems without increasing production. H2, which also states that he has worked for years abroad, explains:

> (...) We need to reach the agricultural and commercial capacity enough to make the people who live in these lands. If we take the straw, even the meat, from the outside, referring to other countries, we can not even look at the university graduate Turk, not Syrians.

Negotiations H4, H6, S3, and S6 advocate that this crisis should be returned to its owner with the sentence "They will go ahead." H5 suggests that the necessary legislative arrangements can contribute to solving this problem and allow the sector to breathe a bit. H1 "It's like this. This is not a solution," he said, saying that, thinks that the industry will not be able to produce solutions to asylum seekers and local people.

Another issue is that all participants agree that reducing the economic burden on employment and reducing worker costs can work to solve the problems of both local people and Syrian asylum seekers. S1 and S6 version would gain from this method, if forced to be profitable in Turkey. H5 "These are people who can not even find bread. There is no house. They say they can sleep better in better places, think more about the job," he argues, improving the conditions of asylum seekers can speed up the solution of the employment problem.

All participants indicate that asylum seekers are engaged in low-quality jobs, competencies and technical capacities are low. H6 "I give up on account, I am not able to give anaother job other than portering people who have problems in collecting and taking out," while affirming this situation, S5, where all his colleagues approve the meticulousness and competence of his neighbors, establishes the following sentences regarding working conditions and competences:

> (…) I hired some of the Syrians who came to me; but it did not. The apprentice comes before your master. The apprentice sometimes comes and sometimes does not. The cover work is a makeup work. When someone does not like the work I do, some people talk badly. After my work goes bad. I did not see Syria in this sensitive job. They probably go abroad too. It downloads tons of merchandise, but I can not take the drawing of the hand as I see it. This requires vocational training. Our people here can not do these things, so they go into construction and field. There is no need to say "knowing Turkish." (Simple sense) There is a lot of business.

All participants declare that Syrian asylum seekers can work with them without complying with social security costs. They say that high costs make them difficult. So much so that H1 and H4 report that they can not cover their own social security costs, not just their employees. It is stated by S2 that the incentives made in this regard should expand to include domestic local people as well. S2 also emphasizes the importance of the controls on the subject by saying, "I have not seen an inspection, an inspector or something to prevent the use of the Syrians." S1 makes an assessment of the issue:

> (…) We can not tell the local people that we have to hire the Syrians. They sometimes react to us and sometimes to the Syrians simply because we think that we are running them because they are cheap. I tried to tell you in the beginning, but then I quit. When I have to choose one of the two unskilled laborers, I prefer the Syrians, of course, considering my extraction. (…) Ha (By referring to local people) our people do not like working too much or it is a separate issue. Indeed, they are giving a plaque to the local employee (in the sense of a prize)! Is this the chamber of tradesmen and artisans tell TRT, now someone needs to tell. So we do not do it. This is the issue of supply and demand. (…)

The H2, which declared earlier that it worked abroad, is still striking.

> There are heavy workers. Everywhere is risky. While I was abroad, they smelled 'smiling' and taught the most important signs, numbers. Later I joined German courses and made up for it. The Syrians in our country do not know exactly what is prohibited or dangerous than what they do not know. The other day, the carbide was exploded as a kind of chemical used to ripen the fruits. If it was vocational training, no one here would be

nosy. 3 people were hospitalized all of them are Syrian. "I think that it is imperative to stay in here to learn Turkish (…)

19.8 DISCUSSION, RESULTS, AND RECOMMENDATIONS

Syrians and the effects of migration that took place in Turkey have reached disturbing levels. This crisis deeply affecting the border region where the population and population density in Turkey is also deeply affect the country's economy naturally. It is obvious that one of the most important ways to overcome these widespread effects of the problem is the training activities. Education is an effective way of ensuring that immigrant individuals are integrated with the migrated community (Karakütük and Kavak, 2017). The depth of the neighborship relationship between the two communities is at a level that can overcome the social problems that will arise (ORSAM, 2015a).

The employment of such a population of more than a mass of refugees is a difficult task (Koç et al., 2015). However, the most effective force on this issue could be training activities. Unqualified workmanship, which can reach significant dimensions—such as the lack of qualified and exploitative workmanship—can be traced back to schooling and vocational training before the problem.

The proportion of Syrian asylum seekers who agree to work in the workplace is remarkable (Kavlak, 2011). Parallel data have also been obtained in the study carried out. It is possible to encounter findings show that Syrian asylum seekers are contributing positively to production by providing cheap labor (HKU, 2013). However, this situation has not been observed in the present study.

Working is also the right of Syrian asylum seekers (Çopur and Demirel, 2017). Syrian refugees in Turkey to participate in economic activities, it is inevitable inclusion (TEDMEM, 2016). However, most of the employees are negative about the employment of Syrian asylum seekers (Kaygısız, 2017). In this study, the related result is reached. It should be remembered that Syrian asylum seekers' own wins will cause the share of the asylum seeker state budget to fall. It should not be forgotten that this situation may also have a negative impact on rising unemployment.

Labor force can be cheaper in countries that receive immigration. The result of this study confirms this general acceptance. According to research results in terms of the quality of working life of the population who migrated to Turkey it seems to be low. Participants have also made a separate critique of the low performance. As stated in other studies, the pressures on the job of Syrian asylum seekers are affecting those with high school and below graduation degrees (i.e., the lower income group, which is already high in unemployment). (Deniz, 2014; Kirişçi, 2014; Koç et al., 2015; Paksoy et al., 2015; ORSAM, 2015b; Uzun, 2015; Sayın et al., 2016; Kaygısız, 2017). Again, according to the results of the research, it is not the case for the professionals and the university educated individuals.

As a result of the research, the following information is presented, taking advantage of the information provided by the participants: (1) establishing and developing a data bank showing the demographic status of the Syrians, continuing to record and organizing the domestic migration/distribution through needs, (2) compulsory language and job security training for all working Syrian individuals, (3) inclusion of asylum seekers in vocational training, (4) adoption of the necessity of providing Syrian individuals with their own livelihoods and the employment of Syrian individuals in local businesses is a work done in this context, (5) reduction of social security burden on SMEs, which constitutes a significant part of tax and workload.

KEYWORDS

- **demography**
- **economics**
- **education**
- **Syrian refugees**
- **migration**

REFERENCES

AFAD. *Türkiye'deki suriyeli sığınmacılar, saha araştırması sonuçları*. Başbakanlık Afet ve Acil Durum Yönetimi Başkanlığı. 2013. https://www.afad.gov.tr/upload/Node/3925/xfiles/syrian-refugees-in-turkey-2013_baski_30_12_2013_tr.pdf (accessed Dec 9, 2017).

AFAD. *Suriye'den türkiye'ye nüfus hareketleri: kardeş topraklarında misafirlik*. Ankara: Başbakanlık Afet ve Acil Durum Yönetimi Başkanlığı. 2014a. https://www.afad.gov.tr/upload/Node/3931/xfiles/webformatisuriyedenturkiyeyenufushareketleri.pdf (accessed Dec 9, 2017).

AFAD. *Türkiye'deki suriyeli kadınlar.* Ankara: Başbakanlık Afet ve Acil Durum Yönetimi Başkanlığı. 2014b. https://www.afad.gov.tr/upload/Node/3932/xfiles/turkiye_deki-suriyeli-kadinlar_-2014.pdf (accessed Dec 9, 2017).

Canbay, T.; ve Gülden, E. Göçün mali etkileri. In *Türk Göçü 2016 Seçilmiş Bildiriler I*; Yazgan, P., Tilbe, F., Eds.; Transnational Press London: London, 2016; pp 94–101.

Çopur, A.; ve Demirel, M. Suriyeli mültecilerin yükseköğrenim sorunu: göç süreci ve sonrası. In *Eğitim bilimlerinde yenilikler ve nitelik arayışı* içinde; Demirel, Ö., ve Dinçer, S., Eds.; Pegema: Ankara, 2017; 13–27.

Deniz, T. Uluslararası Göç Sorunu Perspektifinde Türkiye. *Türkiye Sosyal Araştırmalar Dergisi*, **2014,** *181*, 175–204.

ERG. *Eğitim izleme raporu: 2016–2017*; Eğitim Reformu Girişimi: Ankara, 2016.

Göç İdaresi Genel Müdürlüğü. *2016 Türkiye göç raporu*; İçişleri Bakanlığı, Göç İdaresi Genel Müdürlüğü: Ankara, 2016.

Güven, E. Suriyeli Sığınmacıların Türkiye'deki Çalışma Durumları ve Türkiye Ekonomisine Etkileri. 2013. http://www.ilem.org.tr/blog/wp-content/uploads/2016/07/Bilgi-Notu-Esma-Gu%CC%88ven.pdf (accessed Dec 9, 2017).

HKU (Aralık, 2013). *Suriye mültecileri sorununun olası ekonomik etkileri ve tedbir stratejileri*; Hasan Kalyoncu Üniversitesi: Gaziantep, 2013.

Karakütük, K.; ve Kavak, Y. Göç Olgusu ve Eğitim Planlaması. 2017. https://pictes.meb.gov.tr/meb_iys_dosyalar/2017_05/22222022_GYY_OLGUSU_VE_EYYTYM_PLANLAMASI_KARAKYTYK_KAVAK_22.05.2017.pdf (accessed Dec 9, 2017).

Kaygısız, İ. (Ağustos, 2017). Suriyeli mültecilerin türk işgücü piyasasına etkileri. *Dünyadan.* http://www.fes-tuerkei.org/media/pdf/D%C3%BCnyadan/2017/Du308nyadan%20-%20Suriyeli%20Mu308ltecilerin%20Tu308rkiye%20I307s327gu308cu308%20Piyasasina%20Etkileri%20.pdf (accessed Dec 9, 2017).

Khalaf, S.; ve Ilgar, R. Suriyeli Mülteciler ve Türkiye'de Mülteci Sorunu. *Türk Sosyal Bilimler Araştırmaları Dergisi* **2017,** *2* (2), 40–52.

Kocabaş, İ.; Aladağ, S.; ve Yavuzalp, N. Eğitim sistemimizdeki okullaşma oranlarının analizi. XIII. Ulusal Eğitim Bilimleri Kurultayı, 6–9 Temmuz 2004, Malatya, 2004.

Koç, M.; Görücü, İ.; ve Akbıyık, N. Suriyeli Sığınmacılar ve İstihdam Problemleri. *Birey ve Toplum* **2015,** *5* (9), 63–93.

KTO. *Suriyelilerin türk işgücü piyasasına entegrasyonu sorunlar-öneriler*; Konya Ticaret Odası: Konya, 2016.

ORSAM. *Suriyeli sığınmacıların türkiye'ye etkileri* (Rapor No.: 195); Ortadoğu Stratejik Araştırmalar Merkezi: Ankara, 2015a.

ORSAM. *Suriyeli mültecilerin ekonomik etkileri: sentetik bir modelleme* (Rapor No.: 196); Ortadoğu Stratejik Araştırmalar Merkezi: Ankara, 2015b.

Paksoy, H. M.; Koçarslan, H.; Kılınç, E.; ve Tunç, A. Suriyelilerin Ekonomik Etkisi: Kilis İli Örneği. *Birey ve Toplum* **2015,** *5* (9), 143–173.

Reçber, K.; ve Ayhan, V. Türkiye İle Suriye Arasındaki Krizin Hatay Bölgesi Üzerindeki Etkileri. *Alternatif Politika* **2013,** *5* (3), 324–340.

Sayın, Y.; Usanmaz, A.; ve Aslangiri, F. Uluslararası Göç Olgusu ve Yol Açtığı Etkiler: Suriye Göçü Örneği. *KMÜ Sosyal ve Ekonomik Araştırmalar Dergisi* **2016,** *18* (31), 1–13.

Seydi, A. R. Türkiye'nin Suriyeli Sığınmacıların Eğitim Sorununun Çözümüne Yönelik İzlediği Politikalar. *SDÜ Fen Edebiyat Fakültesi Sosyal Bilimler Dergisi* **2014,** *31*, 267–305.

TEDMEM. *2016 eğitim değerlendirme raporu* (Değerlendirme Dizisi: 3). Türk Eğitim Derneği Yayınları: Ankara, 2016.

Tunç A. Ş. Mülteci Davranışı ve Toplumsal Etkileri: Türkiye'deki Suriyelilere İlişkin Bir Değerlendirme. *Tesam Akademi Dergisi* **2015,** *2* (2), 29–63.

TÜİK. *Adrese dayalı nüfus kayıt sistemi sonuçları, 2016* (Sayı: 24638). Nüfus ve Vatandaşlık İşleri Genel Müdürlüğü, Türkiye İstatistik Kurumu: Ankara, 2017.

Türüt, F. Suriyeli göçmenlerin toplumsal yapı üzerindeki etkileri. Yüksek tezi, Karabük Üniversitesi, Karabük, 2015.

Uzun, A. Günümüzün Sosyal ve Ekonomik Sorunu Olan Suriyelilerin Mülteci ve Ekonomi Hukuku Bakımından Değerlendirilmesi. *Ankara Barosu Dergisi* **2015,** *1*, 105–120.

CHAPTER 20

The World and Turkey in 2030: The Importance of Seeing the Future Today

NİYAZİ ERDOĞAN*

International Trade Department, Faculty of Economics and Administrative Sciences, Ufuk University, Ankara, Turkey

**E-mail: drnierdogan@gmail.com*

ABSTRACT

Economic, political, and geopolitical conditions as well as people or strategic organizations who can add speculative jurisdiction to imagination, write about what kind of world awaits us in the 2030s, based on economic and political tendencies that clearly show them day to day in the past. Experts have been trying to make a definite chart as close as possible to our near future. In this chapter some of the futurism studies are given and they are aimed to draw attention to how societies are prepared for the future. This chapter is about future of the societies and countries from economics pointof view.

20.1 INTRODUCTION

In recent years, the globalizing world has evidently expressed the need to anticipate the changing process of interaction; the long-term future of the societies and countries, "futurism" has become almost fashionable. Economic, political, and geopolitical conditions as well as people or strategic organizations who can add speculative jurisdiction to imagination, write about what kind of world awaits us in the 2030s, based on economic and political tendencies that clearly show them day to day in the past. Experts have been trying to make a definite chart as close as possible to our near future. I have also compiled from various sources; my viewpoint is that the common point of the world 2030s, Turkey will show considerable power in the future. Some of these studies are given below and they are aimed to draw attention to how societies are prepared for the future.

20.2 WORLD AND TURKEY IN 2030

I, along with a large number of senior managers of Turkey Chambers and Stock Exchanges Union's (TOOB) Foreign Economic Relations Board (DEIK), participated in a discussion "World and Turkey in 2030s" by Bahçeşehir Economic and Social Research Director Professor Dr. Seyfettin Gürsel, held on February 11, 2010, in which following opinions were voiced:

- It should be said that we are heading toward a world of great differences, complexities, and imbalances. Such a trend should be expected to cause conflict and instability. The key economic trends show that there will be substantial relative price changes between goods and services prices and natural resource prices over the next two decades.
- Over the next 20 years, China, India, and more than 10 countries with population of 40 million will have grown rapidly and will become increasingly similar to the standard consumer society. This analogy means that a population exceeding 4 billion has been approaching the consumption level of today's developed rich societies. Such an approach will create tremendous pressure on energy resources, so natural resource rents will be scattered too much. The relative price revolution can have very different effects from country to country. As a result, this complexity will lead to large technological progresses (the United States and followers),

1. natural resource leaseholders (oil),
2. those who produce competitive goods (rising surplus),
3. consumer goods producers (rising current account deficit), and
4. marginalized (those who will not be mentioned in the world).

In the coming years, two fundamental economic dynamics will play a particular role in the restructuring of the world. The first is the relative price revolution, while the second is the population movements. Over the last 10 years, energy and commodity prices have increased relatively rapidly compared to commodity prices. The prices of industrial goods are far below the average rate of inflation, and a few units of goods are needed to buy one unit of energy. The cheapness of industry is due to the fact that cheap labor trapped in agriculture in China, especially in Asia, is introduced and technological progress, especially the increase in productivity provided by the information revolution, is easily reflected to the producers of goods.

In the 2030s, increased energy and commodity relative price, domestic demand, and real exchange rate will be overcome with the current surplus enrichment. Financial reserves reaching huge amounts of money will be used for natural resource supply security.

The Rant owners will obtain 35 trillion USD of oil and an estimated 30–40 trillion USD rent in 20 years. Most of annuity will be taken by the Gulf countries, Iraq, Iran, Russia, and Central Asia.

By the way, there will be problems due to tensions such as the rantians' excessive reserve accumulation, the preservation of the value of these reserves, supply constraint (multiple equilibriums), energy supply and energy security power unity, and cooperation of Russia–EU, West–Iran, Russia–China.

Two possible scenarios for Turkey in 2030 are as follows:

Scenario 1: Growth will be 4% (potential), gross domestic product (GDP) US$ 1.5 trillion, and per capita income of $18,000. The conditions are:

- exports weighted growth;
- tax, workforce, education, and public spending reforms; and
- a current account deficit that can be financed under control despite price problems.

Scenario 2: Growth 2%, GDP US$ 1.1 trillion, and per capita income US$ 13,000. In this case, Turkey will be a place struggling with social issues.

The opportunity in Turkey's hands is to enrich by using as leverage the rich natural annuity neighbors. In addition, Turkey's attitude on issues such as European Union (EU) membership and pluralistic democratic regime will determine the future.

In the "Global Trends 2030" report prepared by the participation of some 40 experts with the initiative of the European Parliament in the beginning of May 2012, it is expected that Ankara will become a rising power, while the most likely countries to play a more effective role in 2030 ranked as Indonesia, South Africa, and Turkey. In the report, experts' scenario for 2030s is that:

- in 2030, the world will be multipolar and no country will be able to hegemony alone,

- a more democratic Arab world could emerge over the next 20 years,
- the United States of America and China will be the most powerful actors,
- as India continues to grow, Russia and Japan will lose their "great power" status in the 21st century,
- China, which will increase its global economic share to 19% in 2030, will leave the United States behind and become the world's largest economy,
- the United States will probably continue to be the largest military power position,
- medium-sized forces will play a critical role in international relations in the coming years, primarily at the regional level and increasingly at the global level,
- these are the beginning of Indonesia, South Africa, and Turkey; it will strengthen the democracy of these countries, where strong indications show that they will deepen social and economic modernity,
- democratic medium-sized countries such as Indonesia and Turkey's "soft power" to regional and even global level will grow.

Also, the report states that,

- the future of the EU as an international actor is balanced: to find solutions to the current debt crisis, to increase the speed and quality of economic growth, and to solve the aging population problem, the Union will strengthen the international profile and increase its influence in the world.
- Failure in resolving the economic crisis and declining living standards will lead to concentration of the right and left populist movements and possibly result in an introverted and defensive Europe.
- As globalization reaches a new and more competitive level by avoiding the migration of immigrant incursions to the safety axis of the debate in this area, to attract qualified migrants and to find solutions to the threats stemming from differences in society is critical to the welfare of the EU; in addition, preserving the economic growth and social model so that the EU can protect its soft power, rising populism, xenophobia, and anti-immigrant struggle will be determinant.

By 2030, the Balkan countries are likely to have entered the EU; otherwise there will be many problems and border disputes in the region, especially in Bosnia and Herzegovina.

- Cyprus can be alleviated with the division of rapprochement between Turkey and the EU, this situation might contribute to the peaceful solution of the existing problems.
- The thesis of conflict between civilizations will lose its popularity in the next 20 years with democratization and the widespread use of the internet
- In the next 20 years, social development that weakens nationalism and patriarchy by values of human rights and rule of law will lead to a more democratic Arab world.
- Turkey's global role is connected with the ability to manage the interests of international groups and partners; Turkey's membership in international structures such as North Atlantic Treaty Organization (NATO) and the Organization of Islamic Cooperation is likely to strengthen the role of international.

In the book titled "Bountiful Years: Why Germany Wants Brilliant Years," which includes predictions of world economies in 2030 written by Bert Rürup and Dirk Heillman, one of Germany's most important economists said that:

- Turkey and Poland would be one of the two most important bridges between Asia and Europe.
- After 20 years, the United States will create the world's largest GDP with $21.2 trillion, Japan will be number three, India will be number four, Germany, the world's fourth in 2012, will fall to sixth place after Brazil
- In 2030, Germany will represent Europe alone in the riches club; in the case of the reestablishment of the group of seven, the four Western countries will be left to the BRIC countries, industrialized countries are also in place.
- The economic future of Europe is full of uncertainties. Because of the failure of the debt crisis and the disintegration of the euro area, the economic power of Europe will take a heavy blow; the recovery will take decades and will adversely affect the world economy.
- Africa, which grew by 5% to 7% per annum, should prevent population explosion. The development of these countries owes to the richness of the raw materials; these rich raw materials can also lead to malpractice and conflicts (economists call this "curse of resources").
- South Africa and Indonesia will be the new production centers of the world.

Turkey's foreign trade with the member countries of the Islamic Conference 10 years ago was 10%, while the share now reaches 30%.

- The number one trading partner in the area of Turkey traditionally is Germany and how it changed in recent years, Russia has closed the gap by coming from behind.
- In Turkey, where the door waiting for years on the EU's membership, increasing the strengthening of the predictions that no need for EU among the people, this situation is reflected in Turkey's Middle East policy, in mid-2011, problems experienced by Turkey with its most important partner, Israel, lead to its rapprochement with the new democracies in the north of Africa.
- Major differences in the quality of education in Turkey and girls in front of the continuity in the development of the country on the issue of education is the biggest obstacle; despite all things, the growth rate between 2021 and 2030 will be around 4.2%.

In the "Turkey Economy: Pearl of the Bosphorus" report dated June 2012 that was prepared for 400th years of the diplomatic relations between Turkey and the Netherlands by Rob Rühl who is the manager of the Department of Economics ING Group in Amsterdam stated that:

- Turkey's economy will exceed Netherlands in 2014;
- the reforms carried out after the 2001 crisis showed that the very young population of 5 million and their strategic position was influential;
- Turkey will grow by an average of over 5% between the years 2011 and 2020;
- thanks to the relative growth advantages, the Turkish economy will be bigger than Spain in 2030;
- the Turkish economy will come in fifth place in Europe after Italy;
- in the next 10 years, the share of agriculture, livestock, fisheries, and mining in GDP will decline;
- there will be a strong performance in the construction sector in the same period;
- intermediate goods production in the manufacturing industry will leave behind other industrial sectors;
- the most rapid growth in the category of consumer goods will be in the production of durable consumer goods;
- Turkey will continue its way in becoming one of Europe's major automotive production locations;

- the prosperity of economic growth tends to shift toward more middle-aged consumers;
- this situation is also caused by changes in consumption structures by increasing private consumption;
- the number of middle-aged adults between the ages of 40 and 60, which is the group with the highest purchasing power among age groups, will increase by 19.4 billion as of 2011, which will lead to the increase in services, consumption goods, health, and luxury consumption;
- it is still necessary for Turkey to continue to develop its infrastructure in the area which is open to development;
- Turkey, which has the rating of 71 in the World Bank's ease of doing business among 181 countries, will increase its rating;
- in general, more credibility should be established in the judicial system; and
- the need for commitment to external financing needs to be addressed.

PricewaterhouseCoopers (PwC), which published a report in May 2012, "In 2041, Turkey: Looking to the Future" on Turkey's economic outlook for the year 2041 stated the following:

- the increasing number of qualified labor force;
- the growth of rich natural resources in the food and beverage sector and the emerging export markets;
- direct investment can help to develop a growing agricultural R&D sector;
- there will be an increase in the value chain within the next 30 years; foreign investors may want to take advantage of Turkey's rich renewable energy resources to be effective if there is a transition toward alternative energy supported, as Turkey has the potential to turn into an international center of expertise;
- Turkey's labor market has a potential to create a robust growth momentum for the country and in 2040 is expected to reach 90 million. Compared to today, this population will be younger and more educated. A well-structured university system can help educate future generations for even higher quality jobs;
- with the positive developments in the labor market, the high value-added sectors will support the movement; in this case, until the year 2041 Turkey will create the potential for the emergence of the international center of expertise in some nations;

- the number of foreign companies investing in the Turkish food and beverage sector has increased from 257 in 2005 to 421 in 2009, while in the food and beverage sector, foreign investors will play an important role in supporting the growth; and
- Turkey by using the geographical proximity to Europe, Central Asia, and the Middle East will serve the needs of this market. The Aegean, Black Sea, and Central Anatolia regions are very suitable for the production of organic foods for health conscious consumers in Europe.

In 2030s, there will also be research on food products for diabetes and other related diseases. Frost and Sullivan carried out throughout the world "Diabetes Drugs and Devices Markets titled" research (Milliyet, 16 January, 2012), in the number of diabetic patients, including in Turkey, which is a rapid increase, with $41 billion in 2009, the global diabetes drugs and devices market growth of 12% per annum due to increased demand. According to this research results:

- There are 5 million diabetic patients in Turkey. Obesity-induced diabetes is showing a rapid rise and there is 7% prevalence of diabetes in general and is particularly common in the female population.
- Parallel to the development of diabetes medicines and devices, diagnosis and treatment of diabetes mellitus will grow in the world. By 2030, the number of people with diabetes worldwide will reach 438 million. The solution of the disease will cause fluctuations in the special foods, test equipment, and the drug delivery market which will grow in the market.
- Products that increase the quality of life in the coming period will gain importance.

According to the Times of India, a study conducted by the Public Health School of Tulane University in the United States examined the rising obesity rates between 2005 and 2030. Researcher Tanika Kelly claims that in 2005, 23.2% of the population was overweight and 9.8% was in the obese class. By 2030, the number of obese people would be 1.12 billion and the number of overweight people would be 2 billion.

Scientists state that in 2030, 58% of the world's population will get obesity due to the increasingly sedentary lifestyle in cities and unnecessary food consumption. This suggests that food producers should attach importance to preparing for the future, developing products that will not cause disease, and fighting against obesity, which is the world's first, especially in the healthcare sector.

20.3 CONCLUSION

The butterfly effect is a term used by Edward Lorenz in his 1963 study, which mentions that the seagull flapping of a wing will change the weather forever. At subsequent conferences, Lorenz changed the seagull into a more romantic "butterfly," according to the theory he put forward, "a winged butterfly in Africa will create a storm in America." To sum up, it refers to the effect that a very small change in the world can cause changes in unexpected dimensions in the form of chained events.

Today, developments that are underestimated in the global economy can cause a perfect storm with "butterfly effect." For example, improvements in the ordinary and simple political or financial markets in China, Brazil, Russia, and Turkey may cause storms on a global basis when it comes to day. Just like recent Southeast Asian and Latin American crises. Throughout the history, the world has been experiencing many economic, political, and humanity crises. The most beautiful scenario we want for our future is that all the countries and people in the world live in peace, abundance, and prosperity.

KEYWORDS

- **2030s**
- **2040s**
- **future**
- **Turkey**
- **world**

REFERENCES

EU Parliament. *Global Trends 2030*; May, 2012.

PwC Türkiye. *2041'de Türkiye: Geleceğe Bakış*; 2012. https://www.pwc.com.tr/tr/publications/arastirmalar/pdf/2041deturkiye.pdf. (accessed Feb 2, 2017).

Rühl, R. *Turkey an Economic Pearl on Bosphorus*; 2012. http://www.hollandturkey.com/2012/pdf/Turkey-an-economic-pearl-on-the-Bosphorus.pdf. accessed date 15.01.2017
Rürup, B.; Heillman, D. *Fette Jahre*; Hansen Verlag: München, 2012.
Türkiye Odalar ve Borsalar Birliği (TOBB). *2030'lu Yıllarda Dünya ve Türkiye*; Türkiye Odalar ve Borsalar Birliği (TOBB) Dış Ekonomik İlişkiler Konseyi (DEİK), 2010.
http://www.gazeteler.org/milliyet-gazetesi/2012-ocak-16/.
http://www.timesofindia.com.

PART 3

Third World Economies

CHAPTER 21

A Discussion on Corruption and Political Risks in Latin American Countries[1]: A General View Using the Chaos Theory[2]

JOSÉ ANTÓNIO FILIPE* and MANUEL ALBERTO M. FERREIRA

Lisbon University Institute ISCTE-IUL, BRU-IUL, ISTAR-IUL, Lisboa, Portugal

**Corresponding author. E-mail: jose.filipe@iscte-iul.pt*

ABSTRACT

Along this chapter, a general view over political risks and corruption is given, analyzing the way these phenomena distort the economies. Chaos Theory is used to show social instability and disorganization effects in economies development. The consequences in developing countries socio-economic structures are complex as much as there are dynamic interactions with a complex effects net in systems. This issue is studied having by reference Latin American countries, where corruption is easily understood as a phenomenon that much distorts these countries development. Historically, in general, these countries have great potential to make business and to receive foreign investments, being potentially attractive to companies. Anyway political risks and corruption are seen as serious obstacles when investments are studied in terms of business attractiveness. In recent years, corruption is perceived as being increasing, creating a brake to development and to stability on these socioeconomic systems. In literature, chaos theory has

[1]Latin America is a set of sovereign states and several territories and dependencies which cover an area located from the northern border of Mexico to the southern tip of South America, including some Caribbean territories, comprising nearly 13% of the Earth's total land surface area.

[2]In Filipe, Ferreira and Coelho (2012) a study was made involving corruption and political risks. In Ferreira and Filipe (2012), Filipe and Ferreira (2013a,b) and Ferreira et al. (2014) the "drop of honey effect" in chaos theory was presented. Along this work is intended to join both subjects and integrate them in a unified perspective of analysis, considering the problem of corruption and political risks as phenomena explained by dynamical systems theories, particularly involving chaos.

been applied recently also, particularly, in the politics context. A reflection on chaos theory in dynamical systems is made to understanding political issues and corruption. Some events in the political context in Latin American countries are considered. Macro-strategic ideas of states positioning in the international stage are also taken in account. Following this argumentation the "Drop of Honey Effect" metaphor is applied, somewhat parallel to that of "Butterfly Effect." "Drop of Honey Effect" is considered to be more suited to portray social phenomena and in this particular case to political phenomena, than the "Butterfly Effect".

21.1 INTRODUCTION

Corruption is a huge cause of wealth and resources distribution distortion and of social justice unfairness. To combat this problem, mainly ethical and legal issues must be considered. Intervention mechanisms are crucial to detect and even to avoid, for example, situations in which people are positioned to feel unpunishable to improperly either appropriate or influence ownership of third parties resources or to influence other people by illegal means.

Chaos theory outlines are very interesting in explaining possible disturbs in structure, considering inherently disorder and instability in systems. A possible, and with enormous negative consequences due to their importance in structures administration, appearance of structural instability sources may be associated to governments' interventions, which is so much visible in national and international contexts. Disorganization—social, economic, and even cultural—becomes a mode of expressing the chaotic and complex network of inefficient relationships in a system.

Political risks, corruption, violence, and several kinds of crimes destabilize socioeconomic structures as much as impede economic development and companies' investments. These phenomena are severe barriers to the system's stability and to moral principles that are deeply violated. Foreign investment is often kept away from these countries once international companies avoid countries where corruption undermines societies' structures and justice does not work properly.

As mentioned by Shleifer and Vishny (1993), governments' institution's structures and political processes are very important corruption determinants. Usually, when corruption is present, political institutions are undermined, democratic processes are distorted, political instability prevails... Also, social injustice and inequalities increase, due to rules perversion, usually becoming the judicial system particularly inefficient.

There are many corruption modes in modern societies. Corruption provokes strong distortions not only in economic structures but also in the pillars of democracy and in the political forms of institutions' organization. Particularly, political corruption is so powerful that in many countries completely undermines the economy by preventing any hypothesis of the economic sustainable growth and of sustainable socioeconomic development.

The illegal acts executed by political officials, constitute political corruption crimes, particularly dangerous and reprehensible, when they are directly related to their official duties, involving particularly their ability to influence others' decisions. The use of politic influence by government members implies illegitimate private gains, having as the principal motivation the greed of its perpetrators.

Types of corruption are very different in nature. Anyway, activities that constitute corruption, illegal, are differently classified from country to country and from one jurisdiction to another. The same political funding practices, for example, may be considered legal in one country and illegal in another one.

This last case of campaign contributions is an example of a situation that, even legal, may be considered a form of corruption. Often it is said that they are examples of "legalized corruption." They provoke a bias in the electoral process benefiting particular interests. This kind of practices undermines public trust in political institutions. These practices corrupt institutions even if there are individual members who are not corrupt themselves. But just because they are members of a corrupt collective, they suffer the pejorative of corrupt people. Also, similar problems of corruption may happen on organizations in which an institution's manager decisions conflict by considering his/her personal interests and the primary purposes of the institution he/she manages.

Considering these general points about corruption and having the political structures as a reference for international investment, an analysis of the situation of Latin American countries is made so as to understand the current status and prospects for a future conscientiousness of the situation. For that, chaos theory is used depicting the idea that many activities reflect dynamic forms of analysis and very complex and widespread reality, specific of complex systems, which dynamics are very hard to model (either mathematically or physically) and understand. These realities fall within a range of situations integrated in a broader context. This context is not only

[3] The Madoff investment scandal was a major case of stock and securities fraud discovered in late 2008. In December, Bernard Madoff, the former NASDAQ Chairman and founder of the Wall Street firm Bernard L. Madoff Investment Securities LLC, admitted that the wealth management arm of his business was an elaborate Ponzi scheme.

intended to be reproduced in the theory but also integrated in the complex environment of their own dynamics, with complex and often chaotic features into their essence.

International financial and economic relations have shown the weaknesses of national, international, and transnational economic systems, exposing how international finance is dependent on a set of economic networks and sometimes specific events with strong implications on the system's stability. The Madoff case[3] is a particular example. In chaos theory, this shows the "drop of honey effect" working: Madoff's personal and business asset freeze created a chain reaction throughout the world's business and philanthropic community, forcing many organizations to at least temporarily closed.

21.2 POLITICAL RISKS AND CORRUPTION

International companies' investments intentions in foreign countries require a complex analysis considering the conditions of attraction. Political conditions, involving the country's risk and the corruption levels, are determinant in companies' studies and reports. The stability of a country's government is also a determinant when an international company's study is made.

As stated in Filipe et al. (2012), the economic and social developments of developing countries depend often on commercial advantageous investments and on the reinvestment of capitals on these economies. Economies need to grow up and for that, at a large extent, they require foreign capital and foreign investments. The risk involved in the operation is determinant to an economy to be interestingly attractive for investors.

Even though often governments attempt to present policies that are investment's friendly, with management systems and facing a control of economic activities' structures. In international business high levels of corruption are common in many countries, increasing the political risk of any host country that cannot be neglected in agents' decision-making.

Due to the term "political risk" importance to the context of international investment some ideas are left in this chapter about it. The term "political risk" appears often in the international business literature and it means that there is a strong chance of unwanted consequences arising from political activity. However, the precise meaning is far from just that. Political risk is customarily seen as the (usually host) government interference in business operations.

As referred in Moura et al. (2011), many authors define political risk as the government interference with business operations (Carlson, 1969; Greene,

1974; Baglini, 1976; or Lloyd, 1976). Other analysts define political risk in terms of specific events (certain political acts, constraints imposed on firms, and a combination of both). Political risk can be regarded as any political change that alters the expected outcome and value of a given economic action by changing the probability of achieving business objectives (Price Waterhouse Coopers, 2006).

Simon (1982) definition refers to political risk as the governmental and societal actions and policies, originating either within or outside the host country, and negatively affecting either a select group of or the majority of foreign business operations and investments (Simon, 1982; Jessen, 2012). Simon (1982) definition of political risk includes political risks originating from the host-country environment, home-country environment, international environment, and the global environment. It views political risk in the general environment context, whilst differentiating between internal and external causes of political risk, and includes both the country factors as well as the industry-specific concerns (Jessen, 2012).

Political risks may arise from national governments' actions that interfere with or prevent business transactions, or change the agreements terms, or cause the confiscation of wholly or partially foreign-owned business property (Weston and Sorge, 1972). Root (1976), cited in Kobrin (1978), defines political risk in terms of the "...possible occurrence of a political event of any kind (such as war, revolution, *coup d'état*, expropriation, taxation, devaluation, exchange controls and import restrictions), at home or abroad, that can cause a loss of the potential profit and/or of the assets in an international business operation."

Political risk is related to a set of risks on diverse areas as financial and market risks, for example, or others that imply losses for companies and individuals resulting from political decisions, changes or other disruptions.

Surely, there are many factors influencing business and one of them results from a political decisions' basis. Political decisions taken by governments in areas such as taxes, currencies, trade tariffs or barriers, investment regulation, wage levels or wage legislation, labor laws, environmental regulations and development priorities, for example, affect companies and their environment and market conditions as much as their profitability conditions. Also many noneconomic factors affect businesses framework. Political disruptions (such as, for instance, terrorism, riots, coups d´état, civil wars, international wars, and even political elections...) have huge consequences in business and markets, and companies need to consider all these factors on their analysis.

So corruption is determinant in political risks' analysis. Corruption also may involve government institutions. The policies of corrupt governments, in general, potentiate the resources controlled directly by them and have a particular consequence of increasing poverty. According to Brink (2004), corruption implies that a transaction takes place between a corruptor and the corrupted.

There are many definitions for corruption. All of them highlight the pernicious nature of corruption. In consequence of the corruption multiple effects, it is important to understand the path followed by market economies in their socioeconomic development process.

Moura et al. (2011) refer political risk, in general, is much difficult to evaluate quantitatively contrasting with economic or financial variables, which measuring is often easily performed. If it is possible to calculate political risk "scores" or other quantitative-looking benchmarks, it is important to have in mind that the calculus is at last based on qualitative judgments.

An evident corollary of corruption is the markets' poorer performance and accomplishment. As Senior (2006) reminds, corruption means "decomposition; moral deterioration; use of corrupt practices (bribery, etc.); perversion (of language, etc.) in its original state." In the interesting approach to our purposes, corruption may be highlighted through some of the following ideas in terms of the definitions considered next. For example, according to Morris (2004), the term corruption is described as the illegitimate use of public power to benefit a private interest. Senior (2006) defines corruption considering five conditions that must all be satisfied simultaneously. Corruption occurs when a corruptor (1) covertly gives (2) a favor to a corrupted or to a nominee to influence (3) action(s) that (4) benefit the corruptor or a nominee, and for which the corrupted has (5) authority.

Senior (2006) makes a profound literature review on corruption. This author presents, for example, a corruption classification, made according to Heidenheimer (1989), considering the following three categories:

- "black corruption," which indicates that a particular action is such that a majority consensus of both elite and mass opinion would condemn and claim its punishment on grounds of principle;
- "grey corruption," which indicates, in a not at all consensual attitude, that some elements, usual elites, may want to see the action punished, others not, and the majority may well fall in an ambiguous position;
- "white corruption," which signifies that the majority of both elite and mass opinion, in a quite consensual attitude, probably would not vigorously support an attempt to punish a form of corruption that they regard as tolerable.

Senior (2006) also shows a list of forms that corruption can take, according to the Johnson and Sharma (2004) framework, which, in their view, encompass more than bribery, as follows:

- bribery and graft (extortion and kickbacks);
- kleptocracy (stealing and privatizing public funds);
- misappropriation (forgery, embezzlement, and misuse of public funds);
- nonperformance of duties (cronyism);
- influence-peddling (favor-brokering and conflict of interest);
- acceptance of improper gifts ("speed" money);
- protecting maladministration (cover-ups, perjury);
- abuse of power (intimidation and torture);
- manipulation of regulations (bias and election rigging);
- electoral malpractice (vote buying and election rigging);
- rent-seeking (public officials who illegally charge for services after creating an artificial shortage);
- clientelism and patronage (politicians giving material favors in exchange for citizen support);
- illegal campaign contributions (giving unregulated gifts to influence policies and regulations).

Such a list, resulting from a classification effort, with 13 items, synthetizes in a very clear way the various forms corruption can assume, giving also an idea of its potential, and also real, dimension.

As stated by Morris (2004), corruption tends to be higher in countries at lower levels of economic and human development, with lower levels of education, limited political rights, weak or nonexistent political competition, a relatively large state role in the economy, lower levels of economic freedom and openness, ethnolinguistic factionalism, the lack of judicial independence and a free press, low civil service wages, abundant natural resource endowments, low levels of interpersonal trust, and high levels of permissiveness toward corruption, among others (on the causes of corruption, see also Ades and DiTella, 1997; Brunetti and Weder, 1999; Johnston, 2000; Kaufmann and Wei, 1999; LaPort et al, 1999; Mauro, 1995, 1997; Tanzi, 1994, 1998; Triesman, 1999).

Clammer (2012) sees corruption as "a systemic problem in which the whole social systems are implicated, and which is generated in large part by dysfunctions in systems themselves." Indeed, corruption is generally viewed as a society's problem, with consistently negative consequences, that deform society's socioeconomic structures and often the judicial system in a country, undermining the investment's environment. This general setting brings intricate forms of managing the social country's system and makes a very complicated set of interests that discourage the agents' official actions inside the system. The well-known problem of Latin American countries on this subject is much studied in the literature, giving way to many reflections in the course of the past decades. Recently, this phenomenon has increased again in many Latin American countries.

21.3 DYNAMICAL SYSTEMS

An interesting and important term to the sequence of this chapter is the *dynamical system*. Let us see briefly some notions on this term, presented in a constructive way.

Begin to note that a system may be considered as a set of interacting or interdependent component parts representing a complex and intricate whole. It is defined by the concretization of its spatial and temporal boundaries, influenced by its environment, described considering its structure and its purpose and expressed in its operative mode.

Complex systems are highly composite ones, built up from very large numbers of mutually interacting subunits that are often composites themselves, producing a very intricate dynamics.

Chaotic systems do not have necessarily a lot of mutually interacting subunits. But even when very few interacting subunits have, they interact in such a way as to produce very intricate dynamics.

Following Rickles et al. (2007) "a dynamical system is a system whose state (and variables) evolve over time, doing so according to some rule. How a system evolves over time depends both on this rule and on its initial conditions—that is, the system's state at some initial time instant. Feeding this initial state into the rules generates a solution (a trajectory through phase space), which explains how the system will change over time; chaos is generated by feeding solutions back into the rule as a new initial condition. In this way, it is possible to say what state the system will be in at a particular time in the future." Therefore, the term "dynamical" is related to the evolution of the system over time which here is a crucial variable. When the system does not evolve over time it is said to be "static" or "stationary."

Inherently, those authors conclude that "complex and chaotic systems are both examples of *nonlinear dynamical* systems." Here the term "nonlinear" emphasizes the disproportionality of the response of the system when compared to the very pivotal variation of the initial conditions that gives rise to it.

21.4 SOME GENERALITIES ON CHAOS THEORY

Quite important in the analysis intended to be performed here some concepts, generally accepted in chaos theory, are now introduced.

Begin with this phrase: "the hidden orderly patterns in chaotic behavior can be presented in the so-called phase space," that is an abstract mathematical space. It is a set of structured points, representing each one a complete and detailed state which the analyzed system can eventually assume. Thus, the larger the dimension—number of coordinates—of the phase space, the better will be the description of a particular state reached by the system (I Font and Régis, 2006). But, the larger the dimension, the greater the difficulty of the interpretation, in terms of real world of the state meaning.

A trajectory in its phase space portrays the evolution of any particular system. The existence of a trajectory assumes the idea of the existence of an attractor, because any trajectory of a system running on the long-term is somehow "attracted" by some points or some closed, in sets 'language sense, regions within the phase space describing the system in question.

Generally, a chaotic behavior is characterized by its extreme sensitivity to the initial conditions, as stated already in the former section. This sensitivity means, for example, that a very small perturbation of the system in an initial condition may lead it to an exponential type divergent final state. The trajectories may behave in a very different way in neighboring points, approaching and moving away one from the other in a really unpredictable way: consider for illustration the metaphors either of the "butterfly effect" or of the "drop of honey effect."

Also important to refer the bifurcation points, which permanently challenge the trajectory of the system, positioned where the system sensitivity to the initial conditions is stronger. Up to this kind of moments, the trajectory of the system might behave in a quite predictable pattern, but once reached this bifurcation point, the prior order breaks out and the system is driven by patterns of behavior less predictable than ever before. That is: with nonlinear dynamic systems, the bifurcation implies a change in the system's behavior when it is changing from one attractor to a new one (I Font and Régis, 2006).

Phenomena happen over time as at discrete, separate or distinct, intervals[4] or as continuously[5] (Williams, 1997). Discrete intervals can be spaced

[4]Examples are the occurrence of earthquakes, rainstorms, or volcanic eruptions.

[5]Examples are air temperature and humidity or the flow of water in perennial rivers.

evenly in time or irregularly in time. Continuous phenomena might be measured continuously.

Differential equations are often the most accurate mathematical way to describe a smooth continuous evolution. However, some of these equations, although it is known they have solutions, no one achieved to determine them. In contrast, although the tools adequate to solve difference equations are much simpler than the adequate to differential equations, with the exception of linear difference equations and some other exceptional cases, it is much more difficult to achieve analytical solutions. In opposition, it is in general possible to obtain numerical solutions using computational recurrence processes. The problem with these kinds of solutions is that they do not allow us to determine temporal behaviors in a general form. Furthermore, they are often acceptable approximations of differential equations that allow us to obtain approximate numerical solutions for them. Olsen and Degn (1985) state that difference equations are the most powerful vehicle to the understanding of chaos. But note that either linear differential equations or linear difference equations are the most uninteresting from the point of view of chaos theory (see the nonlinearity discussion above).

It follows an example got from Ferreira et al. (2014) for politics, considering the political credibility, modeled through a difference linear equation. Call x_t the political credibility, of a politician or of a party measured, for instance, in the number of votes, or in the number of chamber's members, or even in money, in the year t; and consider b the political credibility increasing rate, $b \geq -1$. It is admissible that in the year $t + 1$, $x_{t+1} = x_t + bx_t$ that is: in a certain year the political credibility is the one of the former year plus, or minus, a part of it. Solving this difference eq 21.1 (see Ferreira and Menezes, 1992) it is obtained[6]:

$$x_{(t)} = x_0 (1 + b)^t,\ b \neq 0 \text{ and } x_{(t)} = x_0,\ \mathrm{b} = 0 \tag{21.1}$$

Then, according to this model, if the political credibility increasing rate is null the political credibility is kept unchanged, assuming always the initial value. If $b > 0$, the political credibility follows an increasing exponential path. If $-1 < b < 0$, the political credibility follows a decreasing exponential path converging to 0. Finally, if $b = -1$, x_t is permanently null. Evidently, values such as $-1 \leq b < 0$ define political credibility paths that may lead to people's chaotic behaviors if, for instance, they mean either an important politician or an important party loss of political credibility. To detect this

[6]Evidently, this is the compound interest capitalization formula, at interest rate b used for financial purposes.

chaotic behavior, it must be defined a new model, constructed over the just presented, translating the idea that the social disturbances are greater the minor is the politicians' credibility. And this is really a challenging question.

Indeed, chaos is extremely complex and difficult to be identified in the real world, using the workable information. But it is sometimes possible to find specific mathematical relationships for problems to be solved. Indeed, the introduction of the idea of nonlinearity into theoretical models allowed the existence of chaos analysis through the models possible. But in any event, it is not possible just grab a nice little set of data, apply a simple test or two, and declare "chaos" or "no chaos" (Williams, 1997).

Chaos occurs in deterministic, nonlinear, dynamical systems and it is something extremely sensitive to the initial conditions. The sensitive dependence on initial conditions shows how a very small change at either a place or a moment in a nonlinear system can result in quite large differences to a later state in the system.

The deterministic chaos, occurring in some nonlinear systems, in general, imposes limitations on the human ability to predict behaviors. But, the exploration of a big number of alternative conditions by a single deterministic result may create the possibility of a prospective outcome in terms of adaptation and evolution. In the environment of artificial life models, this has led to the notion of "life at the edge of chaos" expressing the principle that a delicate possible balance between chaos and order is optimal for successful evolution (Campbell and Mayer-Kress, 1997).

The "drop of honey effect" is proposed in this work, after its presentation in Ferreira and Filipe (2012), Filipe and Ferreira (2013a,b) and Ferreira et al. (2014) to show how punctual events, some of them apparently with no significant importance, have huge consequences either in economic or social or political terms. This effect was constructed from the wonderful tale written by the Armenian poet Tumanian (1869–1923). The "drop of honey effect" may be used instead of the term "butterfly effect" when dealing with social events. The term "butterfly effect" seems more adequate when dealing with physical and natural events, whereas the "drop of honey effect" seems much more adequate and accurate to evidence chaos in social and political situations.

21.5 POLITICAL RISKS IN LATIN AMERICA

At the beginning of the 1990s, many Latin American countries tried to follow a path toward democracy. The transition has driven to the establishment of new political institutions, which has posed challenges for (1) determining the concept of the state and (2) the way in which these institutions should safeguard democracy (Zambrano, 2012).

As of Zambrano (2012) reports, in some cases, the transition to democracy stimulated the development of successful economies with effective political and social control systems. This led to greater governability and increased transparency, being the region's most prominent example Chile. Overthrowing dictatorship began an intense process of economic development, which coupled with good investment decisions, opened up spaces for transparency and social control.

In many other situations, however, institutions were not enough strengthened and corruption went on at all state levels, being the main impediment to reach equality and the development of counties. Nevertheless, in general, Latin America got interesting results and some important achievements were got as:

- income distribution improvement,
- poverty decrease,
- home ownership increase,
- higher literacy, and
- a more stabilized democracy.

As Schumacher and Giovingo (2012) report, the United Nations Economic Commission for Latin America and the Caribbean (ECLAC) show that global Foreign Direct Investment (FDI) grew only 1% in 2010, but FDI in South America was up 56% to over $85 billion. As has been the case in the past, United States was the leading source of FDI in the region followed by the Netherlands, China, Canada, Spain, and the United Kingdom. These figures suggest that the global business community sees attractive opportunities in many countries across the region.

However, companies can generally expect to encounter certain risks in South America, varying from country to country (Schumacher and Giovingo, 2012), as for example:

- Local government officials tend to be paid lower wages, and there may be expectation wages are to be supplemented with "payments" either from individuals or from companies needing their services,
- The attitude toward bribery is generally highly relaxed, and may even be viewed as an accepted part of doing business,
- Local laws are complex and often make it difficult to prove and prosecute bribery,

- Customer relationships built on trust may result in a general reluctance to memorialize business agreements or understandings into formal written contracts.

Yet according to Schumacher and Giovingo (2012), as companies find potential risks within specific countries, they also may find unique interesting challenges. The following realities are usually considered:

- Chile was rated as having "little or no enforcement" of anticorruption laws in Transparency International's 2011 Progress Report,
- In Argentina, the government has an old habitude of establishing short- and long-term trade barriers such as quotas, additional import/ export fees, or licensing and registration requirements to protect local industries,
- In a Transparency International study, Argentina was cited as having certain inadequacies in its legal framework to combat bribery and corruption, including no criminal liability for corporations, inadequate sanctions, and inadequacies in statutes of limitations,
- In Peru, the government is encouraging state-owned energy companies to play a greater role in the economy, similar to Chile and Brazil, so companies in this industry may be dealing with more government entities and officials going forward,
- In Venezuela, nationalization of key industries has resulted in a sharp drop in FDI with increased government interactions for companies continuing to conduct business there,
- Venezuela has the highest level of perceived corruption among South American countries based on Transparency International's 2011 Corruption Perception Index with a score of 1.9,
- Ecuador ranks 130 out of 180 in the World Bank's "Ease of Doing Business" 2012 ranking by scoring poorly on certain subindices including: starting a business, protecting investors, and getting electricity.

Accordingly, the unique business and regulatory environments in many South American countries may require companies to assess corruption risk on a country-by-country basis.

The political climate of Latin America shows the peculiar facet within the international geostrategic relationships. Indeed Latin America incurs serious risks. The international framework is getting a very significant and complex stage in terms of the international relations for Latin American countries, as much as the new dimensions for the international economy, and allows showing the evidence of complex vulnerabilities for which the solutions seem very difficult in the present context (see Filipe et al., 2012).

21.6 CORRUPTION IN LATIN AMERICA

Corruption got deteriorated along the last years. Comparing 2011 Transparency International Corruption index metrics with those of a decade earlier, it can be seen a surprising (or maybe not…) trend of increased corruption across a Latin America great part. Among a wide range of economic, political and cultural factors influencing a nation's level of corruption, many may be found in Latin America (see Morris, 2004). In Brazil, for example, corruption [and human trafficking] has increased greatly over the past few years owing mainly to the high rates of poverty, income differentials, illiteracy, gendered cultural practices, discrimination, and homelessness. All of which have been described by Bales (2005) as critical "push" factors (MPDFT—Ministério Público do Distrito Federal e Territórios, 2010).

For investors, Latin America represents a perplexing dichotomy. On one hand, strong resource prices, an expanding middle class, and rising discretionary consumption present tantalizing opportunities to global companies. But fortune-hunting businesses must contend with increasing corruption, and not just when trying to sell goods and services to governments. Wherever government oversight is needed—to obtain a permit, to import a component, to inspect a facility, to police a district—the specter of corruption is near (Price, 2013).

Organized crime is also an important pillar of corruption in Latin America. While the Colombian, Peruvian, and even Bolivian governments managed to pulverize the power of cocaine cartels, those managing distribution channels for the drug have become the enemies of the state in other countries. In Central America, Mexico, Venezuela, Trinidad and Tobago, Jamaica, Bahamas, the Dominican Republic, Paraguay, Argentina, and Brazil, the organized crime has grown and became more consolidated, fueled by the incredible profits of drug trafficking and distribution. Per capita cocaine consumption in Buenos Aires is now believed to exceed that of most US cities. In Jamaica, the two leading political parties are both accused of taking contributions from rival street gangs who control the nation's drug trade. Venezuela's military leaders are accused by some of colluding with Colombian drug producers (Price, 2013).

Páchon (2010) categorizes, citing Casal and Associates (2004), that there is a set of incentives/weaknesses that encourage corruption, as follows:

- Civil service in the transition toward professionalization and meritocracy.
- Deliberated confused and often redundant legal systems.
- Extremely complex and bureaucratic procedures and regulations.
- Oversight and control weaknesses.
- Inefficient control on the part of the legislative branches.
- Dysfunctional judicial systems.
- Social control weaknesses.
- Limited awareness of the issue, in conjunction with attitudes and opinions that do not reflect the extent of the problem.
- Insufficient political will.

Price (2013) defends that the greatest weapon against corruption, therefore, is austerity. Latin America became a more transparent region in the 1990s (versus the 1970s and 1980s) because low resource prices and aggressive debt servicing obligated governments to manage costs. By privatizing government-owned resources, infrastructure, and manufacturing firms, many of the vehicles of corruption were taken away from their puppet masters. After a decade of prosperity, some Latin American leaders have deluded themselves into thinking that they can manage energy companies, mining operations, steel production, electricity distribution, ports, and highway tolls. The reversal of Latin America's privatization trend is worrying on two fronts—it threatens the region's competitive standing and provides new channels for government corruption.

In fact, corruption is growing considerably again in Latin American countries. As stated by Price (2013), the data presented by Periodistas Frente a la Corrupción (PFC), which publishes a report every year highlighting a considerable number of corruption cases that have actually been indicted, show a set of examples, being some of them pointed out as follows:

- an unfinished highway in Ecuador costs taxpayers \$106 million when the winning quoted \$36 million;
- in Guatemala, the Ministry of Communications awarded \$27 million to fictitious suppliers linked to government officials;

- in Venezuela, the massive public housing program paid close to $800 million to phantom companies as well as builders who never broke ground.

These are just a handful of indicted cases, a drop in the veritable bucket of existent corruption, estimated by some to be anywhere from 5% to 10% of overall government spending and 10% to 30% of infrastructure spending in Latin America (Price, 2013).

There is a trend that shows that the situation on corruption, in general, is getting worse (figures in 2010 already showed this tendency—see Galán Páchon, 2010).

More recently, for instance, in Brazil, the "Lava Jato" operation (car wash operation) is a set of ongoing investigations by the Federal Police of Brazil, involving for example more than 1000 search warrants and seizure, temporary arrest, pretrial detention, aiming to establish a money laundering scheme that moved from 10 to 20 billion *reais* in bribes. Started on March 17, 2014, the operation counted until February 2016 with 38 operating phases, during which more than 100 people had been arrested and convicted.

This operation investigates crimes of active and passive corruption, mismanagement, money laundering, criminal organization, obstruction of Justice, fraudulent exchange operation, and receiving the undue advantage. According to investigations and contributions received by the task force of *Lava Jato* Operation, are involved administrative members of the State oil company Petrobras, Brazil's largest political parties and politicians, including mayors and governors of states, as well as businessmen from major Brazilian companies.

There are also many suspicions on corruption in the Olympic Games in Brazil. In several phases since the attribution of the Games to Brazil until posterior phases, the corruption suspicion accompanied the Brazilian Olympic Games.

21.7 CHAOS AND THEORY APPLIED TO LATIN AMERICAN COUNTRIES

This work deals with the way societies in Latin America have to deal with corruption and political risks to manage structures and promote development.

In this chapter, several examples are given to show how American countries are undermined by corruption and significant political risks, which provoke serious obstructions to socioeconomic development in these countries. Governments are themselves a serious source of structural instability in these countries, provoking socioeconomic disintegration in administrative organizations and a deficient context in companies' structures development. Clammer (2012) refers to the "social disorganization" to express the degeneration of a functional system, what can be easily visible in many Latin American countries.

The "drop of honey effect" metaphor is perfectly adjusted to this study since here is profusely illustrated how simple facts may have a very strong effect in one country's structures and also internationally, in terms of social, political, and economic impacts. Some recent cases in Brazil (being yet some of them in court) are examples of the importance they have to the development and to the course a country and a region takes. The example of Lava Jato operation is illustrative. In June 2015, the arrest of the then-President of Odebrecht, Marcelo Odebrecht, was evaluated by analysts as a serious blow to one of the largest Brazilian multinational, which became one of the icons of Brazil international projection. This situation has resulted in a very serious problem for this multinational company, which had then tremendous implications in Brazilian economic activity and unemployment. Furthermore, a set of individual events has had a recently tremendous impact in the Brazilian economic situation and in its international image, reflected in the Brazilian poor scenario for business and international investment perspective for foreign companies.

The perspective presented in this chapter may be seen as being pessimistic. Countries in Latin America have great potential to get developed and could initiate a strong and impetuous road to progress but they are bound up, losing a set of excellent opportunities they have to develop and not taking advantage from their excellent capabilities in general and also in particular from their large natural resources availability.

Chaotic systems reflect the situation exposed in this chapter considering the situation of these countries. Also, the situation is similar considering their international context. Besides, international transactions are often invisible. However, they show a face exposed from a market conducive to obscure relations and corruption in these countries. Several examples exist in relations between companies that involve even states and multinational companies, which reflect the existence of chaotic systems and generators of unstable relationships.

21.8 FINAL REMARKS

In this chapter, political risks and corruption in Latin American countries are analyzed using chaos theory. The present situation about this topic is analyzed for these countries. The conclusion that, in general, there is a new wave of increased corruption in the Latin American countries gets obvious, understanding it in the light of chaos theory.

It is well recognized that Latin American countries, in general, potentiate an interesting perspective basis for investments, with considerable returns, as historically it is perceived that they have a great potential to develop and that making business may allow interesting results. In truth, these countries may be potentially attractive to companies to establish there but many dangers around the corner constitute obstacles to a real establishment and often companies retract. In fact, political risks are considerable and corruption is always present when, in the international scenario, business attractiveness is studied by the companies considering investment in these countries. In recent years, corruption is perceived to be increasing.

Chaos theory is adequate to analyze social disorders and their consequences. Corruption itself is the mirror of a society dealing with social organization forms propitious to develop several ways of social distortions. Better policing, change of mentalities, changes in organizational structures and cultural changes involving ethics and morality are required. The nature of social disorganization is essential to understand the emergence of phenomena as violent crimes, endemic corruption, and the nature of chaotic socioeconomic systems that are the image of several Latin American countries.

KEYWORDS

- **political risk**
- **corruption**
- **international investment**
- **Latin America**
- **chaos**
- **drop of honey effect**

REFERENCES

Ades, A.; Di Tella, R. The New Economics of Corruption: A Survey and Some New Results. *Polit. Stud.* 1997, *45*, 496–515.

Baglini, N. A. *Risk Management in International Corporations*; Risk Studies Foundation, Inc.: New York, 1976.

Bales, K. *Understanding Global Slavery;* University of California Press: Berkeley, 2005.

Bergé, P. Y.; Pomeau, C. V. *Order Within Chaos*; John Wiley: New York: 1984.

Brink, C. H. *Measuring Political Risk: Risk to Foreign Investment*; Ashgate Publishing: Aldershot, 2004.

Brunetti, A.; Weder, B. *Explaining Corruption*, University of Basel, 1999.

Campbell, D. K., Mayer-Kress, G. Chaos and Politics: Applications of Nonlinear Dynamics to Socio-political Issues. In *The Impact of Chaos on Science and Society;* Grebogi, C., Yorke, J. A., Eds.; United Nations University Press: 1997.

Carlson, S. *International Financial Decisions*; The Institute of Business Studies: Uppsala, 1969.

Clammer, J. Corruption, Development, Chaos and Social Disorganisation: Sociological Reflections on Corruption and Its Social Basis. In *Corruption: Expanding the Focus;* Barcham, M., Hindess, B., Larmour, P., Eds.; ANU Press: 2012; pp 113–132. http://www.jstor.org/stable/j.ctt24hbwc.11

Ferreira, M. A. M.; Filipe, J. A. The Drop of Honey Effect. A Note on Chaos in Economics. *Int. J. Latest Trends Finance Econ. Sci.* **2012,** *2* (4), 350–353.

Ferreira, M. A. M.; Filipe, J. A.; Coelho, M.; Pedro, I. Chaos in World Politics: a Reflection. The Drop of Honey Effect. In *Chaos Theory in Politics;* Banerjee, Santo; Erçetin, Sefica Sule; Tekin, Ali, Ed.; Springer: Netherlands, 2014; Vol. 1, pp 89–104. ISBN: 978-94-017-8690-4. DOI: 10.1007/978-94-017-8691-1_6. Series: Understanding Complex Systems.

Ferreira, M. A. M.; Menezes, R. Equações com Diferenças: Aplicações em problemas de Finanças, Economia, Sociologia e Antropologia. Sílabo. Lisboa. 1992.

Filipe, J. A.; Ferreira, M. A. M.; Coelho, M.; Moura, D. FDI Political Risks: the New International Context. *Int. J. Latest Trends Finance Econ. Sci.* **2012,** *2* (2), 117–124.

Filipe, J. A.; Ferreira, M. A.; Coelho, M. Corruption and Political Risks in Latin American Countries. A General View, Statistical Review. *J. Greek Stat. Assoc.* **2012,** *8* (1-2), 83–96. ISSN: 1791-1745.

Filipe, J. A.; Ferreira, M. A. M. Social and Political Events and Chaos Theory: the Drop of Honey Effect. *Emerg. Issues Nat. Appl. Sci.* **2013a,** *3*, 126–137. DOI: dx.doi.org/10.7813/einas.2013/3-1/10.

Filipe, J. A.; Ferreira, M. A. M. (2013b), Chaos in Humanities and Social Sciences: an Approach. International Conference on Applied Mathematics: Aplimat 2013. Proceedings. Bratislava. Slovakia. February 2013.

Galán Páchon, J. M. (2010). Corruption and Oversight in Latin America. http://gopacnetwork.org/Docs/SenatorGalansPresentation_CorruptionandOversightinLatinaAmerica_EN.pdf. (assessed Jan 3, 2014).

Greene, M. K. The Management of Political Risk, Bests Review (Property/Liability, ed.) 1974, p 75.

I Font, J. P. P., Régis, D. Chaos Theory and its Application in Political Science. (Draft), IPSA – AISP Congress, Fukuoka, 2006.

Jessen, L. Corruption as a Political Risk Factor for Investors in the Oil and Gas Industry, with Specific Emphasis on Nigeria: Identification, Analysis and Measurement, Master Thesis, Stellenbosch University, 2012.

Johnston, M. Corruption and Democracy: Threats to Development, Opportunities for Reform. Hamilton, N.Y.: Colgate University, 2000.

Lloyd, B. *Political Risk Management*; Keith Shipton Developments, Ltd.: London, 1976.

Kaufmann, D.; Wei, S. J. Does Grease Money Speed up the Wheels of Commerce? National Bureau of Economic Research Working Paper 7093, Cambridge, MA, 1999.

Kobrin, S. J. Political Risk: a Review and Reconsideration. Working Paper #998-78, Alfred P. Sloan School of Management, 1978.

Mauro, P. Corruption and Growth. *Quart. J. Econ.* **1995,** 683–712.

Mauro, P. The Effects of Corruption on Growth, Investment, and Government Expenditure: a Cross-Country Analysis. In *Corruption and the Global Economy*; Elliott, K. A., Ed.; Institute for International Economics: Washington, DC, 1997; pp 83–107.

Morris, S. D. Corruption in Latin America: An Empirical Overview, University of South Alabama, 2004. http://w2.mtsu.edu/politicalscience/faculty/documents/crptn_paper.pdf

Moura, D. F.; Filipe, J. A. Political Risks in Latin America: Analyzing International Investments Decisions. *J. Econ. Engineer.* **2012,** *3* (1), 17–19.

MPDFT - Ministério Público do Distrito Federal e Territórios (2010). Corruption and Human Trafficking in Brazil: Findings from a Multi-Modal Approach. 2014-01-22, Brazil.

Olsen, L. F., Degn. H. Chaos in Biological Systems. *Quart. Rev. Biophy.* **1985,** *18* (2).

Price, J. (2013). Corruption is Growing (Again) in Latin America.

CHAPTER 22

Workforce, Higher Education Policies, and Development in Turkey

İLKNUR ŞENTÜRK*, DERYA YILMAZ KILIÇOĞLU, and GÖKHAN KILIÇOĞLU

Faculty of Education, Eskişehir Osmangazi University, Ankara, Turkey

**Corresponding author. E-mail: ilknurkokcu@gmail.com*

ABSTRACT

In this chapter, initially the status of higher education in Turkey is discussed based on various variables and indicators. In this context, the topics considered as priorities in higher education policies and practices with respect to the higher education, quality of manpower, development, national and international trends, and advances in higher education are addressed. An attempt was to include the concepts, principles, and strategies that would increase the contribution of the academic institutions to workforce planning in Turkey in particular and in the international scale in general based on the discussions mentioned above.

22.1 INTRODUCTION

In the face of indisputable need for change in today's globalizing world, universities have displayed internationally market-oriented development with transcending their national identities. Universities in the new world order have entered into a competition-based higher education marketing in the front of increasing demand for higher education and decreasing in the funds transferred from the public sector. This situation has been increased searching for performance, efficiency, and quality in universities which plays a key role in the formation of technology and innovation. Specifically, neoliberal principles in education, limited public funding, increase in higher education demand, costs, and quality forces in higher education drive higher education to keep up with the globalizing higher education processes in Turkey (Tezsürücü and Bursalıoğlu, 2013). Therefore, it is essential to indicate many changes and innovation initiatives in higher education with the influence of globalization such as quality assurance and accreditation, internalization, privatization, and quantitative changes in Turkish higher education such as number of higher education institutions, students, academic staff, enrollment rate, scientific publication performance, and patent rate.

22.2 QUALITY ASSURANCE SYSTEM AND ACCREDITATION

Within the past 10 years, the concepts of quality and quality assurance have become two important concepts for higher education. Indeed, higher education institutions and administrators are closely concerned with quality, and all are looking for how to adapt quality assurance system into their institutions that systematically monitors and evaluates the quality of services in universities (Mok, 2000; Pond, 2002). The quality assurance system in universities provides an internal audit mechanism for the quality of education and is organized to cover management of the institution, all activities, and stakeholders (Özer et al., 2010). The aim of accreditation is to document that the education carried out in higher education institutions is at an acceptable level in national or international standards (Rehber, 2007). Indeed, accreditation is a voluntary quality assurance system that periodically provides higher education institutions with institutional self-assessment as well as giving opportunities to evaluate universities by independent accreditation institutions (Aktan and Gencel, 2007).

In order to increase quality consciousness in higher education, national quality assurance agencies for higher education were established in all developed countries since 1980s. The quality assurance system established in Turkey is based on the Council of Higher Education that has the authority of planning, regulation, management, and supervision of teaching. Two types of accreditation, namely, institutional accreditation and programmatic accreditation have been embraced in Turkish higher education system. The institutional accreditation provides the assessment of administrative, financial, and academic capacity of the higher education institution. The programmatic accreditation allows for the evaluation of higher education programs whether they are in line with the standards of quality and excellence.

Since 1998–1999 academic years, accreditation of teacher education has been started in faculties of education to improve quality in education. At the beginning of the 1990s and 2000s, international quality assurance of engineering programs was provided by evaluation processes of the Accreditation Board for Engineering and Technology (ABET) and Association for Evaluation and Accreditation of Engineering Programs (MÜDEK). In addition, quality development activities performed by the Turkish Council of Higher Education are "Institutional Evaluation Program" supported by the European University Association (EUA) and "Culture of Quality Project." Academic Evaluation and Quality Development Commission in Higher Education Institutions (YÖDEK) was also established in 2004 to draw guideline about development, approval, and recognition of quality levels of institutions (Tezsürücü and Bursalıoğlu, 2013). Besides, other ongoing national accreditation bodies in Turkey are Association for Evaluation and Accreditation of Medical Education Programs (TEPDAD), Architectural Accrediting Board (MİAK), Association for Accreditation of Science, Literature, Science–Literature, Language and History–Geography Faculties (FEDEK), Association for Evaluation and Accreditation of Veterinary Medicine Educational Institutions and Programs (VEDEK), and Association for Evaluation and Accreditation of Nursing Education Programs (HEPDAK) (YÖK, 2017).

22.3 INTERNATIONALIZATION

Internationalization transcends academic interest and transforms into a strategic process that improves institutional capacity and adaptation in higher education in front of global developments. In fact, policies and strategies have been developed within the concept of internalization of higher education in order to meet the demands of globalization (Altbach et al., 2009; Teichler, 2004; Van Der Wende, 2001). Internalization improves institutional capacity and adaptation process across global developments rather than meeting academic interest. That is, internalization has become both the reason and the catalyst for reforms and changes in higher education. Specifically, internalization in Turkish higher education is also becoming an important agenda. However, it seems that this issue has not been systematically addressed as a public policy domain. Universities are often organized in extraordinary periods and structured in response to previous turbulence. Due to the military interventions, higher education in Turkey is distant from international developments (Vural Yılmaz, 2016).

Higher education in Turkey is almost confined to a bureaucratic and centralized structure. However, efforts to break this introverted structure have been increased in recent years. Thus, internationalization has been taken on the agenda within the context of both restructuring of higher education and foreign policy goals in the year of 2010 (Vural Yılmaz, 2016). In this context, a comprehensive internationalization vision has been embraced that involves Bologna Process. Turkey was involved in Bologna Process in 2001. Indeed, the Bologna Process is a reform process aimed at promoting the convergence of higher education systems in Europe, as well as performing harmonization and cooperation between higher education institutions in European countries. In order to establish a balance between diversity and unity, European universities restructure themselves against the dominant role of American universities in teaching (Aktan, 2007; Davoudi and Ellison, 2006; Kuyumcu and Erdoğan, 2008). Under Bologna Process, Turkish higher education has been performed reform initiatives such as establishing national competency framework, national quality assurance system in parallel with European country standards, and comparable academic degree system, implementing diploma supplement, European credit transfer and accumulation system, joint degree programs between countries, staff and student mobility, and lifelong learning approach (Erdoğan, 2010).

Development of national strategies for internationalization in higher education, structuring of new institutions, the arrangement of legislations, and implementation of programs have been also observed. Indeed, it is obvious that it is not yet possible to mention the existence of a strategic approach about shaping of internationalization policies in higher education in Turkey. Unfortunately, internationalization is not considered holistically in both development plans and strategic plan of Council of Higher Education. However, the report titled "Growth, quality, internationalization: A roadmap for higher education in Turkey" which is published in 2014 by Council of Higher Education is expected to be a starting point in this sense (Cetinsaya, 2014). Also, internationalization efforts need to be transformed into a strategic process in order to achieve sustainability and generate concrete results. Even though fundamental characteristic of Turkish higher education is centralization and Council of Higher Education is responsible from decision making in internalization process, strategic goals of the universities regarding internalization focus on student and staff mobility, language proficiency, international activities, and projects. However, these goals may provide short-term changes due to low cost and human resources rather than long-term changes in higher education. Therefore, more challenging and long-term goals need to be set within educational initiatives such as internalized curriculum, joint degree and corporative programs, launching franchising, branch campus, and twinning arrangements (Vural Yılmaz, 2016).

22.4 PRIVATIZATION

The tendency to reduce the role of the state in higher education means to reduce the dependence on public financing to have a more market-oriented, competitive and quality-based structure, and to show effective professional management (Altbach et al., 2009). The equality of opportunities and the higher education services undertaken by the state have gained a market-oriented structure in which the private sector as well as the state will operate (Tezsürücü and Bursalıoğlu, 2013). Specifically, the Council of Higher Education in Turkey gave the right for establishment of private foundations after 1981 depending on the concept of entrepreneur university and modern management techniques (Gürüz, 1995). Bilkent University is the first private foundation in Turkish higher education system established in 1984. Indeed, private foundations in Turkish higher education system have scientific and financial autonomy. They are subject to provisions of the constitution as regards academic studies, teaching staff recruitment, and security (Hirsch, 1950).

22.5 QUANTITATIVE CHANGES IN TURKISH HIGHER EDUCATION SYSTEM

Against rapid changes in the world as regards technological, economic, political, and social aspects, higher education institutions have to keep pace with these changes and need to restructure their educational opportunities and settings. In fact, universities have increased in number, staff and student number have also been increased, enrollment rate of higher education institutions have changed, and scientific research performance and patent rate of universities have changed day by day. Therefore, this section will present a general framework about these aspects in order to provide detailed information about quantitative changes in Turkish higher education system.

22.6 NUMBER OF HIGHER EDUCATION INSTITUTIONS

Higher education institutions in Turkey have shown a numerical increase after the 1980s in republican history, as can be seen in Figure 22.1. It is obvious that state-based universities dominate the Turkish higher education system. With neoliberalization process stating after 1980s, the number of private foundation universities increased and begun to take its place in higher education system. The number of private foundation universities, which increased especially after 2000, was an increase about four times between the years 2000 and 2015. However, some private foundation universities were closed and state-based universities were reduced in number with the coup attempt in 2016, as shown in Figure 22.1.

22.7 NUMBER OF STUDENTS AND ACADEMIC STAFF

The number of students in Turkish higher education has been increasing every year and reached to 6,689,185 according to 2015–2016 academic year statistics, as can be seen in Figure 22.2. Even though the vast majority of the students in higher education system attend state-based universities, the number of students in private foundation universities was in a great increase parallel with the increase in the number of these private foundations within the 2000s. In this direction, the number of students in private foundation universities between years 2000 and 2015 has been increased by 10 times and reached to 552,172.

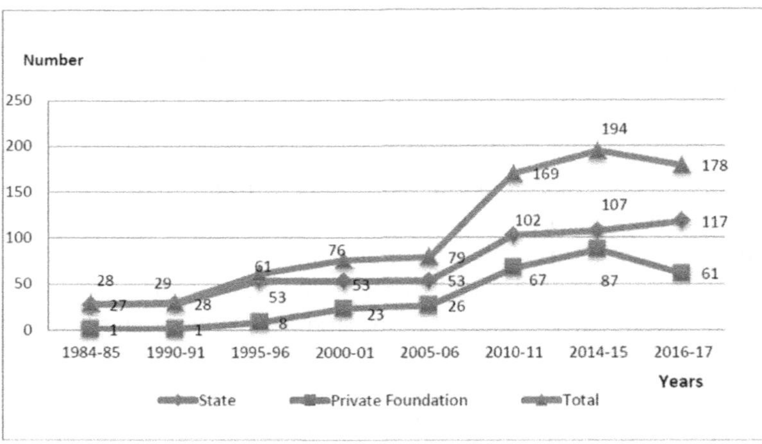

FIGURE 22.1 The number of higher education institutions in Turkey.

Source: Computed and created by the authors using higher education statistics of Measuring, Selection and Placement Center (ÖSYM), reports of Council of Higher Education, and Higher Education Information Management System.

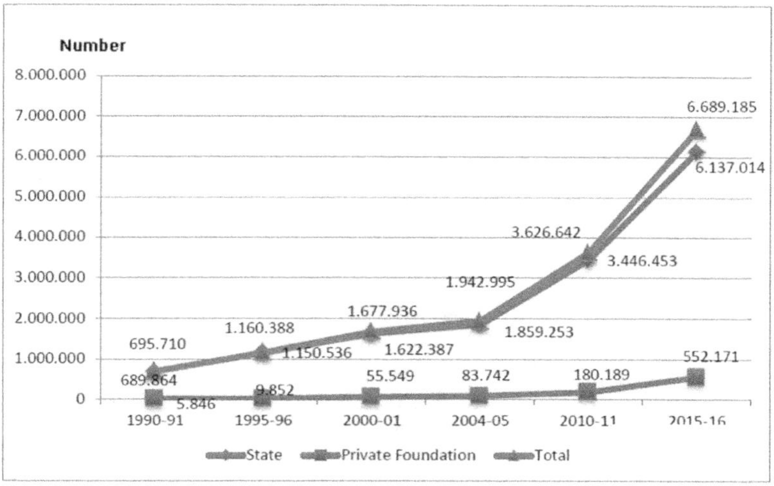

FIGURE 22.2 The number of students in Turkish higher education institutions.

Source: Computed and created by the authors using higher education statistics of Measuring, Selection and Placement Center (ÖSYM), reports of Council of Higher Education, and Higher Education Information Management System.

The number of academic staff in Turkish higher education system has been increasing year by year and reached 156,168 according to 2015–2016 academic year statistics. The great majority of the staff in higher education system works in state-based universities, as can be seen in Figure 22.3. At the beginning of the 1990s, Bilkent University was the only private university served in Turkish higher education. But in 2000s, the number of academic staff in private foundation universities significantly increased as well as the number of private foundations and students. The number of academic staff has increased by six times from the beginning of the 2000s to the present and reached to 23,802.

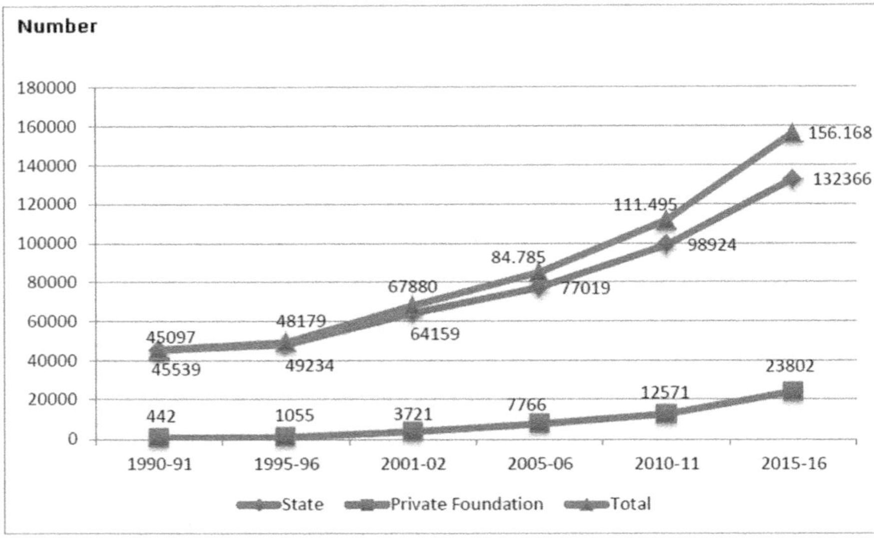

FIGURE 22.3 The number of academic staff in Turkish higher education institutions.

Source: Computed and created by the authors using higher education statistics of Measuring, Selection and Placement Center (ÖSYM), reports of Council of Higher Education, and Higher Education Information Management System.

22.8 ENROLLMENT RATE

Enrollment rate in Turkish higher education system is directly proportional to the number of institutions, students, and academic staff. As can be seen in Figure 22.4, the gross enrollment rate has increased steadily from 2000 to 2015. Gross enrollment rate in formal education increased from 15.06% to 49.98%. Similarly, the enrollment rate in higher education also increased from 22.30% to 94%. Moreover, it is obvious that the fastest increase has been started in 2008 and the enrollment rate increased twice in 2015 (94%) compared to 2008 (40.78%). This serious increase is based on about two and a half times increase in the number of open university students compared to previous years.

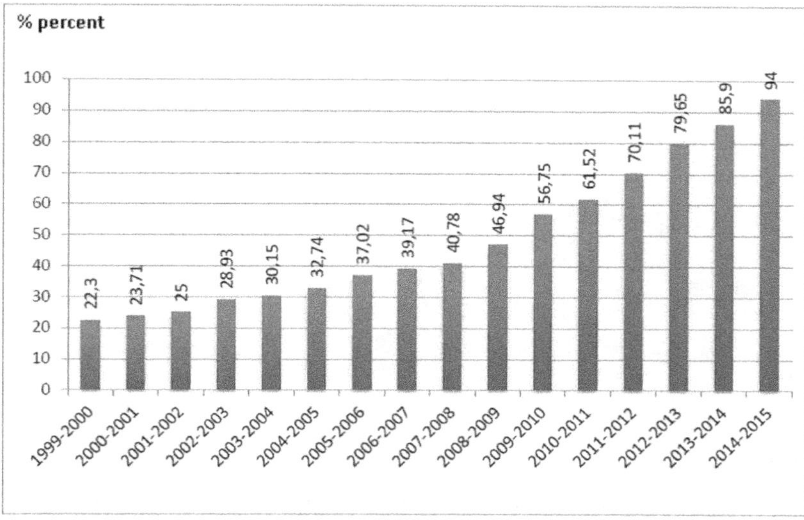

FIGURE 22.4 The gross enrollment rates in higher education in Turkey.

Source: Computed by Günay and Günay (2016) using higher education statistics of Measuring, Selection and Placement Center (ÖSYM) between years 2000 and 2013, Higher Education Information Management System (2014–2015), population census results in 2000 by Turkish Statistical Institute, population projections according to 2000 population census (2001–2006), and address-based population registration system (2007–2015).

There is also a significant increase in the net enrollment rates in higher education, similar to the gross enrollment rate. As can be seen in Figure 22.5, the net enrollment rate in 2000 was 38.93 % (1.99% open education and 9.62% formal education), whereas it was 38.93% in 2015 (7.59% open education and 31.34% formal education). Indeed, net enrollment rates in higher education showed a rapid increase after 2008. However, the fact that net enrollment rates in open education compared to formal education have become stagnant from 2010 onward, and fall into a downward trend is noteworthy. Moreover, the fact that the net enrollment rate is very low as compared to gross enrollment rate shows that there are many enrolled students in higher education in Turkey except for 18–22 age group.

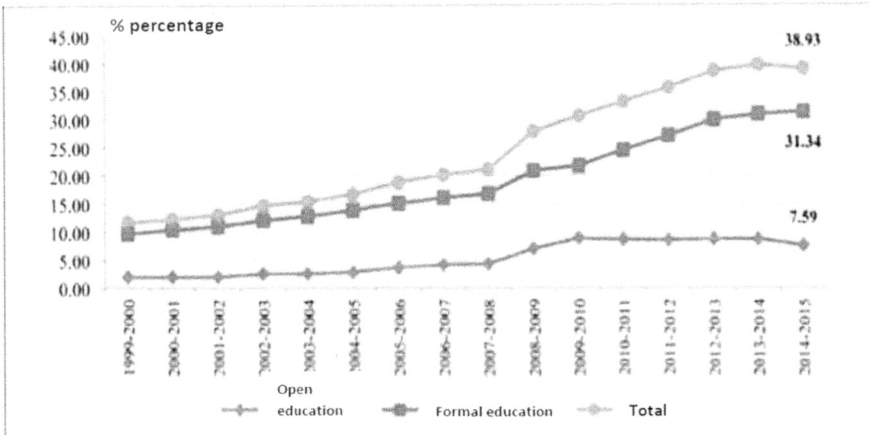

FIGURE 22.5 The net enrollment rates in Turkish higher education based on education type.

Source: Computed by Günay and Günay (2016) using higher education statistics of Measuring, Selection and Placement Center (ÖSYM) between years 2000 and 2013, Higher Education Information Management System (2014–2015), population census results in 2000 by Turkish Statistical Institute, population projections according to 2000 population census (2001–2006), and addressed-based population registration system (2007–2015).

22.9 SCIENTIFIC PUBLICATION PERFORMANCE AND PATENT RATE

When scientific publication performance of Turkey is examined, the number of publications addressed to Turkey was 204,216 between years 2010 and 2015 and the number of citations to these publications was 674,332. The impact value average of publications was 330 while the rate of cited publication was 5134. Besides, the world ranking of Turkey varied between 20 and 18 between years 2006 and 2015, as can be seen in Figure 22.6. The latest rank of Turkey specified in 2015 was 18 (TÜBİTAK, 2017).

In addition, bibliometric study results performed by Al, Sezen and Soydal (2012) showed that Turkish scientists produced 198,687 scientific publications in 2012, but three quarters of these publications have never been cited. Therefore, the quality of the produced publications is necessary to be emphasized scrupulously (Kılıçoğlu and Yılmaz, 2015).

When we look at the patents registered at Turkish Patent Institute, it has been seen that Turkey experienced a decrease between 2000 and 2003 in domestic and foreign patent registration while there has been a linear increase since 2003, as can be seen in Figure 22.7. If domestic and foreign patents are compared in terms of number, foreign patents have significant dominance over the domestic ones. The total number of registered patents was stated as 16,778 in 2016, which is twice the total number of patents in 2010.

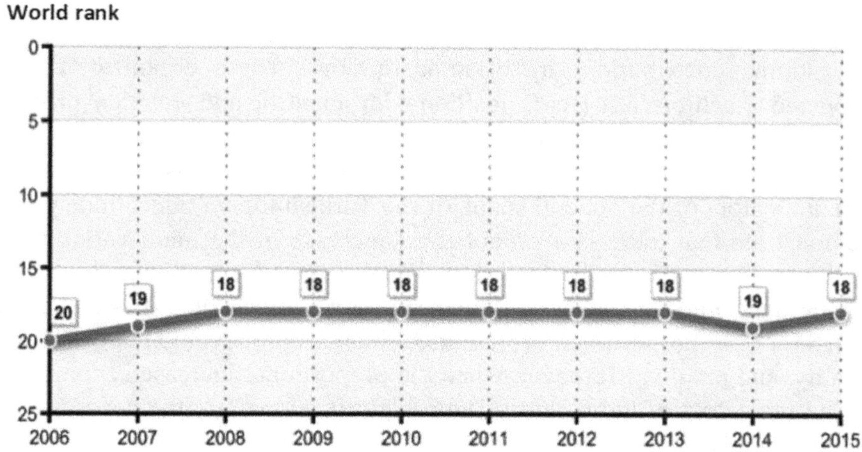

FIGURE 22.6 Scientific publication performances in Turkish higher education.
Source: TUBİTAK, 2017.

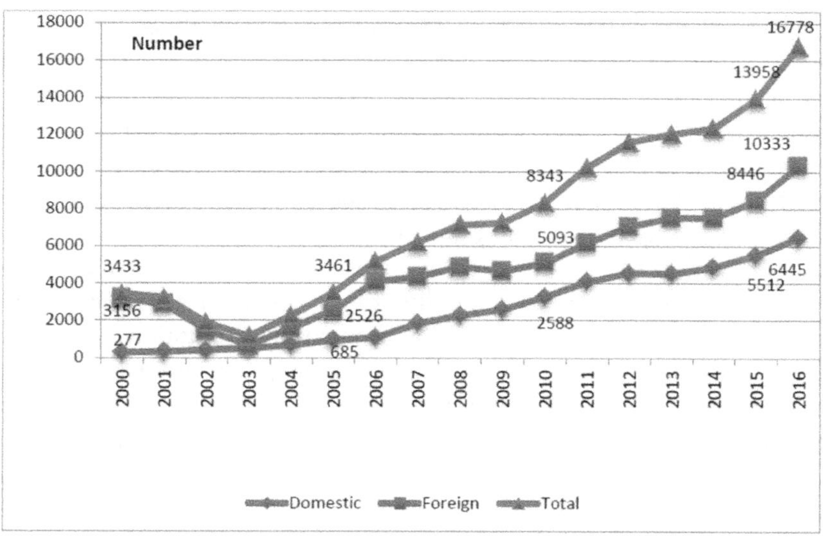

FIGURE 22.7 The number of patents registered at Turkish Patent and Trademark Office.
Source: Turkish Patent and Trademark Office (2017).

22.10 DISCUSSIONS AND RECOMMENDATIONS FOR HIGHER EDUCATION POLICIES, WORKFORCE QUALITY, AND DEVELOPMENT

Academic organizations are areas where human capital needs to be trained with the highest proficiency in the dimensions of social, political, economic, and cultural development with the scientific, technical, institutional, and educational contributions by these institutions. These organizations are expected to achieve a strategic position with scientific and technical production at universal, regional, and local scales. Thus, the focus is on quantitative and qualitative indicators of their positions and productions. The quantitative indicators of the general status of the Turkish higher education system demonstrate that there is a proportional increase in the main variables of scrutiny. However, discussion of the quality of human power in higher education would neither be sufficient nor convincing when it is examined based on only proportional or numerical increase in development, welfare, quality, and quality. Moreover, when the proportional increase is compared to the indicators of the countries with high development, the necessity for Turkey to adapt more vigorous and reformist policies becomes the topic of a new debate. In this section of the chapter, the possible principal, political, and economic variables for Turkey within the context of the higher education system are examined based on the quality indicators.

Özsoy (2008: 32), who found a positive correlation between higher education and economic growth in Turkey, emphasized that developments in higher education affected the national growth performance and Turkey should focus on financing, management, and supervision in higher education to meet competitiveness, information society, and economic growth expectations.

The contribution of higher education institutions to human development results in universal interest, comparison, and competition. Thus, collaborative policies were introduced at the international level. From the perspective of education and development policies, the role of higher education in increasing the quality and productivity of human capital that global policies require is significant. Higher education is one of the institutions that directly influences the quality of manpower. The national and international human capital potentials of the nations are reflected in higher education indicators (Çalışkan et al., 2013, Şimşek and Kadılar, 2010). Schooling rates, the number and quality of academicians, quality indicators, scientific publications, projects, technical and technological production capacities, patent production, and internationalization status are at the focus of development policies. Thus, the initial section of this chapter is devoted to the indicators that demonstrate the general status of higher education in Turkey.

The impact of education on development is determined by its contribution to the change in social, political, and economic dimensions, its effect on employment policies, income distribution, equality, welfare, and quality of life (Taş and Yenilmez, 2008). Quantitative indicators of higher education in the dimensions of manpower, services, production, research, and teaching influence the national economic growth and social development (Paton et al., 2005; Molenaers and Renard, 2003). It was reported that the increase in the number of students who can attend higher education and included in the teaching process causes significant and positive effects on growth (Filho and Pace, 2016, Çalışkan et al,, 2013). It is necessary to think about and address the contributions of higher education institutions to human development in political, social, and cultural fields as well as their financial contributions. The primary vision of the future strategies includes the talent, skill, work capacity, and creativity of human capital based on the information capital. Every decision and action that increases the quality of human capital in the process of planning and policy making introduces the essential factor for development. The reflections of education on social policies such as reducing the risk of unemployment, enhancing income levels, contributing to social welfare, improving income distribution, social justice, and reducing poverty are directly proportional to the increase in the quality of the education process (Bashir, 2007). On the other hand, policies, programs, and plans that can improve the human capital quality in Turkey have a critical significance for the realm of higher education. A holistic approach toward higher education policies should be considered as a necessary component of all public administration policies. Increasing the quality of management processes and approaches, delegation, accountability, transparency, autonomy, cooperation, strategic planning, just, and equal education opportunities should be emphasized (Kavak, 2011).

Neoliberal policies increase their impact on higher education. Competition, focus on quantitative indicators, privatization policies, and search for capital have painted higher education institutions into a corner. These realities that support quantitative indicators and economic reductionism, restrict access to education and areas of employment in education with the concerns of privatization and commercialization, meeting capital demands and expectations. The free and autonomous science, instruction, research, and service activities of academic organizations are suffering under the guidance and dominant role of the capital. As a result of the abovementioned pressure, public welfare is neglected and the institutions are alienated from their raison d'être and the liberating nature of science and education (Aydın, 2015). Absolute focus on financial wisdom and mathematical ratios would demonstrate that the rate of increase in the access of certain groups to education such as women, individuals with different ethnic origins, cultures, mother languages, or religions, easing the conscience. However, when the processes that follow the outcome-oriented thinking are considered, it would be observed that the inequality in various variables and the unequal and discriminatory practices continue to influence the process in Turkey (Başkaya, 2002; Demirer, 2002; Önder, 2002; Özsoy, 2002). Conducting situation analyzes comparatively using access, processes, results, quality and quality indicators, and indicators related to conformity to equality and organizational, social justice, merit, production, universal and academic criteria, and planning the future of institutions strategically have primary significance in training a productive workforce.

It was emphasized that the bureaucratic and centralized administration approaches are predominant in the administrative operations in academic organizations in Turkey (Kasalak et al., 2016). Centralized management, planning, and applications impede the democratization and autonomy of decision-making mechanisms, organizational structure, and operations. This may result in the risk of decreasing the quality of the research and education services and human resources. Furthermore, commercialization and marketization could be considered as a serious obstacle to academic freedom. Social, cultural, and political elements such as diversity, multiculturalism, multiple learning environments, guidance of skills, interests and curiosity, and multiculturalism are the significant components that need to be managed in the realization of freedom and autonomy in academic organizations (Erkan, 2011).

Higher education in Turkey faces the problem of overcoming multidimensional opportunity inequality. This issue is the most fundamental variable in improving the quality of human resources. Furthermore, it also fuels social inequalities. Having increased the enrollment rates in higher education to 80% (www.yok.gov.tr) based on 2016 data could facilitate a solution for the problem for the factor of access to higher education. However, there remains the unsolved problem of access to education that is caused by student admission conditions, which is an extension of centralized policies, and student guidance strategies, which are not activated starting from the lower grade levels. This situation that limits the potential and right of the individual to design his/her future and to attend the training for his/her professional choices could be regarded as the greatest obstacle to the development policies. It would be advisable to focus on a process-oriented planning by constantly monitoring access to higher education and the employment. Without overlooking the social benefits of higher education, the principle of justice should be observed in distribution of public resources and in public spending. The system should also review its relationship with the labor market. Studies that measure the effects of the processes of foundation, privatization, and commercialization on the quality of higher education and on the potential and quality of manpower should be sustained.

Studies that tackle human resources with development indicators and offer international comparisons provide significant perspectives for nations on their developmental strengths and weaknesses. The United Nations Development Programme (UNDP) is one of the organizations that offers this perspective. Human development is a concept that focuses not only on the wealth of economies but also on the enhancement of human life. This approach focuses on people and the options and opportunities offered to people for sustainable development. (http://www.tr.undp.org/content/turkey/en/home/presscenter/pressreleases/2017/03/turkey-ranks-71st-by-human-development-index.html). In the United Nations Development Programme (UNDP) Human Development Index (HDI) report published in 2016, Turkey ranked 71st among 188 countries and regions with 0.767 points. Between 1990 and 2015, the HDI score of Turkey has risen from 0.576 to 0.767. It could be observed that the average education period in Turkey increased 3.7 years and the expected education period increased 5.7 years between 1990 and 2015. It was reported that gross domestic product per capita in Turkey increased by 78.2% between 1990 and 2015. However, when social and economic inequalities are considered, it was emphasized that Turkish HDI score decreased 15.8%. Turkey's HDI score was 0.767 in 2015. However, when the differences of inequality in human development are considered, it was reported that Turkey's score decreased to 0.641 with a loss of 15.8%, based on the disparity in the distribution of dimension indices. Turkey ranked 69th out of 159 countries with a Gender Inequality Index (GIE) score of 0.328 in the 2015 GIE. The rate of female Member of Parliament in the Turkish parliament is 14.9%. Among adult women, the ratio of the graduates of at least the secondary education is 43.5%, compared to 64.8% for men. It was observed that the women's share in the labor market was 30.4%, while the same was 71.4% for men (http://www.tr.undp.org/content/turkey/tr/home/presscenter/pressreleases/2017/03/turkey-ranks-71st-by-human-development-index.html). These indicators demonstrate the kinds of policies that Turkey should address to improve its position on the indicators of human development: Extending the education process as much as possible, organizing access to all levels of education, conducting justice-oriented public policies, removing inequalities and discrimination systematically, providing the rights of labor, life and education to all social classes, and most importantly, increasing specialization and scientific production and prioritizing higher education policies as much as the primary education.

The significant contribution of the universities to human development indicators will be achieved through the transformation of the programs that have lost their functionality today. However, the main logic in this process should be constructed not only in the direction of rationalization of neo-liberal rational and expectations but through the leadership in political, social, and economic transformation (Adserias et al., 2017) led the academia. Because the concrete example of the abovementioned situation is the reduction in the capacity of training human resources and information capital production in higher basic science education in the universities due to the failure of Turkish labor market, enterprises, institutions, and financial policies in creating employment in these fields. The priority for academic organizations is to increase their potential as the driving force of social, political, and economic development. It is to realize organizational models and administrative approaches beyond hierarchical, centralized, and bureaucratic structuring. It entails increasing participation and not compromising the purpose of democratic culture. It is to adapt the principles of self-control and accountability. It means not compromising merits and the science. It is to consider scientific production and freedom of thought as much as profitability and productivity (Şenses, 2007).

The expectations from higher education institutions to shape future societies are related to the improvement in their organizational intelligence and scientific power. These institutions should be able to work in coordination with local, regional, and universal collaborators in instructional processes. This proficiency framework is related to the quality of the academia. However, the dysfunctional quality of policies and practices related to the training of faculty members and the performance of the academicians within the system are among the most fundamental problems in Turkey (Erçetin, 2001).

The main trend in higher education in Turkey is to increase the schooling rate. This trend is considered as an element of the political strategy. Increasing access to tertiary education continues to be the primary strategy since the 1980s. The area of higher education is popularized. Along with popularization, the access of disadvantaged groups can also be facilitated to a certain extent. The cities and regions expect to grow further with each established university, and that the quality of the labor force would rise, the variety in production and consumption would increase. Thus, competition among universities also increases. It is expected that the competition would develop social, economic, and cultural structure. To fulfill these expectations, it is necessary for the universities in Turkey to implement decisions and plans to improve the quantity and quality of financial, infrastructural, material, and human resource inputs. It is necessary to implement efficient use of resources, to differentiate and diversify the curricula, content, and instructional methods and techniques instead of standardization and to adopt internationally accepted competencies. Student admission and placement conditions should be structured within the framework of basic education and access rights, equality, skills, and personal competences. Academic organizations should include their professional relationships with social fields and institutions and the workforce in coordinated plans (Altınsoy, 2011)

Universities should not be left to purely pragmatic approach of the market culture. Functionalism and utilitarian reasoning may not primarily organize the increase in the quality of academic, scientific, and intellectual capital in universities. It may not emphasize the preservation of institutional autonomy (Çelik and Gür, 2014; Erkan, 2011) and productive freedoms. Besides, there is a risk that academic organizations could make strategical plans an instrument of financial control and restraint . It raises the risk of being driven by market conditions. Measuring the strategic objectives only by quantitative indicators and reduction of academic performance to numerical indicators lead to the internalization and legitimization of market values. This would limit the produced knowledge and services with the concepts of pragmatism, market utility, entrepreneurship, competition, and quality, and would reduce these to value criteria. It facilitates monitoring and central accountability of the performance with managerial approach and processes. In short, it results in instrumentalization of the academic production (Aydın, 2013). It could be argued generally that academic production is rapidly instrumentalized in Turkey based on that fact. Quantitative definition and measurement of academic performance strengthen the control mechanisms of planning and auditing activities. Administration focuses its practices on this.

Based on the areas of discussion structured so far, it would be proper to be recommended for Turkish higher education system to accelerate efforts to develop continuous development, coordination, a participative and supportive organizational culture, organizational change management, new theories, and approaches (Ramos et al., 2015). The diversity and quality of the labor force trained by universities could be effective if it focuses on technology production, training producers of knowledge, and increasing global proficiency skills (Burke and Ng, 2006).

It was observed that central policies in university administration have lost their efficiency in managing financial and human resources. Efficiency should not be expected in an organizational structure where authority and responsibilities are not balanced, democratic process and rights are not observed. In centralist and hierarchical structures, top administrators are unable to utilize functions that would maximize the potential of the employees in the organization and cannot recognize the motivational and performance elements that would increase productivity (Calp, 2013).

For higher education to increase the quality of human potential in the society, it is necessary to design and implement administrative and financial autonomy and accountable, democratic, participatory, and pluralistic management processes strategically, legally, and practically in the administrative level in higher education institutions. The principal method to improve the production and service performances of the academia is primarily democratization and autonomy (Yılmaz and Sarpkaya, 2016). Furthermore, academic training and employment policies in Turkey are critical problem based on input quality of the system. Merit, academic freedom, universal qualities and criteria, objectivity, science, and justice-based nonfavorist human resources policies would also increase the quality of the system output. These principles require political commitment (Kavak, 2011; Grifficen and Jong, 2017).

Neoliberal economic political practices have transformed higher education services into a marketable, commercial activity along with constantly increasing global circulation of capital. In this process, internationalization applications, agreements, protocols, joint programs, and accreditation practices are also very effective. These experiences, which can support the quality of human resources based on circulation, access to service, universal common principles and values, also lead the attention of global international society to the issues of equality and justice in access to education and access to similar opportunities. Commercialization and privatization also create a greater risk of exclusion of low-income social classes. Based on development potential, this could affect the social mobility of certain social classes negatively in terms of employment potential.

Although it is a fact that higher education is a high personal gain level economically, it also has a great potential to increase the public interest by transforming the quality of human and information capital (Rana et al., 2015). Thus, development policies in Turkey should primarily resolve the problems of low employment and high unemployment among higher education graduates. To solve this problem, it is necessary to include the quality of institutional output of higher education, especially the human capital, intellectual capital, information capital, and technical skills variables in the development process by coordinating these outputs with the labor and production markets (McCurdy, 2010; Tyurina and Troyaskaya, 2017). Thus, lifelong learning and development (Filho and Pace, 2016) benefits of higher education would be activated. Higher policies are expected to increase resources and to utilize these resources actively and functionally. Distribution, sharing, and efficiency of public expenditures are the indicators of true development. Turkish central decision-making mechanisms should be decisive in policies that would improve the quality of public services.

As Gündüz and Balyer (2013) emphasized, implementation of the idea of reforms that would introduce the dynamics of organizational change from student services to educational, instructional, and research processes in universities, from administrative approaches to information, communication, and education technologies, could be exploratory: Above all, a balanced environment should be established. Laboratories, classrooms, and workshops should be adequate. Scientific and technological developments should be adapted in learning environments rapidly. Employment opportunities for students should be monitored, planned, and developed in coordination with other institutions. Professional development processes and programs that would improve the quality of academic (Griffioen and Jong, 2017) and administrative staff should be designed and implemented. Existing human resources should be planned based on rational roles and duties according to skills, interest, and expectations. The capacity to utilize international databases should be established at the highest level in research, education, and service processes. Coordination and implementation of international studies in projects and research should be improved (Collins, 2015). Academic programs should increase the capacity of the personnel, motivate the personnel according to their talent and skills, and support information capital (Holdsworth and Thomas, 2016). Accomplished knowledge and skills should be integrated into social and economic development plans as well. In knowledge-based economies, training and employment policies that could meet global human resource qualities and expectations are prerequisites for sustainable development. Thus, the initial focus should be on administration and administrator variables. The capacity of the administrator and the administration should be reviewed and merit, equality, scientific principles, and professionalism should become uncompromised principles in administrative approaches (Yıldız and Talih, 2011).

The essential condition for achieving all technical processes and functions mentioned above is to transform the organizational culture with change management. Equalizing the human resource quality in a global scale with the policies and strategies established with the international accreditation process would consist of the first stage of the culture of change (Adgerias et al., 2017). It is possible to make the changing culture a part of the organization through leader-member interactions (Horne et al., 2016). Positioning all unit managers in academic organizations as leaders who improve the potential of human resources based on talent, knowledge and skill, and structuring their administrative processes accordingly will be a strategic step (Gürkan, 2007; Hoşgörür and Hoşgörür, 2011) for higher education in Turkey. Academic talent is supported by the lead of the academic leadership. Every academic organization should be able to create its own praxis within its own potential. Employees in the academic organization expect a work environment that considers the critical perspectives that facilitate the recognition of their skills and qualities for organizational development (Lightcap, 2013) that can be sustained based on their potential. On the other hand, for academic development, academicians should be able to organize and conduct their scientific thoughts and production processes with a vocational value perception that is not conformist or status quoist and their educational and instructional process autonomously. Adapting this universal principle in government, public and institutional policies are a measure for real development and democracy beyond the numerical indicators.

KEYWORDS

- **workforce**
- **higher education policies**
- **development in Turkey**
- **quality assurance system**
- **accreditation**

REFERENCES

Aydın, A. *Eğitim politikaları;* 1. Basım. Pegem Akademi Yayın Evi: Ankara, 2015.

Adserias, R. P.; Charleston, L. J.; Jackson, J. F. L. What Style of Leadership Is Best Suited to Direct Organizational Change to Fuel Institutional Diversity in Higher Education? *Race Ethn. Educ.* **2017,** *20* (3), 315–331. DOI: 10.1080/13613324.2016.1260233.

Aktan, C. C.; Gencel, U. Yükseköğretimde akreditasyon. In *Değişim çağında yükseköğretim,* Aktan, C. Can Ed.; Yaşar Üniversitesi Yayınları: İzmir, 2007.

Aktan, C. C. Yüksek öğretimde değişim: Global trendler ve yeni paradigmalar. In *Değişim çağında yükseköğretim;* C. Can Aktan Ed.; Yaşar Üniversitesi Yayınları: İzmir, 2007.

Altbach, P.; Reisberg, L.; Rumbley, L. *Trends in Global Higher Education, Tracking an Academic Revolution;* UNESCO: Paris, 2009.

Altınsoy, S. Yeni devlet üniversitelerinin gelişimi: sorunlar ve politika önerileri. Development of New Public Universities: Problems, and Policy Recommendations. *Yükseköğretim ve Bilim Dergisi/J. Higher Educ. Sci.* **2011,** *1* (2), 98–104.

Bashir, S. *Trends in International Trade in Higher Education: Implications and Options for Developing Countries;* Working Paper Series. World Bank: Washington DC, USA, 2007.

Brewer, P. D.; Brewer, K. L. Knowledge Management, Human Resource Management, and Higher Education: A Theoretical Model. *J. Educ. Bus.* **2010**, *85* (6), 330–335.

Burke, R. J.; Ng, E. The Changing Nature of Work and Organizations: Implications for Human Resource Management. *Hum. Res. Manage. Rev.* **2006**, *16*, 86–94.

Calp, M. Yükseköğretim örgütlerinde kurumsal değişimi okumada mecazların gücü: Bir Türk devlet üniversitesi örneği. *Kuram ve Uygulamada Eğitim Yönetimi [Educ. Admin.: Theory Pract.]* **2013**, *19* (2), 215–248.

Collins, I. Using International Accreditation in Higher Education to Effect Changes in Organisational Culture: A Case Study from a Turkish University. *J. Res. Int. Educ.* **2015**, *14* (2), 141–154.

Çalışkan, Ş.; Karabacak, M.; Meçik, O. Türkiye'de eğitim-ekonomik büyüme ilişkisi: 1923–2011 , Kantitatif bir yaklaşım. *Yönetim Bilimleri Dergisi* **2013**, *11* (21), 9–48.

Çelik, Z.; Gür, B. S. Yükseköğretim sistemlerinin yönetimi ve üniversite özerkliği: küresel eğilimler ve Türkiye örneği. The Governance of Higher Education Systems and University Autonomy: Global Trends and the Case of Turkey. *Yükseköğretim ve Bilim Dergisi/J. Higher Educ. Sci.* **2014**, *4* (1), 18–27.

Çetinsaya, G. *Büyüme, kalite, uluslararasılaşma: Türkiye yükseköğretimi için bir yol haritası.* Eskişehir: Anadolu Üniversitesi Basımevi, 2014. Retrieved from http://www.yok.gov.tr/documents/10279/2922270/Büyüme+Kalite+Uluslararasılaşma+cetinsaya-19x27-12%2C5forma.pdf/e5681887-1560-4fc3-9bab-0402e7f3ec2b

Davoudi, S.; Ellison, P. *Bologna Survey 2006.* Brokes University: Oxford, 2006.

Erçetin, Ş. Ş. Biz akademisyenler geleceğin yükseköğretim kurumlarını yaratmaya hazır mıyız? *Kuram ve Uygulamada Eğitim Yönetimi* **2001**, *25*, 75–86.

Erdoğan, A. *Yükseköğretimde yeniden yapılanma: 66 soruda Bologna Süreci uygulamaları.* Yükseköğretim Kurulu: Ankara, 2010.

Erkan, S. Change in Higher Education and a Core Value: Academic Freedom. *Kuram ve Uygulamada Eğitim Yönetimi [Educ. Admin.: Theory Prac.]* **2011**, *17* (3), 349–376.

Filho, W. L.; Pace, P. Teaching Education for Sustainable Development: Implications on Learning Programmes at Higher Education. In *Teaching Education for Sustainable Development at University Level;* Filho, W. L., Pace, P. Eds.; Springer, 2016; pp 1–9.

Griffioen, D. M. E.; Jong, U. M. The Influence of Direct Executive Managers on Lecturers' Perceptions on New Organizational Aims in Times of Academic Drift. *Int. J. Leadersh. Educ.* **2017**, *20* (4), 451–467, DOI: 10.1080/13603124.2015.1066867

Gündüz, Y.; Balyer. A. Yükseköğretim örgütlerinde değişim ve yenilemeye ilişkin akademik algılamalar. Perseptions of the Academicians About Change and Innovation at Higher Education Institutions. *Elektronik Sosyal Bilimler Dergisi, Electron. J. Soc. Sci.* **2013**, *12* (43), 47–66.

Günay, D.; Günay, A. Dünyada ve Türkiye'de yükseköğretim okullaşma oranları ve gelişmeler. *J. Higher Educ. Sci.* **2016**, *6* (1), 13–30.

Gürkan, Ç. G. Öğrenen örgütler: Yükseköğretim kurumlarının öğrenen örgüt olması. *Yönetim Bilimleri Dergisi* **2007**, *5* (2), 201–225.

Gürüz, K. *Türkiye'de ve dünya'da yükseköğretim, bilim ve teknoloji raporu üzerine görüşler.* TÜSİAD Yayınları No: T/95, 3-175.

Holdsworth S.; Thomas, I. A Sustainability Education Academic Development Framework (SEAD). *Environ. Educ. Res.* **2016**, *22* (8), 1073–1097, DOI:10.1080/13504622.2015.10 29876

Horne, A. L.; Plessis, Y.; Nkomo, S. Role of Department Heads in Academic Development: A Leader–Member Exchange and Organizational Resource Perspective. *Educ. Manage. Admin. Leadersh.* **2016**, *44* (6), 1021–1041.

Hoşgörür, V.; Hoşgörür, T.. Planlı dönemde yükseköğretimdeki gelişmeler. *Hasan Ali Yücel Eğitim Fakültesi Dergisi* **2011**, *15* (1), 111–131.

Hirsch, E. E. *Dünya üniversiteleri ve Türkiye'de üniversitelerin gelişmesi,* Ankara Üniversitesi Yayınları: İstanbul, 1950.

Kasalak, G.; Günbayı, İ.; Özçetin, S. Öğretim elemanlarının yönetsel hizmetlere ilişkin görüşleri: Bir durum çalışması/*The Views of Academic Staff on Administrative Services: A Case Study Educational Administration. Kuram ve Uygulamada Eğitim Yönetimi/Theory Pract.* **2016**, *22* (4), 507–536. DOI: 10.14527.

Kavak, Y. Türkiye'de yükseköğretimin görünümü ve geleceğe bakış. Overview of Higher Education in Turkey and Future Directions. *Yükseköğretim ve Bilim Dergisi/J. Higher Educ. Sci.* **2011**, *1* (2), 55–58.

Kılıçoğlu, G.; Yılmaz, D. Türkiye'de bilim ve teknoloji politikaları. In *Türkiye'de yükseköğretim: Alanı, kapsamı ve uygulamaları*; Aypay, A. Ed.; Pegem A Yayıncılık: Ankara, 2015.

Kuyumcu, A.; Erdoğan, T. Yükseköğetimin toplumsal değişmeye etkisi. *Hacettepe Üniversitesi Eğitim Fakültesi Dergisi* **2008**, *35*, 240–250.

Lightcap, S. J. *Organizational Commitment and Job Satisfaction: Leadership Perspectives on the Effectiveness of Interinstitutional Collaborations.* Doctorate Thesis University of Pennsylvania, 2013.

McCurdy, J. W. *Organizational Commitment and Job Satisfaction: An Examination of Staff in Higher Education*; Doctoral Thesis. Claremont Graduate University, 2014.

Mok, K. Impact of Globalization: A Study of Quality Assurance Systems of Higher Education in Hong Kong and Singapore. *Comp. Educ. Rev.* **2000**, *44* (2), 148–174.

Molenaers, N.; Renard, R. University Development Co-Operations and Good Practices: Discussions. In *University Development Co-operations and Good Practice.* Beneitone, P.; Höivik, S.; Molenaers, N.; Obrecht, A.; Renard, R. Eds.; University of Deusto: Bilboa, 2003; pp 173–186.

Öğrenci Seçme ve Yerleştirme Merkezi [ÖSYM]. *1995–1996 Öğretim yılı yükseköğretim istatistikleri,* 1996. http://www.osym.gov.tr/TR,8388/1995-1996-ogretim-yili-yuksekogretim-istatistikleri-kitabi.html

Öğrenci Seçme ve Yerleştirme Merkezi [ÖSYM]. *2000–2001 Öğretim yılı yükseköğretim istatistikleri,* 2001. http://www.osym.gov.tr/TR,7152/2000-2001-ogretim-yili-yuksekogretim-istatistikleri-kitabi.html

Öğrenci Seçme ve Yerleştirme Merkezi [ÖSYM]. *2001–2002 Öğretim yılı yükseköğretim istatistikleri,* 2002. http://www.osym.gov.tr/TR,7252/2001-2002-ogretim-yili-yuksekogretim-istatistikleri-kitabi.html

Öğrenci Seçme ve Yerleştirme Merkezi [ÖSYM]. *2004–2005 Öğretim yılı yükseköğretim istatistikleri,* 2005. http://www.osym.gov.tr/TR,7625/2004-2005-ogretim-yili-yuksekogretim-istatistikleri-kitabi.html

Öğrenci Seçme ve Yerleştirme Merkezi [ÖSYM]. *2005–2006 Öğretim yılı yükseköğretim istatistikleri,* 2006. http://www.osym.gov.tr/TR,7769/2005-2006-ogretim-yili-yuksekogretim-istatistikleri-kitabi.html

Öğrenci Seçme ve Yerleştirme Merkezi [ÖSYM]. *2010–2011 Öğretim yılı yükseköğretim istatistikleri,* 2011. http://www.osym.gov.tr/TR,8260/2010-2011-ogretim-yili-yuksekogretim-istatistikleri.html

Özer, M.; Gür, B. S.; Küçükcan, T. *Yükseköğretimde kalite güvencesi.* SETA Yayınları: Ankara, 2010.

Özsoy, C. Türk yükseköğretim sisteminin durumu ve iktisadi büyüme performansına katkısı. *Nigde Üniversitesi İktisadi ve İdari Bilimler Fakültesi Dergisi* **2008**, *1* (2), 31–48.

Paton, R.; Peters, G.; Storey, J.; Taylor, S. Corporate Universities as Strategic Learning Initiatives. In *Handbook of Coporate University Development. Managing Strategic Learning Inıtiatives in Public and Private Domains*; Paton, R.; Peters, G.; Storey, J.; Taylor, S. Eds.; Gower Publishing Company: USA, 2005; pp 7–17.

Pond , W. K. Distributed Education in the 21st Century Implications for Quality Assurance. *Online J. Distance Learn. Admin.* **2002**, *5* (2), 1–7.

Ramos, T. B.; Caeiro, S.; Hoof, B.; Lozano, R.; Huising, D.; Ceulemans, K. Experiences from the Implementation of Sustainable Development in Higher Education Institutions: Environmental Management for Sustainable Universities. *Journal of Cleaner Production* **2015**, *106*, 3–10.

Rana, S.; Ardichvili, A.; Taing, S. K. National Human Resource Development in a Transitioning Society: The Case of Cambodia. *Human Res. Dev. Int.* **2017**, *20* (2), 127–146. DOI: 10.1080/13678868.2016.1243937

Rehber, E. Dünyada değişen yükseköğretim ve kalite anlayışı. In *Değişim çağında yükseköğretim*; Can Aktan, C. Ed.; Yaşar Üniversitesi Yayınları: İzmir, 2007.

Şen, Z. Türkiye'de yükseköğretim eleştirileri ve öneriler. Critics and Suggestions for Higher Education System in Turkey. *Yükseköğretim Dergisi* **2012**, *2* (1), 1–9.

Şenses, F. Uluslararası gelişmeler ışığında Türkiye yükseköğretim sistemi: temel eğilimler, sorunlar, çelişkiler ve öneriler, 2007. ERC Working Papers in Economics 07/05 September 2007 http://erc.metu.edu/menu/series07/0705.pdf

Şimşek, M.; Kadılar, C. Türkiye'de beşeri sermaye, ihracat ve ekonomik büyüme arasındaki ilişkinin nedensellik analizi. *C.Ü. İktisadi ve İdari Bilimler Dergisi* **2010**, *11* (1), 115–132.

Taş, U.; Yenilmez, F. Türkiye'de eğitimin kalkınma üzerindeki rolü ve eğitim yatırımlarının geri dönüş oranı. *Eskişehir Osmangazi Üniversitesi Sosyal Bilimler Dergisi* **2008**, *9* (1), 155–186.

Teichler, U. International Student Mobility in Europe in the Context of the Bologna Process. *J. Int. Educ. Leadership* **2012**, *2* (1), 1–13.

Tezsürücü, D.; Bursalıoğlu, S. A. Yükseköğretimde değişim: Kalite arayışları. *KSU J. Soc. Sci.* **2013**, *10* (2), 97–108.

Turan, Ş. A. Türk üniversitelerinde yeni yönetim uygulamaları ve ortaya çıkan yönetsel kontrol anlayışına dair bir inceleme. *İnsan ve Toplum Bilimleri Araştırmaları Dergisi* **2013**, *2* (1), 158–189.

Turkish Patent and Trademark Office. *İstatistikler,* 2017. Retrieved from http://www.turkpatent.gov.tr/TurkPatent/statistics

Tyurina,Y.; Troyanskaya, M. Perspectives of Increase of University Education Effectiveness: Use of Private Educational Resources. *Int. J. Ed. Manag.* **2017**, *31* (3), 396–403, DOI: 10.1108/IJEM-02-2016-0026.

Üsdiken, B.; Topaler, B.; Koçak, Ö. Yasa, piyasa ve örgüt tiplerinde çeşitlilik: 1981 sonrasında Türkiye'de üniversiteler. *Ankara Üniversitesi SBF Dergisi* **2013**, *68* (3), 191–227.

Türkiye Bilimsel ve Teknolojik Araştırma Kurumu [TUBİTAK]. *Bibliyometrik analiz*, 2017. Retrieved from http://cabim.ulakbim.gov.tr/bibliyometrik-analiz/turkiye-bilimsel-yayin-performans-raporlari

Van Der Wende, M.C. Internationalization Policies: About New Trends and Contrasting Paradigms. *Higher Educ. Policy* **2001**, *14* (3), 249–259.

Vural Yılmaz, D. Türkiye'deki devlet üniversitelerinde uluslararasılaşma süreci: Stratejik planlar üzerinden bir değerlendirme. Suleyman Demirel University. *J. Faculty Econ. Admin. Sci.* **2016**, *21* (4), 1191–1212.

Yıldız, E. Üniversitelerin kalkınmadaki rolü: Babaeski meslek yüksekokulu örneği. *Girişimcilik ve Kalkınma Dergisi* **2011**, *6* (2), 269–287.

Yılmaz, T.; Sarpkaya, R. *Eğitim ekonomisi*. Eleştirel bir yaklaşım.1. Basım. Anı Yayıncılık, Ankara, 2016.

Yükseköğretim Kurulu [YÖK]. *Türk yükseköğretiminde 10 yıl (1981–1991): 1981 reformu ve sonuçları*. Yükseköğretim Kurulu: Ankara, 1991. http://www.yok.gov.tr/documents/10279/30217/TÜRK+YÜKSEKÖĞRETİMİNDE+ON+YIL+1981-1991.pdf/05f9f18c-a406-45e5-b24b-0bfde59e3bbb

Yükseköğretim Kurulu [YÖK]. *Türk yükseköğretiminde gelişmeler*. Yükseköğretim Kurulu: Ankara, 1994. http://www.yok.gov.tr/documents/10279/30217/turk_yuksekogretiminde_gelismeler.pdf/ab0b683e-0a96-4248-8587-ca2d49877f48

Yükseköğretim Kurulu [YÖK]. *Türk yükseköğretiminin bugünkü durumu*. Yükseköğretim Kurulu: Ankara, 1996. http://www.yok.gov.tr/documents/10279/30217/TÜRK+YÜKSEKÖĞRETİMİNİN+BUGÜNKÜ%20DURUMU+MART+1996.pdf/0ff0f9ec-a524-4fff-b170-b701528086dd

Yükseköğretim Kurulu [YÖK]. *Türk yükseköğretiminin bugünkü durumu*. Yükseköğretim Kurulu: Ankara, 1997. http://www.yok.gov.tr/documents/10279/30217/TÜRK+YÜKSEKÖĞRETİMİN+BUGÜNKÜ%20DURUMU+1997.pdf/fcfda3a8-a1e7-496c-abd8-df55d1f38c1b

Yükseköğretim Kurulu [YÖK]. *Türk yükseköğretim sistemi 1994–1995'den 1995–1996'ya gelişmeler*. Yükseköğretim Kurulu: Ankara, 1997. http://www.yok.gov.tr/documents/10279/30217/TÜRK+YÜKSEKÖGRETİM+SİSTEMİ%201994-1994+ve+1996+%27%20ya+Gelişmeler+1997.pdf/f170ba97-738a-4f2d-be05-97b6c49d7507

Yükseköğretim Kurulu [YÖK]. *Türk yükseköğretiminin bugünkü durumu*.Yükseköğretim Kurulu: Ankara, 1998. http://www.yok.gov.tr/documents/10279/30217/TÜRK+YÜKSEK+ÖĞRETİMİNİN+BUGÜNKÜ%20DURUMU+MART+1998.pdf/0594f7ca-9bd6-4751-8bb2-2cd8aa1bee89

Yükseköğretim Kurulu [YÖK]. *Türk yükseköğretiminin bugünkü durumu*. Yükseköğretim Kurulu: Ankara, 1999. http://www.yok.gov.tr/documents/10279/30217/turk_yuksekogretimin_bugunku_durumu_mart_1999.pdf/b150b17a-1794-4e09-bb65-4a5013d0cf8b

Yükseköğretim Kurulu [YÖK]. *Türk yükseköğretiminin bugünkü durumu*. Yükseköğretim Kurulu: Ankara, 2000. http://www.yok.gov.tr/documents/10279/30217/turk_yuksek_ogretimin_bugunku_durumu_mart_2000.pdf/acfbb960-bbb3-44fe-871e-ee960e11e30e

Yükseköğretim Kurulu [YÖK]. *Türk yükseköğretiminin bugünkü durumu*. Yükseköğretim Kurulu:Ankara,2003. http://www.yok.gov.tr/documents/10279/30217/turk_yuksekogretim_bugunku_durumu_mart_2003.pdf/61b573f7-3675-4063-8889-dd133d74177c

Yükseköğretim Kurulu [YÖK]. *Türk yükseköğretiminin bugünkü durumu*. Yükseköğretim Kurulu: Ankara, 2004. http://www.yok.gov.tr/documents/10279/30217/turk_yuksekogretimin_bugunku_durumu_kasim_2004.pdf/b2c1f7dc-8ff7-4841-8cf0-577ce4a3d0f4

Yükseköğretim Kurulu [YÖK]. *Türk yükseköğretiminin bugünkü durumu*. Yükseköğretim Kurulu: Ankara, 2005. http://www.yok.gov.tr/documents/10279/30217/turk_yuksekogretimin_bugunku_durumu_kasim_2005.pdf/eec6a5a4-c0ee-4bca-ac3c-735dda4199cf

Yükseköğretim Kurulu [YÖK]. *Türkiye'nin yükseköğretim stratejisi*. Yükseköğretim Kurulu: Ankara, 2007. http://www.yok.gov.tr/documents/10279/30217/yok_strateji_kitabi/27077070-cb13-4870-aba1-6742db37696b

Yükseköğretim Kurulu [YÖK]. *Vakıf üniversiteleri raporu*. Yükseköğretim Kurulu: Ankara, 2007. http://www.referansgazetesi.com/UPLOAD/haber/dokuman/2007/12/yök%20vakif_rap.pdf

Yükseköğretim Kurulu [YÖK]. *Yükseköğretim bilgi yönetim sistemi*, 2017. https://istatistik.yok.gov.tr

Yükseköğretim Kurulu [YÖK]. *Yükseköğretim kalite kurulu*, 2017. http://www.yok.gov.tr/web/kalitekurulu/tescil-suresi-devam-edenler.

CHAPTER 23

Understanding Tourism Demography Through Guests and Hosts in the Himalayan State Sikkim

DEBASISH BATABYAL*

Teaching Faculty, Amity Institute of Travel & Tourism, Amity University, Kolkata, West Bengal, India

**E-mail: debasishbatabyal@gmail.com*

ABSTRACT

Since time immemorial, tourism in hilly areas has been considered and confined to be leisure centric and recreation oriented. This demand-led orientation was a mere perception of the tourists and supply means and ways were largely ignored. Tourism in hilly areas is believed to have been a mean of spending from disposable and discretionary income mostly for nonessential activities. On the other hand, it has been empirically proved that the degree of income from tourism largely influences the degree of acceptability and reliability of tourism in a region along with the changes and adoption of new orientation of sustainable development. This article has dealt with a relationship between income and tourism with other issues addressed in destination development literature that are essentially striving to achieve an equilibrium between demand (or marketing) and supply (or destination) parameters. More specifically, this chapter has shown how tourism phenomenon is influencing the community income and vice versa.

23.1 INTRODUCTION

Sikkim is a small Indian state, bounded by vast stretches of Tibetan plateau in the North, the Chumbi Valley and the kingdom of Bhutan in the East, the kingdom of Nepal in the west, and Darjeeling (West Bengal) in the South. Its latitude is 27°30' North and the longitude is 88°30' East. Sikkim is famous for scenic valleys, forest, snow clad mountains, magnificent Buddhist culture and heritage, and hospitable peace-loving people. Though small, the environmental , social, and cultural diversities are not so. Some scholars believe that the word Sikkim has evolved from the Limbu language Su-Khim which implies "*new home*." According to the Nepalese dialect, it refers to a "*now place*" or the term has been derived from a Sanskrit word which means a "*mountain crest*." The people of Sikkim have ethnic diversity. The Bhutias came from Tibet, the Lepchas were the aboriginal community, and the Nepalese came from Nepal. Sikkim was an independent state and it experienced many invasions by its neighboring countries like Nepal and Bhutan. The king took the help of the Britishers in India and later gifted some of its region including Darjeeling to the British India. Now this 22nd Indian State (joined Indian Union in 1975) has over 81% of the total geographical area under the administrative managerial control of the Ministry of Environment and Forest, Government of India. Over 45% of the total geographical area of the state is under tree cover and nearly 34% of the geographical area is set aside as protected area network in the form of national park and wildlife sanctuary. The basic statistics of flora and fauna are as follows (Table 23.1):

TABLE 23.1 Flora, Fauna, and Avifauna of Sikkim (2001–2011).

Flora/fauna/avifauna	2001	2011
Mammals	144 species	125
Birds	550 species	574
Butterflies and moths	650 species	689
Reptiles	33 species	88
Frogs	16 species	Amphibians 50
Orchids	550 species, 95 Genera	527
Rhododendrons	36 species, 45 varieties	38 species
Flowering plants	Over 4000 species	4458, Primulas 58
Ferns and allies	300 species	480, tree ferns 9, bamboo 11
Conifers	9 species	16
Medicinal plants	Not enumerated	700

Source: Economic survey 2010–2011, Govt. of Sikkim.

The maximum summer temperature is 28°C and minimum winter temperature is 0°C. Sikkim has different types of mineral resources including coal, limestone, iron ore, graphite, pyrite, etc. The temperature in the bottom of the valleys (up to 600 m) situated at lower elevations, particularly during summers, is similar to the monsoon type of climate. The temperature starts falling between 600 m and 2000 m above sea level and one enjoys cool temperature climatic conditions and further up (between 2000 m and 3000 m) it is cold temperate climate. Arctic type of climate is found above 5000 m. Sikkim is ecologically a fragile area. The state has the responsibility to conserve its rich ecological and biological diversity that includes coexistence and protection of over *5000 species of angiosperm (one third of the total national angiosperms)*. Again this place has multi-ethnic communities (Table 23.2).

TABLE 23.2 Demographic Features of Sikkim's Districts, 2011.

	Population	% Urban	Population density (Persons per km^2)
East	286,583	43	295
South	146,850	14	196
West	136,435	4	117
North	43,709	11	10
Sikkim	610,577	25	86

Source: Census data, India, 2011.

Sikkim, India's least populous state, reported a population of 610,577 in 2011. It is also the second smallest state in India in terms of land area—after Goa. Population is unevenly distributed across the state over 7096 km^2. As a result, population density varies enormously across the state. For instance, North Sikkim reports a population density of only 10 persons per km^2, as its population of 43,709 people is spread over 4226 km^2. On the other hand, East Sikkim reports a population density of 297, as 46% of the state's population resides in this district which is spread over 954 km^2. Seventy five percent of the state's population resides in rural areas, with the rural proportion being as high as 96% in Sikkim's West District. Increased accessibility by roadways and air transport, rapid socioeconomic development, and competitive advantage both from the point of view of the destination and geographical proximity to tourist generating states contributed to the development of tourism in Sikkim. Recognizing the increased tourist arrivals, accommodation units were set up in Gangtok and a few towns mostly by outsiders without proper land use planning and architectural design. As per the 2011 census, Hindus are majority in Sikkim state. Hinduism constitutes 57.76% of

Sikkim's population. Buddhists play an important role in electoral of Sikkim state forming significant 27.39% of total population. Buddhism is followed in majority of districts. From years 2012 to 2016, it has been noticed that there has been an increase of 29,000 in the population of Sikkim. Therefore, it has been seen that every year the population increases by 5800.

Ever increasing tourist arrival and increasing dependence on tourism as a powerful industry and employment generating source, restoration of peace and harmony, hospitable people, plenty of diversified natural and cultural resources, and a typical interest of the people of India to discover the unknown hidden Sikkim brought about a new dimension for the development and marketing of tourism in Sikkim. Development and adoption of marketing approach extensively include a demand–supply equilibrium and as such takes into account tourism system. Destination marketing is unlike a marketing of essential consumables as it considers and moves toward the characteristics of services. There are very few destinations well managed by scientific research and background analysis. Sikkim has poor quality database and only in few cases the available data have been scientifically analyzed and interpreted as majority of destination planners and government officials are not from tourism background. So, an understanding of the demand for Sikkim as a destination and the capacity levels (physical, biological, social, psychological, and financial) largely contributes to the overall marketing strategy. The government of Sikkim is one of the very few Indian states trying to optimize benefits from tourism for their local people. The recently adopted policy to project the state as an Ultimate Ecotourism Destination was really a committed responsibility toward sustainable development. Next to this is rural tourism and adventure tourism. Almost all these tourism facilities will contribute to the alternative tourism development in the state and the changes in types and forms of tourism will automatically have an impact on the activities of the tourists, duration of stay, the number of tourist arrivals, and the tourism industry as a whole. So a proactive environment friendly approach has already been adopted in the marketing mix and Unique Selling Proposition (USP) of Sikkim though there are many things still remaining. The major task for the government is how to coordinate and control interest of all participants and infuse the idea of sustainable practices in the state. The priority area is really confusing as there is a traditional clash between development and conservation. The destination marketing should not only increase the arrival of tourists but also be proactive in selecting target groups and introduce sustainable practices including mass awareness for environmental conservation.

23.2 LITERATURE REVIEW

In the article *Demography*, Hall (2006) said although the ageing of the population will be a dominant demographic factor in tourism trends in the future, other factors will also be significant. Although Hall (2000) said that the breakdown of the traditional nuclear family in many western countries marked by increased single parenting, people marrying later, more people never marrying, and many people never having children, not only influences demographic characteristics but will also affect the pattern of holiday taking and leisure travel. In addition, the growth of large migrant populations in many developed countries, particularly as labor may be required to sustain economic growth within an ageing population, is also expected to contribute to the development of new transnational tourism patterns as migrant groups seek to retain relationships to "home" (Williams and Hall, 2002). In fact, one of the greatest predictors of future mobility patterns is previous mobility (Hall, 2005). The implications of these predicted changes for tourism are incredibly complex and need to be seen in relation to other factors discussed in this book, particularly with respect to global environmental change. Even assuming that the world's environmental systems can sustain such population growth, the economic and political instability that may result as a result of resource insecurity may severely damage tourism growth in the less developed world.

In the article "The impact of socio-demographics on tourist behavior – analyzing segments of cultural tourists visiting Coimbra," Elisabeth et al. (2004) identified different cultural tourist segments based on a set of socio-demographic variables and to further analyze differences existing between these groups in terms of travel behavior. Results exhibited a clear distinction between a wealthier, older group of tourists, with a higher level of education, contrasting with a poorer, younger, and less educated group. These clusters or culture tourist segments were further analyzed with regard to other aspects of their socio-demographic profile and their travel behavior, revealing significant differences between groups. Results may help understand the culture tourist market and contribute to an improved target marketing to each group, simultaneously understanding the actual and potential role of each group for the destination area in question.

In another article "The Future of Family Tourism," Heike and Ian (2015) have discussed the current knowledge on family tourism with pointers to the future by drawing on multidisciplinary contributions to a recent consolidative book. They said that the research in family tourism is limited, fragmented, and individualized, thus not accounting for the diversity of family forms, sociality of family groups, and multidimensionality of family experiences. Integration of thinking about and understanding the changing nature of families in tourism is needed. The key concepts of change are captured through a cognitive mapping perspective. This research process allows an exploration and systemic tool for constructivism interpretation. The cognitive model resulted in nine main themes that represent emerging and changing trends in family tourism. The key purpose here is the identification of the slow moving changes and trends that are occurring in family tourism based upon cognitive modeling of current research and their significance for predicting the future. This demonstrates connectivity between concepts providing a more holistic and critical interpretation of the future of families and tourism.

Tourism planning process is designed to produce goals, strategies, and objectives for the destination area related to tourism development marketing, industry organization, and awareness, and to other support services and activities while tourism development plan normally provided overall guidelines for development, outlines broad development concepts, and identified individual development opportunities worthy of in-depth analysis through feasibility studies and/or cost–benefit analyses (Mill and Morrison, 1986). While moving with tourism development the destination area would first find it necessary to establish overall development guidelines to ensure that when development takes place, it complies with the area's economic, environmental, social, and cultural policies and goals, and as the destination going for development are needed to draft more specific guidelines describing the basic characteristics of development that it wishes to encourage. Accordingly, the interpretations of strategy and models for strategy formulation in tourism are found to be prescriptive and there is a need for further empirical research about patterns and processes that accompany tourism strategies other than planning (Mintzberg, 1994). The best destination development plans are worked out in cooperation with private, public, and nonprofit sector (Müristaja, 2003). Following the opinion of the author, one could win through cooperation and improvement and not through inner competition and division (Godfrey and Clarke, 2000). Organizing and coordinating were the activities with the key meaning of development. An ideal variant would be if someone took the whole responsibility for tourism development, but still a better solution would be the formation of a holding enterprise that would represent and coordinate different local groups.

According to Tourism Policy (1997), this belief was based on an information gap that systematically constrained the development of tourism in India over the years. National Action Plan for Tourism presented in the Parliament on May 5, 1992. It proposed to achieve different types of tourism products, continual growth of tourism infrastructure, effective marketing and promotional efforts in the overseas markets, and removal of all impediments of tourism. The Action Plan could not specify the infrastructural requirements and the investments needed to meet the targets and the sources of funding for the same. The Tourism Synergy Program consisting the activities and infrastructural components to be provided by the various agencies including the private sector and State Governments was thus prepared in 1993. It was further modified and converted into a "National Strategy for the Development of Tourism" during 1996. These documents were evident for realizing the importance of tourism and ensuring a consensus on the developmental needs. It also contributed positively to all the infrastructural departments in a coordinated manner and ensures higher plan allocation and introduction of new innovative schemes for accelerated development and growth of tourism.

"Tourism strategy can take the three forms." These were: (1) lower costs where production efficiency led to value added to both consumer and producer matching market prices with greater profitability or being able to offer lower prices while maintaining average profits. (2) Differentiation, where additional value was created for the consumers through adding value that allowed a greater price to be commended, for example, branding, product features or additional benefits or support. (3) Focus, relating to the breath of market, scope, focuses referring to targeting specific niche markets or product ranges (Porter, 1980). It is also empirically found that the essence of strategy formulation is coping with competition where competition in an industry is rooted in its underlying and competitive forces.

The article, "Environmental Economics of the Khangchendzonga National Park in the Sikkim Himalaya, India," shows the application and relevance of contingent valuation method (CVM) to measure willing to pay (WTP) important part of polluter pays principles (PPP). Using a random survey, the average WTP was 8.84 USD for foreign visitors per visit, followed by 6.20 USD per household per year, and 1.91 USD per domestic visitor. The WTP was strongly influenced by age, education and income (Maharana et al., 2000).

In the article, "Prevalence and socio-demographic correlates of substance use in a rural community in Sikkim, North East India: Results from a pilot population survey," Goel and Chakrabarti (2010) have marked that alcohol use is higher than national prevalence rates in India requiring targeted intervention in Sikkim. Prevalence of substance use has been 76.7% (only alcohol 55.3%, cannabis 13.6%, and opioid 5.8%). Most drug users belonged to 15–44 years age group with 74.6% having onset in the 15–20 years age range, but only 3.9% had a history of migration. Knowledge about AIDS was significantly low at 36.7% compared with the nondrug using population (62.5%).

23.3 OBJECTIVES AND METHODOLOGY

Tourism in Sikkim is traditionally for leisure and recreation and found to be isolated from the mainstream important economic activity that contributes to the common people and expands thereby. The prime objective of this study was to identify whether tourism is really an effective tool or activity of the rich people for the other rich people from outside. So, the only hypothesis tested here is given below.

H0: Tourism in Sikkim is the activity for rich people and by rich people mainly for leisure and recreation and not for developing the demographic profile of the local community of Sikkim.

A total of 475 customers were contacted, and the overall response rate was 42.32% (201 completed, usable questionnaires). A total of 585 community members of the different areas of Sikkim were contacted and the overall response rate was 16.92% (99 completed, usable questionnaires).

To analyze the primary data Kendall's Coefficient of Concordance, a nonparametric test is used. It determines the degree of association among several (k) sets of ranking of N objects or individuals. It is considered an appropriate measure of studying the degree of association among three or more sets of rankings. This descriptive measure of the agreement has special applications in providing a standard method of ordering objects according to consensus when we do not have an objective order of the objects. The basis of this test is to imagine how the given data would look if there were no agreement among the several sets of rankings, and then to imagine how it would look if there were perfect agreement among all the sets. In our study, the consistency of ranks (maximum disagreement = 0 and minimum is 1) were tested several times with the appropriate P value. While computing Kendall's Coefficient of Concordance (W), the mean of rank (Rj) was taken into note of with the appropriate consistency.

Here, $s = \sum (Rj - \overline{Rj})^2$;

$$W = \frac{s}{\frac{1k^2(N^3 - N)}{12}},$$

where k = no. of sets of rankings, that is, the number of respondents; N = number of objects ranked and

$\frac{1}{12k^2(N^3 - N)}$ = maximum possible sum of square deviations, that is, the sum s which would occur with perfect agreement among k rankings.

Apart from this, the collected data were analyzed by using different statistical tools like percentage, means, etc. and the computer packages like SPSS and Excel were also used.

23.4 DATA ANALYSIS AND DISCUSSION

To raise the supply side aspect from within the destination, one of the major demographic factors with respect to monthly income and monthly income from tourism among the people of Sikkim is taken into consideration. The economic viability is measured to notify which section/s is/are earning as well as depending more on tourism, households in Sikkim were divided into seven important or viable slabs keeping in view their already available demographic profile. Thirty-eight percent (38%) of all respondents were found to be associated with tourism business in Sikkim and the range of monthly average income was less than 9500 to above 34,500 while the same was from Rs. 500 to more than 9500.

Figure 23.1 shows the distribution of mean income and mean income from tourism in the State. Seven important slabs are set along the x-axis while the mean income and mean income from tourism are measured along the y-axis. The result exhibited was interesting as the percentage increase in mean income was always less than that of the mean income from tourism. The mean incomes from tourism have experienced a depression unlike the ever increasing trend of mean income. These two types of mean incomes indicate that the income groups belonging to the fifth and sixth slabs are earning less from tourism though the trend is increasing thereafter for higher income groups in Sikkim. As an important aspect of tourism demand a set of questions are asked to tourists. They are requested to rank simply for 12 infrastructural and/or superstructural[1] variables in which one was considered to be the most important variable. A nonparametric Kendall's W rank test was conducted to test whether ranks put by all respondents were consistent (H0) or not (H1). Here, ranks were found to be consistent as the P value was greater than 0.10 and mean of ranks ($\bar{R}_j$) for each variable was used and compared for further ranking to measure importance. Following was the mean of ranks showing the importance of different variables here under (Table 23.3).

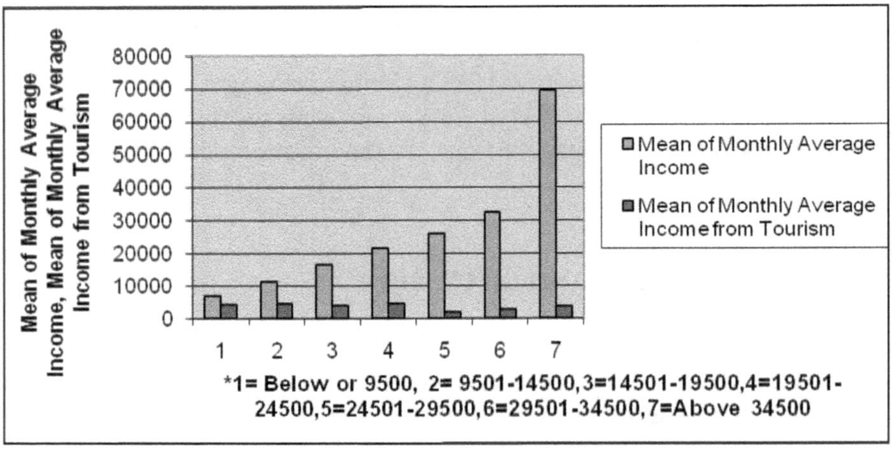

FIGURE 23.1 **(See color insert.)** Mean of monthly average income and mean of monthly average income from tourism of the local people involved in tourism business in Sikkim.

Source: Field survey 2015–2016.

Note: Statistical analysis has been made using Excel.

Majority of tourists visiting Sikkim are price sensitive except their foreign counterparts. Foreign tourists did not prioritize "better income opportunity" to be the second important criterion for future travel to Sikkim unlike their domestic counterparts. Most of the foreigners travelling to Sikkim are alternative/adventuresome/special interest tourists but advocated to or emphasized on basic infrastructure such as "availability of accessible vehicles for travelling in or around Sikkim" (ranked two), availability of effective communication aids like better telecommunication facility, postal services, cyber café, information kiosk, etc. (ranked three) and ramps and elevators at attractions (ranked five). Instead of diluted superstructural or tourism related infrastructural facilities they expected little more infrastructural facilities along with the moderate fulfillment of their associated derived demand for tourism products and services.

TABLE 23.3 Mean of Rank for Infrastructural and/or Superstructural Variables.

Factors	All tourists		Domestic tourists		Foreign tourists	
	Mean of ranks	**Ranks**	**Mean of ranks**	**Ranks**	**Mean of ranks**	**Ranks**
Better income opportunities	3.16	1	3.38	1	4.38	4
Reasonably priced tourist products	3.67	2	3.72	2	3.84	2
Change in attitude of local people toward tourists	8.59	10	8.32	10	9.72	12
Trained staff and their friendly treatment	6.82	7	6.80	7	7.66	9
Information about availability of aids/equipment at attractions	6.05	5	5.95	5	6.58	6
Safety support system	6.64	6	6.66	6	6.97	7
Availability of accessible vehicles for travel	4.03	3	3.93	3	3.59	1
Accessible pathways at attraction premise	7.66	9	7.60	9	7.56	8
Clear signage at all major points	9.27	11	9.33	11	8.50	10
Ramp and elevators at attractions	7.20	8	7.32	8	6.31	5
Effective communication aids	5.19	4	5.37	4	3.88	3
Easily accessible tour operators	9.72	12	9.63	12	9.02	11

Source: Field Survey 2015–2016.

Note: Statistical analysis has been made using SPSS statistical package.

23.5 CONCLUSION

The preoccupied notion of thinking that tourism can only be an elitist tag is found to be perception of hosts as clients or tourists are richer than the common local people ever. The study has promulgated the further scope for analyzing the socioeconomic conditions of guests and hosts and their combined effects on ever changing tourism marketing environment of Sikkim. Again, majority of tourists visiting Sikkim are found to be price sensitive, seeking more income opportunities for future visit with more spending, and asking for better communication. Except price sensitivity, that is evident among all leisure tourists, almost all are asking for basic and moderate infrastructure and little and important superstructural facilities oriented particularly for specialized tourists. It can also be said that tourism development in Sikkim has rightly started following important principles of sustainable tourism development. So, tourism development strategies should broadly incorporate more sustainable orientation for the future by offering alternatives tourism, alternative tourists, and alternative infrastructural development.

KEYWORDS

- **tourism**
- **demography**
- **guests**
- **hosts**
- **Himalayan State Sikkim**

REFERENCES

Chettri, N.; Sharma, E. Assessment of Natural Resources Use Pattern: A Case Study Along a Trekking Corridor of Sikkim Himalaya, India. *Tata Energy Research Institute* **2006,** *3* (1), 21–34. http://search.proquest.com/docviews (accessed Mar 15, 2011).

Elisabeth, K.; Maria, J. C., et al.. The Impact of Socio-Demographics on Tourist Behavior – Analyzing Segments of Cultural Tourists Visiting Coimbra. University of Aveiro, Department of Economy, Management and Industrial Engineering. Project by Association for Tourism and Leisure Education, 2004. http://www.tram-research.com/atlas/Aveiro.pdf (accessed Apr 6, 2017).

Goel, A.; Chakrabarti, A. Prevalence and Socio-demographic Correlates of Substance Use in a Rural Community in Sikkim, North East India: Results from a Pilot Population Survey. *J. Substance Use*. **2010,** *15* (1), 13–23.

Hall, M. Demography. In *Tourism Management Dynamics: Trends, Management and Tools*; Buhalis, D., Costa, C. Eds.; Elsevier Butterworth-Heinemann: London, 2006, 10–18.

Heike, A. S.; Ian, Y. The Future of Family Tourism. *Tour. Recr. Res.* **2014,** *39* (3), 343–360.

Maharana, I.; Rai, S. C ; Sharma, E. Environmental Economics of the Khangchendzonga National Park in the Sikkim Himalaya, India. *Geo J.* **2000,** *50*, 329–337. http://search.proquest.com/docviews (accessed Mar 15, 2011).

Maharana, I.; Rai, S. C.; Sharma, E. Valuing Ecotourism in a Sacred Lake of the Sikkim Himalaya, India. *Environ. Conserv.* **2000,** *27* (3), 269–277. http://search.proquest.com/docviews (accessed Mar 15, 2011).

Mill, R. C.; Morrison, A. M. *The Tourism System: An Introductory Text*; Prentice Hall Publication: Englewood Cliffs, NJ; 1986.

Moscardo, G. Community Capacity Building: An Emerging Challenge for Tourism Development. In *Building Community Capacity for Tourism Development*; Moscardo G., Ed.; CABI: UK, 2008.

Moutinho, L. Positioning Strategies. In *Tourism Marketing and Management Handbook*; Witt, S., Moutinho, L., Eds.; Prentice-Hall International: Hertfordshire, 1994; pp 332–336.

Müristaja, H. Development Trends and the Association of Stakeholders in Tourism Development Process in the Case of Pärnu County (Estonia). University of Tartu, Pärnu College, Estonia, 2009.

Nicolau, J. L.; Mas, F. J. Micro Segmentation by Individual Tastes on Attributes of Tourist Destinations. In *Tourism Management: New Research*; Terry V. L., Ed.; Nova Science Publishers: New York, 2006; pp 1–122.

Porter , M. E. From Competitive Advantage to Corporate Strategy. *Harvard Business Review*, May 1987, pp 43–59.

Walker, J. R.; Walker, J. T. *Tourism: Concepts and Practices*; Dorling Kindersley (Authorized by Pearson Education): New Delhi, 2011; pp 197–229.

ADDITIONAL REFERENCES

Batabyal, D. Changing Aspects of Destination and its Marketing: A Case Study of Sikkim. *Tour. Theory Pract Mark. Issues* **2010,** *8* (1), 175–185. (Kolkata).

Batabyal, D. Destination Development: Strategy for Tourism in Sikkim. Doctoral Dissertation, University of Burdwan, W. B., 2013.

Maheshwari, B. *Decision Styles and Organizational Effectiveness.* Vikas: New Delhi, India, 1980.

Rahman, S. A. *The Beautiful India-Sikkim*; Reference Press: New Delhi, 2006; pp 23–56.

Robbins, S. P.; Sanghi, S. *Organizational Behavior*; Pearson Education: New Delhi, India, 2007.

Wang, C. *Managerial Decision Making and Leadership*; John Wiley & Sons: Singapore, 2010.

CHAPTER 24

Measurement of the Economic Vulnerability Index in the Third World and Emerging Economies: A Comparative Analysis Using Fuzzy Analytical Network Process and Technique for Order Preference by Similarity to an Ideal Solution

HASAN DİNÇER*, ÜMIT HACIOĞLU, and SERHAT YÜKSEL

School of Business and Management, İstanbul Medipol University, Beykoz 34810, Istanbul, Turkey

**Corresponding author. E-mail: hdincer@medipol.edu.tr*

ABSTRACT

The aim of this study is to rank the third world and emerging economies according to the economic vulnerability index. For this purpose, economic vulnerability data of United Nations for the year of 2015 is taken into the consideration. In the analysis process, 10 least developed economies which have high GDP growth and 10 selected emerging economies are evaluated by using fuzzy ANP and TOPSIS methods. It was concluded that Senegal is the most vulnerable country whereas Uganda has the best position among least developed countries. Furthermore, it was also identified that China is the most vulnerable emerging economy while Turkey has the best situation in this category. While ranking least developed and emerging economies together, it was defined that all least developed economies have the better vulnerability conditions than China, India and Egypt.

24.1 INTRODUCTION

Economic vulnerability is a significant problem for the economies. It shows the size of the volatility and uncertainty in the market. Therefore, when the economy of a country is considered vulnerable, investors will be reluctant to go into the market of this country. In other words, vulnerability influences economic growth of the country negatively. Hence, the reasons for economic vulnerability should be identified and necessary actions should be implemented to avoid this problem (Chauvet and Guillaumont, 2003).

Emerging and the third world economies are especially subject to the economic vulnerability problem. The main reason is that they try to provide economic growth by attracting foreign direct investments. Hence, they are influenced by the changes in interest rate and currency exchange rates (Kaplan, 2006). That is to say, this situation increases the vulnerability of these countries. Thus, when there is an economic crisis, it will affect these countries much more negatively than the others.

Economic vulnerability index (EDI) was developed by the United Nations to understand the economic problems of a country (Guillaumont, 2009). It measures the vulnerability of the countries against economic and environmental shocks. By looking at this index, it is possible to compare the economic vulnerability conditions of different countries. The EDI is used to select the countries that need financial aid. The data for this index are issued once in 3 years.

This index has eight indicators, which are size of the population, remoteness, export concentration, share of agriculture, forestry and fisheries in gross domestic product, homelessness due to natural disasters, instability of agricultural production, instability of exports of goods and services, and the percentage of population living in low-elevated coastal zone (Wang, 2013). As it can be seen, all of these criteria negatively affect economy. In other words, they increase economic vulnerability.

In this study, it is aimed to rank the third world and emerging economies regarding EDI. Within this scope, 10 least developed economies that have high gross domestic product (GDP) growth and 10 selected emerging economies are taken into the consideration. Additionally, economic vulnerability data of the United Nations for the year 2015 are evaluated by using fuzzy analytical network process (ANP) and technique for order preference by similarity to an ideal solution (TOPSIS) method to reach this objective. As a result, it will be possible to identify the economic vulnerability levels of these countries.

24.2 LITERATURE REVIEW

Some of the similar studies in the literature are presented in Table 24.1.

Table 24.1 shows that most of the studies are related to defining the factors that affect economic vulnerability. For example, Bussiere and Mulder (1999) made a study by considering 23 different countries with regression analysis. They defined that political stability increases economic vulnerability. Also, Cordina (2004) and Briguglio et al. (2009) emphasized the importance of higher economic growth to decrease vulnerability. Furthermore, Whelan and Maître (2010) concluded that the consumption level of European Union countries is a significant indicator of economic vulnerability by using logit analysis.

In addition to those studies, Kaplan and Yapraklı (2014) conducted a study that contains 12 emerging economies. As a result of regression analysis, they reached a conclusion that current account deficit and short-term debt increase vulnerability of the economies. Rocha and Moreira (2010), Greene and Liu (2015), Kaplan (2006) and Shin and Kim (2015) identified that volatility in the financial market stimulates vulnerability of the countries by using the same methodology. Also, Tunay (2009) underlined the importance of the level of completion for the vulnerability of the Turkish banking sector.

On the other side, there are also some other studies that focus on EDI. For instance, Guillaumont (2009) analyzed 65 least developed countries in his study. With the help of correlation analysis, it was determined that the EDI is very helpful to allocate aids between developing countries. Chauvet and Guillaumont (2003) emphasized a similar conclusion by using the same method. Additionally, Cariolle and Goujon (2013) analyzed the calculation process of this index and made some recommendation. Similarly, Wang (2013) concluded that the weights of the components in EVI should be changed to reach more accurate results.

Moreover, it can also be seen that some studies tried to analyze the level of vulnerability for some countries. Briguglio and Galea (2003) considered 117 different countries and identified that small island developing states are more vulnerable than other countries. Furthermore, Kerschner et al. (2013) presented a vulnerability map of US economy regarding oil prices. Also, Melkonyan (2014) defined that the agriculture sector in Armenia is vulnerable to any shocks by using correlation analysis. In addition to this type of studies, Gnangnon (2017) and McGillivray and Pham (2017) focused on the relationship between EDI and aid allocation.

TABLE 24.1 Featured Studies in the Literature.

Author	Scope	Methodology	Result
Bussiere and Mulder (1999)	23 different countries	Regression	Political stability is a very significant issue for economic vulnerability
Briguglio and Galea (2003)	117 different countries	Descriptive statistics	Small island developing states are more vulnerable than other countries
Chauvet and Guillaumont (2003)	59 developing countries	Generalized method of moments	Economic vulnerability is a significant aspect for enhancing aid effectiveness
Cordina (2004)	Literature review	Descriptive statistics	Vulnerable economies tend to have higher economic growth and lower consumption levels
Kaplan (2006)	China	Descriptive statistics	Undervalued exchange rate in Chine increases economic vulnerability
Guillaumont (2009)	65 least developed countries	Correlation analysis	Economic vulnerability index is very helpful to allocate aids between developing countries
Tunay (2009)	Turkey	Logit	Competition in Turkish banking sector increases vulnerability
Briguglio et al. (2009)	86 different countries	Regression	Countries, which have high economic growth, are not vulnerable
Rocha and Moreira (2010)	23 emerging countries	Regression	Financial liberalization is a significant aspect to decrease economic vulnerability
Whelan and Maître (2010)	European Union countries	Logit	Consumption level of the countries is a key aspect that changes the vulnerability of them
Cariolle and Goujon (2013)	130 developing countries	Correlation analysis	They analyzed the calculation process of economic vulnerability index
Kerschner et al. (2013)	The United States	Descriptive statistics	They created a vulnerability map of US economy about oil prices
Wang (2013)	184 different countries	Regression	He concluded that the weights of the components in Economic Vulnerability Index should be changed to reach more accurate results

TABLE 24.1 *(Continued)*

Author	Scope	Methodology	Result
Melkonyan (2014)	Armenia	Correlation analysis	He concluded that agriculture sector in Armenia is vulnerable to any shocks
Kaplan and Yapraklı (2014)	12 emerging economies	Regression	Current account deficit and short-term debt increase vulnerability of the economies
Shin and Kim (2015)	Korea	Regression	Uncertainty in the financial markets causes economic vulnerability
Greene and Liu (2015)	The United States	Monte Carlo	Volatility in oil prices increases the vulnerability of US economy
Gnangnon (2017)	145 different countries	Regression	In case of higher economic vulnerability, there is an increase in the financial aid to least developed countries
McGillivray and Pham (2017)	Literature review	Optimization analysis	Development aid allocation should be improved by giving more importance to economic vulnerability of the countries

As it can be understood from Table 24.1, the subject of economic vulnerability attracts the attention of many researchers. Therefore, there are lots of studies in the literature that consider the vulnerability of the countries and the meaning of EDI. It is also identified that regression and logit analyses are mainly used in these studies to reach the conclusion. Hence, it can be said that there is a need for a new study in the literature that focus on this subject by using a new and original methodology.

24.3 METHODOLOGY

24.3.1 FUZZY ANP

Saaty (1996) developed a fuzzy ANP method to make a decision in the complex situation. In the process of ANP, first of all, the purpose and clusters are explained. Next, supermatrix and the weights of the elements in this matrix are identified. Finally, the best alternative is selected. Nevertheless, fuzzy ANP is preferred instead of ANP because it reflects the real values much more effectively. The steps of fuzzy ANP are explained below.

Step 1: Fuzzy synthetic extent value is identified.

$$S_i = \sum_{j}^{m} M_{g_i}^{j} \otimes \left[\sum_{i=1}^{n} \sum_{j=1}^{m} M_{g_i}^{j} \right]^{-1} \tag{24.1}$$

In eq (1), $M_{g_i}^{j}$ shows triangular fuzzy numbers and G = $\{g_1, g_2, \ldots, g_m\}$ refers to the goal set. Moreover, $\sum_{j=1}^{m} M_{g_i}^{j}$ can be obtained by using following equation:

$$\sum_{j=1}^{m} M_{g_i}^{j} = \left(\sum_{j=1}^{m} l_j, \sum_{j=1}^{m} m_j, \sum_{j=1}^{m} u_j \right) \tag{24.2}$$

Furthermore, to get $\left[\sum_{i=1}^{n} \sum_{j=1}^{m} M_{g_i}^{j} \right]^{-1}$ the following two equations are used:

$$\sum_{i=1}^{n} \sum_{j=1}^{m} M_{g_i}^{j} = \left(\sum_{i=1}^{n} l_i, \sum_{i=1}^{n} m_i, \sum_{i=1}^{n} u_i \right) \tag{24.3}$$

$$\left[\sum_{i=1}^{n} \sum_{j=1}^{m} M_{g_i}^{j} \right]^{-1} = \left(\frac{1}{\sum_{i=1}^{n} l_i}, \frac{1}{\sum_{i=1}^{n} m_i}, \frac{1}{\sum_{i=1}^{n} u_i} \right) \tag{24.4}$$

Step 2: The degree of the possibility of $M_2 = (l_2, m_2, u_2) \geq M_1 = (l_1, m_1, u_1)$ is shown on the equation 5.

$$V(M_2 \geq M_1) = hgt(M_1 \cap M_2) = (d) = \begin{cases} 1, & if\ m_2 \geq m_1 \\ 0, & if\ l_1 \geq u_2 \\ \dfrac{l_1 - u_2}{(m_2 - u_2) - (m_1 - l_1}, & otherwise \end{cases} \quad (24.5)$$

Step 3: The degree of the possibility for a convex fuzzy number greater than k convex fuzzy numbers is shown in equation (6).

$$V(M \geq M_1, M_2, \ldots, M_k) = min\ V(M \geq M_i),\ i = 1,2, \ldots, k \quad (24.6)$$

In equation (6), "M" explains convex fuzzy numbers. Additionally, by considering the equations (6) and (7), weighted factor is determined in equation (8).

$$d'(A_i) = min\ V(S_i \geq S_k) \text{ where } k \neq i \quad (24.7)$$

$$W' = (d'(A_1), d'(A_2), \ldots, d'(A_n))^T \text{ where } A_i\ (i = 1,2, \ldots n) \quad (24.8)$$

Step 4: Following normalization weight vectors are calculated.

$$W = (d(A_1), d(A_2), \ldots, d(A_n))^T \quad (24.9)$$

Fuzzy ANP methodology is used in various studies in the literature. Nilashi et al. (2016) tried to analyze hospital information system, Altıntaş et al. (2016) determined the criteria of Six Sigma projects and Tavana et al. (2013) used this method for social media platform selection. In addition to these studies, Uygun et al. (2015) and Dargi et al. (2014) selected the best outsourcing provider with the help of fuzzy ANP technique. Similarly, Tadic et al. (2014) considered this method to select city logistic concept.

24.3.2 TECHNIQUE FOR ORDER PREFERENCE BY SIMILARITY TO AN IDEAL SOLUTION

Chen and Hwang (1992) developed the TOPSIS method as a multicriteria decision-making approach. While using this method, it can be possible to compare the values of different alternatives. In the first step of TOPSIS method, an evaluation matrix is created that contains m alternatives and n criteria. After that, this matrix is normalized by using the following equation:

$$r_{ij} = \frac{x_{ij}}{\sqrt{\sum_{i=1}^{m} x_{ij}^{2}}} \tag{24.10}$$

In eq (10), x_{ij} represents the jth objective value of the ith alternative. Next, the elements of this matrix are weighted as shown in equation (11).

$$T = (w_j r_{ij})_{mxn} \tag{24.11}$$

Furthermore, the best (A^+) and the worst (A^-) alternatives are determined. They are defined in equation (12).

$$A^{+} = t_{wj} \text{ and } A^{-} = t_{bj} \tag{24.12}$$

Just then, the distances to the best (d_{ib}) and worst (d_{iw}) alternatives from the target alternative are identified. They are shown in eq 24.13.

$$d_{ib} = \sqrt{\sum_{j=1}^{n} \left(t_{ij} - t_{bj}\right)^{2}} \text{ and } d_{iw} = \sqrt{\sum \sqrt{_{j=1}^{n} (t_{ij} - t_{wj})^{2}}} \tag{24.13}$$

In the final stage of TOPSIS method, the relative closeness to the ideal solution is determined. Within this scope, when this relative closeness is equal to 1, it will be the best alternative (Shojaeefard and Zare, 2016). TOPSIS is a very popular method in the literature. Mousavi-Nasab and Sotoudeh-Anvari used this method for material selection problems, whereas Karahalios (2017) tried to evaluate ballast water treatment systems. Additionally, Wang et al. (2017) benchmarked the performance of the energy firms with the help of TOPSIS method. Similarly, Akbaş and Bilgen (2017) aimed to choose the ideal gas fuel by using this methodology.

24.4 AN APPLICATION FOR LEAST DEVELOPED AND EMERGING ECONOMIES

24.4.1 DATA AND SCOPE

Economic vulnerability data of the United Nations for the year 2015 are used in this study. In the analysis process, 10 least developed economies that have high GDP growth and 10 selected emerging economies are considered. These countries are given in Table 24.2.

TABLE 24.2 Countries Used in the Study.

Least developed economies	Emerging economies
Angola	Bangladesh
Chad	Brazil
Democratic Republic of Congo	China
Ethiopia	Egypt
Ghana	India
Senegal	Indonesia
Sudan	Mexico
Tanzania	Nigeria
Uganda	South Africa
Zambia	Turkey

Determinants of the economic vulnerability index are described in Table 24.3.

Exogenous economic and environmental shocks are focal points of the economic vulnerability factors. That's why the exposure factors such as location and economic structure as well as environmental factors such as the trade and natural shocks are the main pillars of the economic vulnerability determinants.

TABLE 24.3 Economic Vulnerability Factors.

Criteria	Definition
Population (C1)	A variety of natural shocks
Remoteness (C2)	The geographic distance between the capitals or major cities
Share of population in low elevated coastal zones (C3)	The coastal impacts associated with climate change
Export concentration (C4)	The trade shocks related to the merchandise export activities
Shares of agriculture, forestry, and fisheries in GDP (C5)	The share of gross value added in the agriculture, forestry and fishery sectors in GDP
Victims of natural disasters (C6)	The human impact of weather, climate, etc.

Agricultural instability (C7)	The annual changes of agricultural production
Export instability (C8)	The changes of export earnings of goods and services

Source: United Nations, Development Policy and Analysis Division.

24.5 ANALYSIS RESULTS

The integrated model has been applied to measure the determinants of the EDI. The first phase includes the fuzzy ANP approach for computing the relative importance of each criterion. The second phase defines the TOPSIS method for ranking the performance results of the selected economies.

The fuzzy ANP results present the weights of the criteria for the EDI in Table 24.4.

TABLE 24.4 Weights of Economic Vulnerability Index.

Criteria	Weights
Population (C1)	0.135
Remoteness (C2)	0.130
Share of [opulation in low elevated coastal zones (C3)	0.128
Export concentration (C4)	0.141
Shares of agriculture, forestry, and fisheries in GDP (C5)	0.123
Victims of natural disasters (C6)	0.110
Agricultural instability (C7)	0.115
Export instability (C8)	0.117

Weights of EDI can be listed as export concentration (C4), population (C1), remoteness (C2), share of population in low elevated coastal zones (C3), share of agriculture, forestry and fisheries in GDP (C5), export instability (C8), agricultural instability (C7), and victims of natural disasters (C6), respectively. According to the expert choices, export concentration (C4) has the best importance in the determinants of the EDI while victims of natural disasters (C6) are the weakest factor in the index.

In the following phase, selected least developed and emerging economies have been ranked by the TOPSIS method. Weights of the EDI in Table 24.4 have been considered to compute the final evaluation results of the countries. Table 24.5 represents the comparative analysis results of EDI for the least developed economies.

TABLE 24.5 Ranking Results of the Least Developed Economies.

Least developed economies	Closeness coefficient	Ranking
Angola	0.329	6
Chad	0.355	4
Democratic Republic of Congo	0.290	7
Ethiopia	0.410	2
Ghana	0.255	9
Senegal	0.512	1
Sudan	0.379	3
Tanzania	0.270	8
Uganda	0.199	10
Zambia	0.338	5

Senegal is the first with the value of 0.512 and Ethiopia follows with the value of 0.410. Uganda has a last place in the economic vulnerability as seen in Table 24.5. In other words, it can be said that Senegal is the most vulnerable country while Uganda has the best position.

Table 24.6 illustrates the ranking of economic vulnerability in the emerging economies. According to the closeness coefficient values, China has the first position in the economic vulnerability, while Turkey is placed in the last. That is to say, China is the most vulnerable emerging economy, but Turkey is the most successful country according to the analysis results.

Table 24.7 presents the overall ranking results including least developed and emerging economies together.

TABLE 24.6 Ranking Results of the Emerging Economies.

Emerging economies	Closeness coefficient	Ranking
Bangladesh	0.379	5th
Brazil	0.206	8th
China	0.476	1st
Egypt	0.403	4th

Emerging economies	Closeness coefficient	Ranking
India	0.413	3rd
Indonesia	0.277	6th
Mexico	0.161	9th
Nigeria	0.474	2nd
South Africa	0.237	7th
Turkey	0.139	10th

TABLE 24.7 Overall Ranking Results.

Countries	Closeness coefficient	Ranking
China	0.491	1
India	0.403	2
Egypt	0.384	3
Sudan	0.347	4
Senegal	0.341	5
Chad	0.317	6
Angola	0.312	7
Bangladesh	0.305	8
Nigeria	0.299	9
Zambia	0.295	10
Ethiopia	0.284	11
Tanzania	0.227	12
Indonesia	0.224	13
Democratic Republic of Congo	0.223	14
Ghana	0,219	15
Uganda	0.188	16
South Africa	0.179	17
Brazil	0.171	18
Mexico	0.132	19
Turkey	0.084	20

When both set of countries considered in the analysis, the results demonstrate that China is always first in the economic vulnerability evaluation. However, Turkey has the last place in ordering of EDI within the emerging and the least developed economies. It can also be understood that China, India, and Egypt are more vulnerable countries than all least developed economies.

24.6 CONCLUSION

In this study, it was aimed to rank the third world and emerging economies according to the EDI. Within this context, economic vulnerability data of the United Nations for the year 2015 is taken into consideration. In the analysis process, 10 least developed economies that have high GDP growth and 10 selected emerging economies are analyzed by using fuzzy ANP and TOPSIS methods.

In the first stage of the analysis, the criteria of EDI are weighted by using fuzzy ANP method. According to these results, export concentration has the highest weight, which means it is the most important criterion. On the other side, the criterion of victims of natural disasters has the lowest weight. This situation shows that this criterion is the weakest factor in the index.

After determining the weights of the criteria, countries were ranked regarding EDI with the help of TOPSIS method. With respect to the ranking results of the least developed countries, it was identified that Senegal is the first with the value of 0.512, whereas Uganda has the last place. This result explains that Senegal is the most vulnerable country among least developed countries. On the other hand, it can also be said that Uganda has the best position for this concept.

Moreover, while considering the ranking results of the emerging economies, it was defined that China has the first position in the economic vulnerability while Turkey is placed in the last. In other words, it was concluded that China is the most vulnerable emerging economy. Also, Turkey is the most successful emerging economy with respect to the economic vulnerability. Additionally, Mexico has the second best position among all the emerging economies.

In addition to these analyses, least developed and emerging economies are also ranked together. It was determined that China is the first economy, whereas Turkey is the last economy regarding economic vulnerability. Moreover, India is the second worst and Mexico is the second best country. Another important result of this analysis is that all the least developed economies have better vulnerability conditions than China, India, and Egypt.

KEYWORDS

- **economic vulnerability index**
- **third world economies**
- **emerging economies**
- **comparative analysis**
- **fuzzy analytical network process**
- **technique for order preference by similarity to an ideal solution**

REFERENCES

Akbaş, H.; Bilgen, B. An Integrated Fuzzy QFD and TOPSIS Methodology for Choosing the Ideal Gas Fuel at WWTPs. *Energy* **2017,** *125*, 484–497.

Altintas, M., Erginel, N.; Kucuk, G. Determining the Criteria and Evaluating Six Sigma Projects via Fuzzy ANP Method in Group Decision. *IFAC-PapersOnLine* **2016,** *49* (12), 1850–1855.

Briguglio, L.; Galea, W. Updating and Augmenting the Economic Vulnerability Index. *Occasional Paper, University of Malta,* 2003.

Briguglio, L., Cordina, G., Farrugia, N.; Vella, S. Economic Vulnerability and Resilience: Concepts and Measurements. *Oxford Develop. Stud.* **2009,** *37* (3), 229–224.

Bussiere, M.; Mulder, C. Political Instability and Economic Vulnerability. IMF Working Paper No. WP/99/46, 1999.

Cariolle, J.; Goujon, M. A Retrospective Economic Vulnerability Index, 1990–2011 Using the 2012 UN-CDP Definitions. *Development*. **2013,** *17*.

Chauvet, L.; Guillaumont, P. Aid and Growth Revisited: Policy, Economic Vulnerability and Political Instability. *ABCDE* **2003,** *95*.

Chen, S. J.; Hwang, C. L. Fuzzy Multiple Attribute Decision Making Methods. In *Fuzzy Multiple Attribute Decision Making*; Springer: Berlin, Heidelberg, 1992; p. 289–486.

Cordina, G. Economic Vulnerability and Economic Growth: Some Results from a Neo-classical Growth Modelling Approach. *J. Econ. Develop.* **2004,** *29* (2), 21–39.

Dargi, A.; Anjomshoae, A.; Galankashi, M. R.; Memari, A.; Tap, M. B. M. Supplier Selection: A Fuzzy-ANP Approach. *Procedia Comput. Sci.* **2014,** *31*, 691–700.

Gnangnon, S. K. Structural Economic Vulnerability, Openness and Bilateral Development Aid Flows. *Econ. Anal. Policy* **2017,** *53*, 77–95.

Greene, D. L.; Liu, C. US Oil Dependence 2014: Is Energy Independence in Sight? *Energy Policy* **2015,** *85*, 126–137.

Guillaumont, P. An Economic Vulnerability Index: Its Design and use for International Development Policy. *Oxford Develop. Stud.* **2009,** *37* (3), 193–228.

Kaplan, F.; Yapraklı, S. Ekonomik Kırılganlık Endeksi Göstergelerinin Döviz Kuru Üzerindeki Etkileri: Kırılgan 12 Ülke Üzerine Panel Veri Analizi. *Uluslararası Alanya İşletme Fakültesi Dergisi* **2014,** *6* (3).

Kaplan, S. B. The Political Obstacles to Greater Exchange Rate Flexibility in China. *World Develop*. **2006,** *34* (7), 1182–1200.

Karahalios, H. The Application of the AHP-TOPSIS for Evaluating Ballast Water Treatment Systems by Ship Operators. *Transp. Res. Part D: Transp. Environ.* **2017,** *52*, 172–184.

Kerschner, C.; Prell, C.; Feng, K.; Hubacek, K. Economic Vulnerability to Peak Oil. *Global Environ. Change* **2013,** *23* (6), 1424–1433.

McGillivray, M.; Pham, T. K. C. Reforming Performance-based Aid Allocation Practice. *World Develop.* **2017,** *90*, 1–5.

Melkonyan, A. Environmental and Socio-economic Vulnerability of Agricultural Sector in Armenia. *Sci Total Environ.* **2014,** *488*, 333–342.

Mousavi-Nasab, S. H.; Sotoudeh-Anvari, A. A Comprehensive MCDM-based Approach Using TOPSIS, COPRAS and DEA as an Auxiliary Tool for Material Selection Problems. *Mater. Design.* **2017,** *121*, 237–253.

Nilashi, M.; Ahmadi, H.; Ahani, A.; Ravangard, R.; Bin Ibrahim, O. Determining the Importance of Hospital Information System Adoption Factors Using Fuzzy Analytic Network Process (ANP). *Technol. Forecast. Soc. Change* **2016,** *111*, 244–264.

Rocha, K.; Moreira, A. The Role of Domestic Fundamentals on the Economic Vulnerability of Emerging Markets. *Emerg. Markets Rev.* **2010,** *11* (2), 173–182.

Saaty, T. L. *The Analytic Network Process*. RWS Publications: Pittsburgh, PA; 1996.

Shin, D.; Kim, B. Liquidity and Credit Risk Before and After the Global Financial Crisis: Evidence from the Korean Corporate Bond Market. *Pacific-Basin Finance J.* **2015,** *33*, 38–61.

Shojaeefard, M. H.; Zare, J. Modeling and Combined Application of the Modified NSGA-II and TOPSIS to Optimize a Refrigerant-to-air Multi-pass Louvered Fin-and-flat Tube Condenser. *Appl. Thermal Eng.* **2016,** *103*, 212–225.

Tadić, S.; Zečević, S.; Krstić, M. A Novel Hybrid MCDM Model Based on Fuzzy DEMATEL, Fuzzy ANP and Fuzzy VIKOR for City Logistics Concept Selection. *Expert Syst. Appl.* **2014,** *41* (18), 8112–8128.

Tavana, M.; Momeni, E.; Rezaeiniya, N.; Mirhedayatian, S. M.; Rezaeiniya, H. A Novel Hybrid Social Media Platform Selection Model Using Fuzzy ANP and COPRAS-G. *Expert Syst. Appl.* **2013,** *40* (14), 5694–5702.

Tunay, B. Türk Bankacılık Sektöründe Rekabet ve Kırılganlık. *Bankacılar Dergisi*. **2009,** *68*, 30–54.

Uygun, Ö.; Kaçamak, H.; Kahraman, Ü. A. An Integrated DEMATEL and Fuzzy ANP Techniques for Evaluation and Selection of Outsourcing Provider for a Telecommunication Company. *Comput. Ind. Eng.* **2015,** *86*, 137–146.

Wang, C. C. Reconsidering the Economic Vulnerability Index of the United Nations. *Canadian J. Develop. Studies/Revue canadienne d'études du développement* **2013,** *34* (4), 553–568.

Wang, E.; Alp, N.; Shi, J.; Wang, C.; Zhang, X.; Chen, H. Multi-Criteria Building Energy Performance Benchmarking Through Variable Clustering Based Compromise TOPSIS with Objective Entropy Weighting. *Energy* **2017,** *125*, 197–210.

Whelan, C. T.; Maître, B. Welfare Regime and Social Class Variation in Poverty and Economic Vulnerability in Europe: An Analysis of EU-SILC. *J. Eur. Social Policy* **2010,** *20* (4), 316–332.

CHAPTER 25

Happiness Conundrum in the Third World

DEBOJYOTI ROY*

Department of Business Management, Adamas University, Kolkata, West Bengal, India

**E-mail: debojyotiroy3333@gmail.com*

ABSTRACT

The chapter tried to take a look at the different determinants of happiness. We did a comparison between countries of first and third world. Although literature review found that money cannot only buy happiness but the results shows otherwise. The area of economy to happiness need further review. Generosity came out as more significant determinant than anticipated. Future research should investigate more in the area of trust in society to explore the relation between trust and happiness.

25.1 INTRODUCTION

It may not be easy to assume people's happiness score. Income, education, health levels, and other factors determine the level of happiness. But because it is such difficult to find how much people are happy, it is becoming a favorite subject for economist and politician. If politicians and economist analyze the happiness score it becomes useful for them to formulate policies. Most nations along with the gross domestic product (GDP), export, and unemployment reports track happiness report.

According to Adrian White, a psychologist from the University of Leicester, a nation's level of happiness is closely related to wealth. In the level of significance, wealth is followed by education and health. But the findings of White are not in line with conventional ideas like capitalist nation are unhappier or people from nations with strong collective identities are happier than countries where society is individualistic in nature. Third world countries where income is low where people are perceived to have a sense of collective identity perform badly in the index of happiness. In his survey, it has been found that countries that are economically better are more worried about happiness than being wealthy.

United nations have principally agreed that for sustained development both in economic and social front wellbeing of common mass are a critical component. People all over the world are demanding that policies should be aligned to the things that they care for.

25.2 WHAT IS HAPPINESS?

Is it not surprising that some people are content with less, while others need a whole lot to be satisfied? Some reports do suggest that mental health is one of the most important reasons to determine if a person is happy with life or not. The question is do people make them unsatisfied and unhappy? What differentiate people?

If we ask people what they want in life, the most common answer is happiness. Our most common aspiration from life is to find satisfaction in our deeds. There is an old Chinese story that defined happiness as "something to do, someone to love, and something to hope for." What most of the people aspires from life are freedom from the ordeal, thriving in their daily duties, plenteousness, feeling well about one's way of life, and being aware of the real feeling of bliss (Barrow, 1980; Argyle, 1999, 2001; Boniwell and David, 2014).

According to Aristotle, "happiness should be viewed as a consolidation of 'hedonia' (immediate pleasures) and 'eudaimonia' (a life well lived)." In other words, happiness should be as mixtures of short- and long-term desire, that is, what one wants currently and what they need in future to be happy. The fortunate one feels positive feelings for a longer period, negative feelings for short durations, and high level of satisfaction with life (Diener et al., 1999, 2009; Diener, 2008).

Particularly, the word "happiness" can have different meanings as per the context. Particularly for this chapter, the construct is applied to illustrate what makes one satisfied with life. Veenhoven (1984) defined the construct happiness "*as the degree to which someone evaluates the overall quality of his or her present life-as-a-whole positively.*"

Happiness is a debatable hypothetical experience (Bradburn, 1969; Tatarkiewicz, 1976; Diener and Seligman, 2002; Diener and Biswas-Diener, 2008; Eid and Larsen, 2008; Seligman, 2011), which is a state of mind. In the colloquy of Marcus Aurelius, "the happiness of your life depends on the quality of your thoughts," while Seneca put it more directly by saying that "A man's as miserable as he thinks he is." The decision is on oneself how one is going to face a life crisis or incidence (i.e., optimism or pessimism).

The perception of being happy can be defined in a couple of theories. The first theory propagates the idea that life events that are of significance to one may bring transformation in their outlook, and thus may also have enduring effects on happiness. Actual life circumstances do have cognizable effects on happiness. The second idea of the Set-point theory is also used to define happiness (Lykken and Tellegen, 1996). It is generally believed that people have an established set-point that flags the reason for their happiness. The set points are determined by the inborn attributes and psyche. People may initially react to sudden life-changing events but will ultimately "adapt" to the challenging situations. There may be an increase or decrease in emotion due to sudden change in lifestyle but eventually, those dwindling emotions are replaced by neutral feelings (Frederick and Loewenstein, 1999). The third theory of happiness is about the theory of expectation. The theory states that happiness is not dependent on material accomplishment but more related to the gap between what one achieved and what they aspired for. As people achieve more in life, aspirations do rise. With achievement, happiness does increase but with the rise of aspiration, there will be a feeling of un-fulfillment. As a result of the above theory amount of net happiness remains same (Easterlin, 1974).

Therefore, we can conclude that people's aspiration and thereby their happiness is shaped by two key mechanisms—social correlation and adaptation. Social contrast is a psychological phenomenon that exhibits one's penchant to compare their lifestyle with their peer group's (specifically earnings and social standings). If people comparing themselves to others doing better than them, they grow distaste toward their own life situation. Adaptation refers to the extent of capability how individuals acclimatize to changes of life scenarios. Thereby, a better outlook of life circumstances will give an individual a sense of happiness although the feeling can be transient in nature. This infers that people's analogous rather than outright circumstances are the main cause of their individual content. This may help us in explaining the phenomenon why happiness levels will remain solid, irrespective of the timeline, even when circumstances that may have caused happiness have changed (Easterlin, 1974, 2001).

25.3 BUT WHY IS HAPPINESS IS SO MUCH EMPHASIZED RECENTLY?

It is because happy people live longer, are more effective, earn more, and are better residents. Reports have been published that has considered factors other than wealth, health, and education. The factors are, namely, GDP/capita, healthy life expectancy, social support, freedom, generosity, and absence of corruption in a country.

25.4 HEALTH AND HAPPINESS

Poignant states such as happiness and outlook toward life can lead to different levels of feelings to life events such as stress and anxiety. Various findings in medicine have exhibited that one's sentimental reactions to individual life scenarios affect one's physiology in a pattern that likely can be destructive or favorable for health (Collet et al., 1997). Studies also found that happiness can foresee durability and physical well-being (Lyubomirsky et al., 2005; Bjørnskov, 2008). "Positive thinking people," who proclaim that they are satisfied with life shows better health conditions (Stewart et al., 2007). Good mental and physical health is a precursor of satisfaction and content. People with good health find it easier to bounce back from difficult events of life. Healthier people can cope better with stress and trauma. Research has shown that happy people have a better immune system (Veenhoven, 2008; Diener and Chan, 2011). Happy people are prone to act healthy, who generally avoid unhealthy behavior. They get embroiled in sports more often and are more inclined to watch weights (Rasciute and Downward, 2010). Therefore, we can argue that a positive state of health can lead to a positive lifestyle and thereby ensures happiness.

25.5 FAMILY AND HAPPINESS

People who have a cordial family reported to be happier. They have exhibited higher positive fervors and less negative emotions. Family not only gives us love but also provides physical, spiritual, economic, and emotional support. The family gives us the opportunity to associate with society. It has an important role to play in the development of conversation, problem solving, behavior control, and emotional response. Those who have a strong family to support, their life gratification and self-efficacy are comparatively higher (Shi et al., 2006) People with strong family values and better relationships are happy. They tend to have a rich social network with a lot of friends. Although having a lot of friends and a steady family do not guarantee happiness, having a positive family life helps in the pursuit of happiness (Stack and Eshelman, 1998; Steel and Ones, 2002).

25.6 GENEROSITY AND HAPPINESS

A number of studies have observed that although wealthy people are happier than people with less money, it is how the money is spent matters most. Past studies which explored the correlational statistics of variables show that people report greater happiness when they have spent money on others (Dunn et. al., 2017). Having money and spending money on others are reported to be equally important (Dunn and Norton, 2013). Aknin et al. (2013) recognized from a study of 136 countries that there is a correlation between charitable spending and happiness. In 120 out of those 136 countries, there was a positive accord between charity and happiness (the researcher controlled for income and other demographic variables). This strength of the relationship between charities to happiness is significant in a majority of those 136 countries. Although the fortitude of the relationship varied between countries, specially those between first world countries to third world countries, residents from both poor and rich countries alike reported that happiness goes north if money is spent on charity.

Action-oriented people have a philanthropic beam (Ricard, 2006; Dunn et al., 2008). They find that wealth can buy happiness but only by giving it away. Wang and Graddy (2008) proposed that people who are happy are emotionally more capable. They have the desire to help others and demonstrate better confident identities, upholding altruistic giving behavior. Konow and Earley (2008) also contended that happier people provide more to the needy because they are driven by emotions, which are predominantly positive in nature. Their study further suggests that people who are happy do indeed aid more at every opportunity. Harbaugh et al. (2007) observed by performing functional magnetic resonance imaging (fMRI) on the brain that when giving money to charity brain shows a similar response to the experience of pleasure and reward. Experimental works by Field et al. (1998) hint at a spontaneous relationship between giving and happiness. Meier and

Stutzer (2008) found that the more one performs volunteer work, the more they experience life satisfaction. People who are happy live more in the moment, their values and behavior are closely aligned, and their satisfaction with life remains quite significant.

25.7 FREEDOM AND HAPPINESS

When an individual has the ability to make major life choices it can be termed as "Freedom." The "scope" implies that for an individual, there should be a "convenience" to choose. The presence of "preference" requires that an individual has the "scope" to choose (Veenhoven, 2000). Freedom can be classified in two broad categories: actual freedom and perceived freedom (Bay, 1970)

Actual freedom can be defined in broadly three categories—*social, psychological, and potential.* Social freedom talks about ease of adoption and points how people can act without any hindrance from people or existent system. Psychological freedom discusses about one's ability to accept things as they come without any inner strictures. Potential freedom is a construct that defines the presence of possible choice and knowledge of all foreign opportunities (Veenohoven, 2000).

Perceived freedom is a related concept which can veer. One thinks one is free while in actuality it is not, or one may think one is bonded in spite of extensive choice. Each of the concept of actual freedom and perceived freedom can affect happiness, possibly independently (Veenohoven, 2000).

Social freedom can be further classified in three categories—*economic, political, and personal.*

Economic freedom is the ability to do business without any restriction and political freedom is to get involved in any political process without any fear or restriction. Personal freedom is the absence of any restriction in the personal sphere of life (Veenohoven, 2000).

A study by Brule and Veenohoven (2014) found that freedom and happiness are conjoint in nature. The strength of the association between happiness and psychological freedom is modest, whereas the relation between potential freedom and happiness is very high. Both social and potential freedom are attained with the due process of societal evolution, whereas psychological freedom is an independent phenomenon or can be tied to societal advancement (albeit negatively). The strongest relation is between happiness and perceived freedom among different nation. Hence, we can conclude that all kinds of freedoms are the root cause of happiness across different nations.

Since different categories of freedom are intercorrelated, the change in one affects happiness through the other.

25.8 ECONOMY AND HAPPINESS

"Those who say that money can't buy happiness don't know where to shop."—Anon Easterlin (1974) posed a fundamental question, "Will raising the incomes of all increase the happiness of all?" Though most researchers have steadfastly assumed that the answer is "yes," theories of relative predilection predict that the answer may be "no."

Wealth should be seen as a relative concept. It is a natural tendency of even wealthy people to compare their wealth with others, especially who are superior to them. As civil creature, the sense of social worth and happiness is borrowed in part from comparison with other groups and individuals that we aspire to emulate. Therefore, countries where the class divide in terms of wealth is higher, the happiness is significantly low and vice versa (Lucas and Schimmack, 2009; Wilkinson and Pickett, 2009). A similar study of 21 countries has been conducted by Hagerty and Veenhoven where they found of the 21 countries, 7 now show a compelling positive corollary of income growth, and only one shows a negative coefficient. Higher income countries show smaller effects of unadulterated income than lower income countries, persistent with diminishing the minimal utility of money (Hagerty and Veenhoven, 2003).

Some studies have shown that people those who are wealthy are not happier than people on a tight budget (Hagerty and Veenhoven, 2003). Even a number of bystanders who are socially active have pointed out that the humongous increases in income in countries which are democratic in nature have not seen any significant difference in happiness level. In a seminal paper, Easterlin (1974) asked a question like "Are you happy?" and collected "happiness data," where he exhibited that individual's livelihood in a singular year was positively correlated with happiness responses. But over a timespan, the moderate happiness feedback remains unchanged although there is a pointed increase in average wage levels. It can be concluded that once the basic obligations are fulfilled, riches is not directly commensurate to satisfaction. Scitovsky (1976) is of the viewpoint that, beyond a certain level of earthly contentment, further abundance does not add to well-being—and may even depreciate from it—unless it is escorted by satisfying social grids and intellectually exhilarating recreational activities. Personal happiness does not imperatively increase with personal fortune as earthly happiness is commonly short lived (Lane, 2000; Kasser, 2002).

25.9 TRUST IN GOVERNMENT AND HAPPINESS

In the composition of social science, fraud is one of the important issues that recently have gained prominence. In a significant factual study conducted on more than 70 countries where the data are primarily based on the World Value Survey and WHO, Helliwell and Huang interpreted that "the effects of good government remain as the single most important variable explaining international differences 17 in life satisfaction in the full global sample, while international differences in per capita incomes are frequently insignificant" (Helliwell and Huang 2008a, p. 617). "Quality of government," is a construct which is deliberated to the extent of adhering to governmental doctrines such as the rule of law, probity in the civil service, management of corruption, and institutional efficacy, and "has an independent impact on the life satisfaction of people in rich as well as in poor countries" even when controlled for variables such as piety, the extent of democracy, life expectancy, and postmaterialist codes (Samanni and Holmberg, 2010, 11).

Why living under corrupt, nepotic, and prejudiced government institutions would produce unhappiness is a million dollar answer. The answer may lie in the antithetic philosophy of global systems of provisions of social aids and welfares. The phenomenon can be interpreted with a quotation from a research analysis done by the United Nations Human Development program about the bearings in Bosnia and Herzegovina in the year 2002. The survey conducted in the report came to a conclusion that about 70% of the residents in the country contemplated that their provincial governmental authorities and international aid organizations working in place (including the UN organizations) were "severely corrupt." In addition, the report also states: "For the average citizen, therefore, it seems that corruption has broken down all barriers and dictates the rules of life. That is not very different from saying that they interpret life in terms of corruption" (UNDP 2002, 77). Therefore, it can be argued with reason that residents who "interpret life in terms of corruption" are not likely to be happy with their current situation.

Truancy of awareness about corruption in government leads to social trust. The research on social capital and subsequently social trust is quite prevalent from mid-1990 onwards. The overwhelming interest in the area of social trust is because empirical exploration portrays that having a high degree of social faith is allied to a number of compelling elements at the personal level, such as resilience toward minorities, aid in civic activities, literacy, hardiness, and subjective well-being. Although conceptualizing social trust is not easy, it is defined by Uslaner as "social trust is an expression of optimism about the future and the possibility that one has control over one's destiny in life" (Uslaner, 2002). According to Deely and Newton, most people can be trusted, when they act otherwise, they are actually making a righteous appraisal of the society in which they live (Delhey and Newton, 2003). In the presence of social trust, people are assured about the bright and prosperous future and expect that people will demonstrate moral character. A number of surveys confirm that life gratification interacts very strongly with trusts which are organizational in nature (i.e., trust in coworkers and trust management) even if the wages are considered (Helliwell and Huang, 2005). The connection can be explained with the argument that, work-life experiences are important and one of the key indicators for happiness.

25.10 DATA ANALYSIS

We have collected data for a period of 2 years. We have asked respondents from all walks of life questions how happy they are in general in their life. We asked how happy they are with the available freedom in their country. Do generosity in the society makes them happy? Do having a strong social network give them moral support and help them to face challenges in life? Are they earning enough and happy with their standard of living? Are people happy with their overall health and are they really bothered about corruption done by elected officials and bureaucrats?

We observed that there is a high degree of correlation between happiness and the economy, family, freedom, and health. Less important is trust in government (although at 40% certainly not something to ignore) and generosity.

Health, family, freedom, and wealth seem to correlate well with overall happiness, while trust in government and generosity are less important. The result is in line with the study of Frey and Stutzer (2008), where they observed that people tend to value extrinsic desire which is directly related to income or social status and prestige, over intrinsic ones, such as spending time with family and friends or leisure activities.

In the following results, we find when people have enough wealth to lead a comfortable life they are happiest. There is an old saying that "Money can't buy happiness." But in our data, we find the antecedent of that and it is proved that money really can buy happiness. From our current dataset, we can conclude that money is directly proportional to happiness because of strong correlational coefficient. We can also see that economy and health are very strongly correlated.

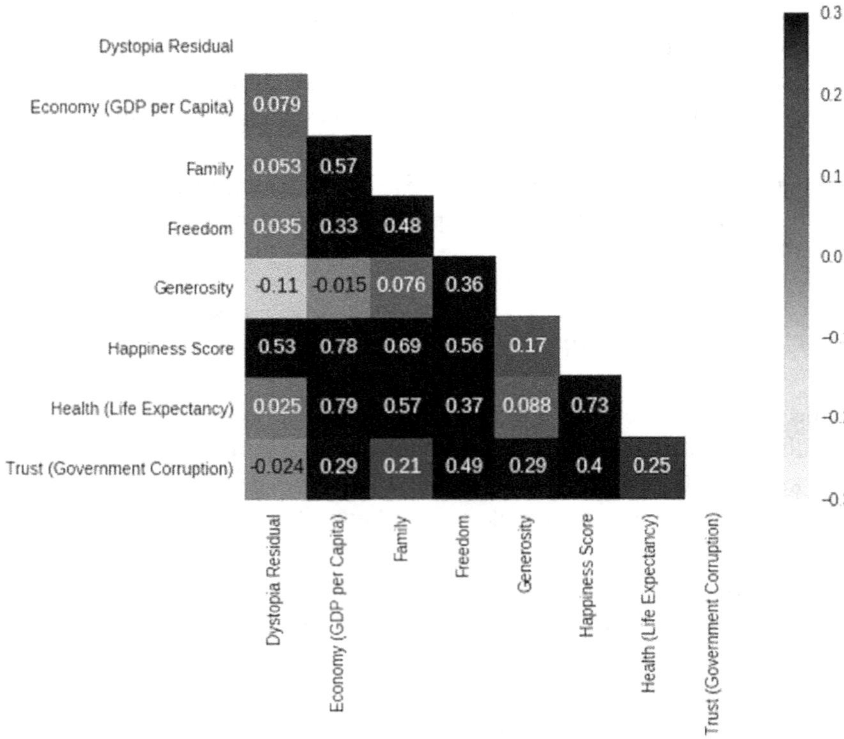

FIGURE 25.1 Correlation between determinants of happiness.

Hence, we can conclude that if one has wealth, they can guarantee better healthcare, and therefore they can be really happy in their life.

Now we need to check the lag plots that are used to check if a dataset or time series is random. Random data should not exhibit any structure in the lag plot. Nonrandom structure implies that the underlying data are not random. Except family and economy there is no pattern in the data, and it can be said that from the lag plots that economy and happy family are directly proportional to happiness.

Ensembling all determinants for happiness, we can find from the following graph that.

Sub-Saharan Africa places give more importance on family and economic success relative to the other variables. Middle Eastern countries place more importance on health. Countries of southern Asia place importance on generosity, health, and trust. Eastern Asia places more importance on generosity and health and less regard for government corruption.

Overall, we have a few countries with decidedly unusual preferences for generosity, a good number of countries from around the world that are more heavily influenced by health considerations, and a strong overall tendency toward the importance of freedom and family. It is also interesting to note that trust in government is not decisive for anybody, apparently. Or, more accurately, that nobody has a high enough trust in government for that factor to be one of the most important ones.

Now we will take a look by region. Americas and Tasmanian countries which are traditionally looked as First World countries are happier. They have a high cumulative GDP per capita and going by previous results, it can be termed as one of the strong factors for happiness. On the other hand, countries from the Saharan region and southeastern Asia which have countries such as Bangladesh and Pakistan where GDP per capita is significantly lower are unhappy. In the span of 2 years when the survey was done it is observed that the overall happiness score remain same all across the region but in almost all regions GDP per capita has increased. It is also to be observed that except GDP, all other determinants of happiness such as family, freedom, health, and trust in government are significantly reduced in the following year than the base year. But generosity has either remained same or has increased across the region. We can conclude from the results that money really made people happy, and they shared the happiness by giving more.

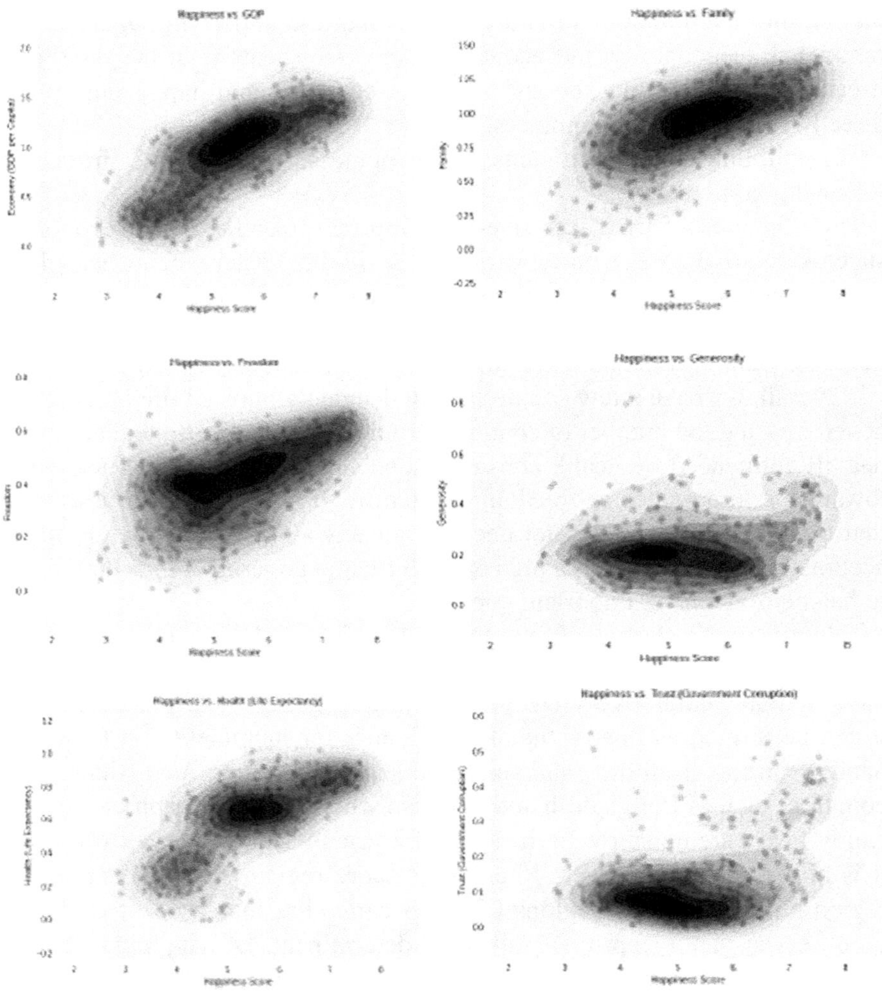

FIGURE 25.2 Heat map happiness versus determinants.

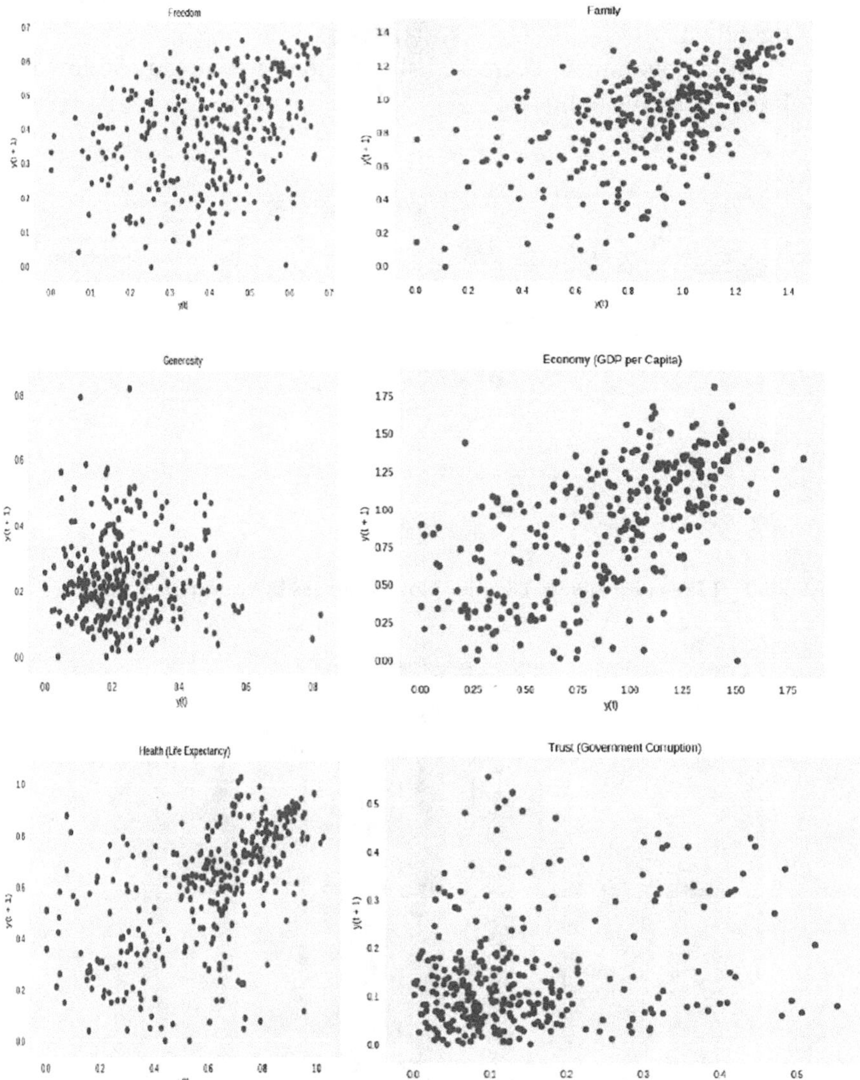

FIGURE 25.3 Scatter plot of happiness determinants.

In general, we can say that in 2015–2016:

1. Generosity and happiness were effectively unchanged.
2. GDP rose.

3. Life expectancy, family, and freedom declined pretty much across the board.
4. Trust in government decreased slightly, but was already so low that it hardly seems to matter.

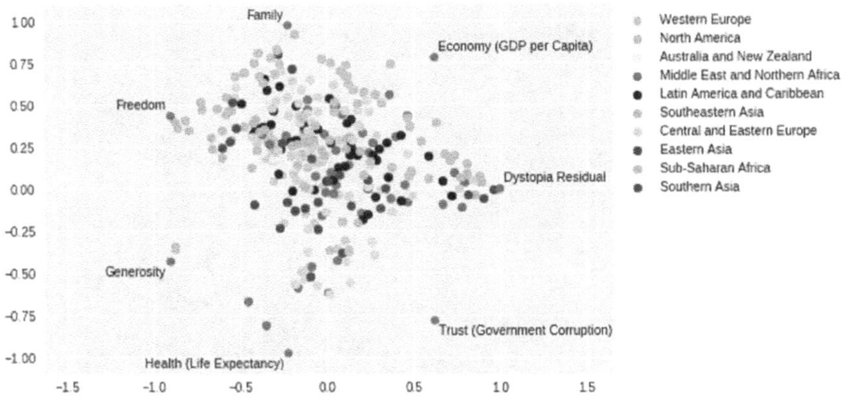

FIGURE 25.4 (See color insert.) Scatter plot happiness by countries.

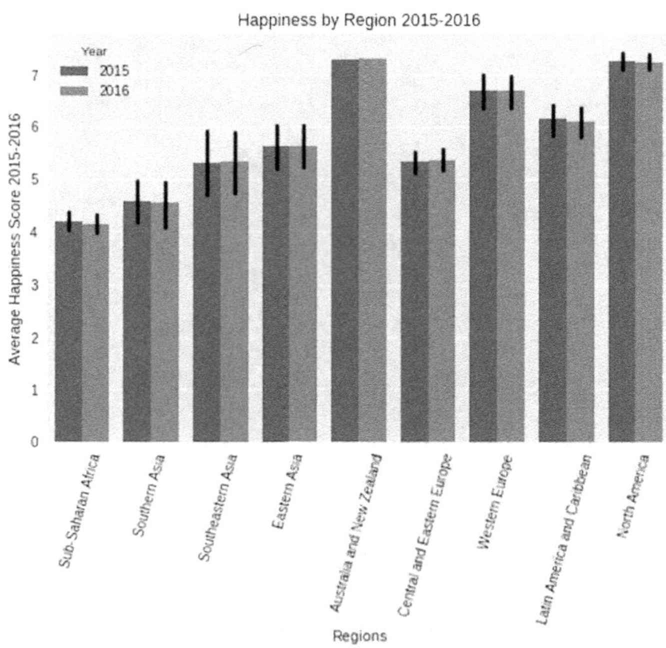

FIGURE 25.5 Bar-plot happiness by region.

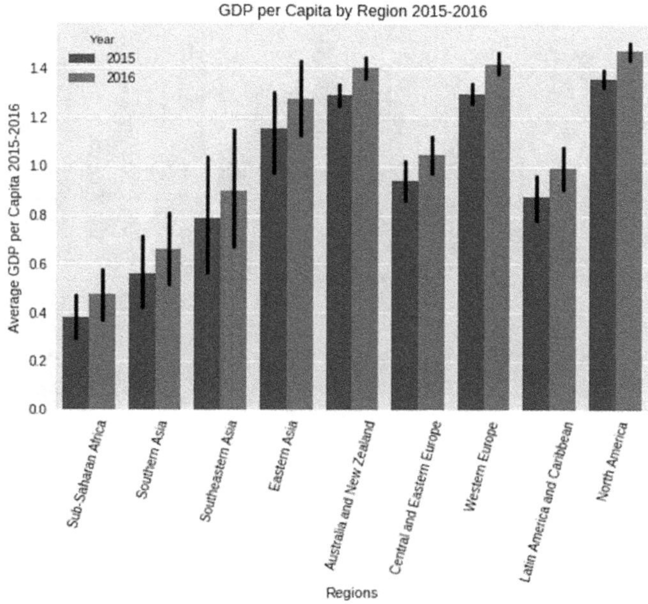

FIGURE 25.6 GDP per capita by region.

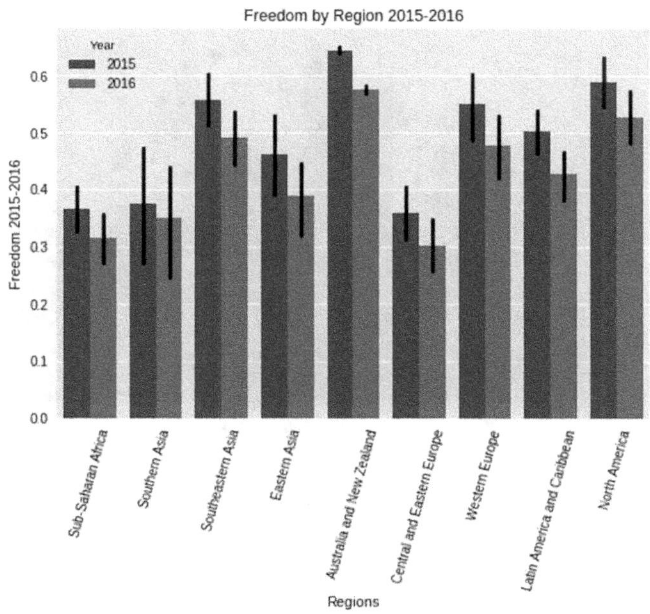

FIGURE 25.7 Stacked plot freedom by region.

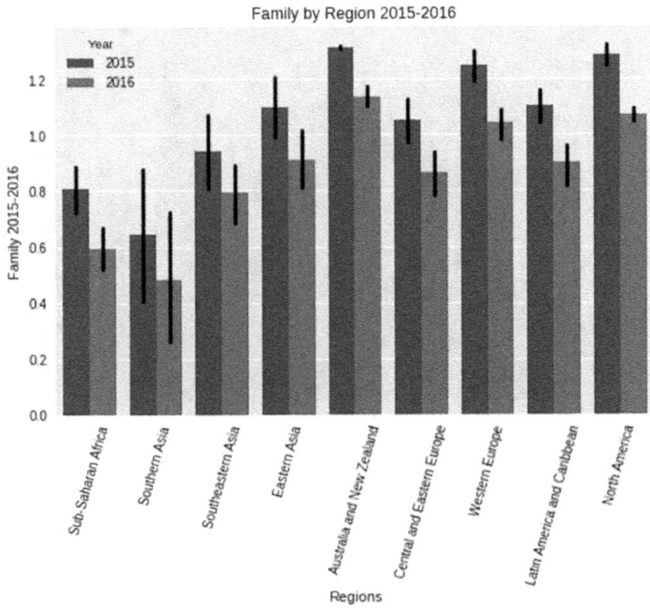

FIGURE 25.8 Family by region.

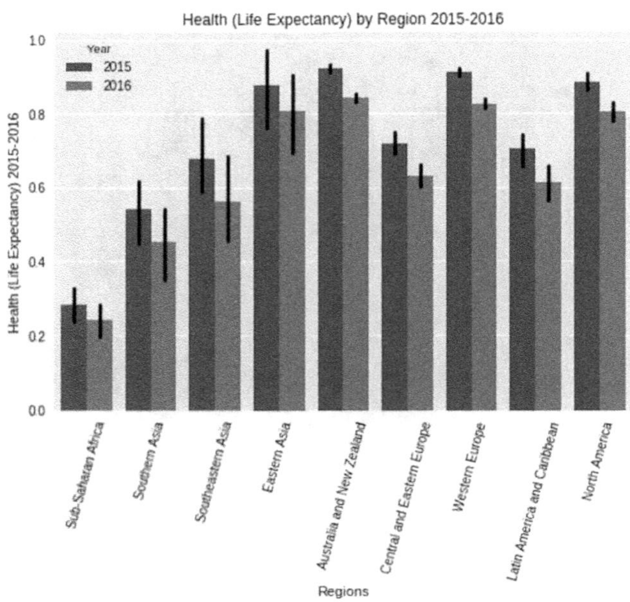

FIGURE 25.9 Life expectancy by region.

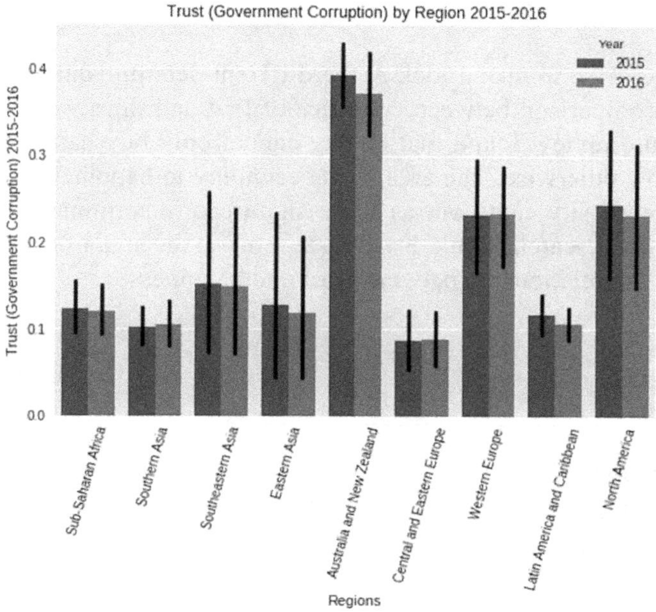

FIGURE 25.10 Government corruption by region.

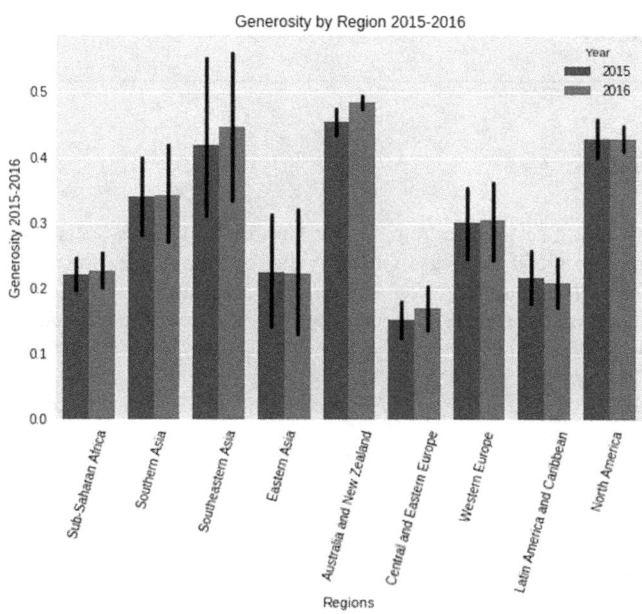

FIGURE 25.11 Generosity by region.

25.11 CONCLUSION

The chapter tried to take a look at the different determinants of happiness. We did a comparison between countries of first and third world. Although the literature review found that money cannot only buy happiness but the results show otherwise. The area of the economy to happiness need further review. Generosity came out as more significant determinant than anticipated. Future research should investigate more in the area of trust in society to explore the relationship between trust and happiness.

KEYWORDS

- **happiness**
- **happiness conundrum**
- **third world**
- **health**
- **family**
- **generosity**
- **economy**

REFERENCES

Aknin, L. B.; Dunn, E. W.; Sandstrom, G. M.; Norton, M. I. Does Social Connection Turn Good Deeds into Good Feelings? On the Value of Putting the "Social" into Prosocial Spending. *Int. J. Happiness Develop.* **2013,** *1* (2), 155–171.

Argyle, M. Causes and Correlates of Happiness. In *Well-Being: The Foundations of Hedonic Psychology;* Kahneman, D., Diener, E., Schwarz, N. Eds.; Russell Sage Foundation: New York, NY, 1999; pp 3–25.

Argyle, M. (2001). *The Psychology of Happiness*, 2nd ed.; Barrow, R. Routledge: London, UK, 1980.

Boniwell, I.; David, S. (Eds.) *The Oxford Handbook of Happiness*. Oxford University Press: Oxford, UK; 2014.

Bjørnskov, C. Healthy and Happy in Europe? On the Association Between Happiness and Life Expectancy Over Time. *Social Sci Med.* **2008,** *66*, 1750–1759.

Bradburn, N. M. *The Structure of Psychological Wellbeing*. Aldine Publishing: Chicago, IL, 1969.

Collet, C.; Vernet-Maury, E.; Delhomme, G.; Dittmar, A. Autonomic Nervous System Response Patterns Specificity to Basic Emotions. *J. Autonom. Nerv. Syst.* **1997,** *62* (1–2), 45–57.

Delhey, J.; Newton, K. Who Trusts? The Origins of Social Trust in Seven Societies. *Eur. Soc.* **2003,** *5*, 1, 93–137.

Diener, E.; Chan, M. Y. Happy People Live Longer: Subjective Well-Being Contributes to Health and Longevity. *Appl. Psychol. Health Wellbeing* **2011,** *3* (1), 1–43.

Diener, E. Myths in the Science of Happiness, and Directions for Future Research. In *The Science of Subjective Well-Being*; M. Eid, R. J. Larsen, Eds.; Guilford Press: New York, 2008; pp 493–514.

Diener, E.; Biswas-Diener, R. *Happiness: Unlocking the Mysteries of Psychological Wealth*; Blackwell: Malden, MA, 2008.

Dunn, E.; Norton, M. *Happy Money: The Science of Smarter Spending*; Simon & Schuster: New York, 2013.

Easterlin, R. A. Does Economic Growth Improve the Human Lot? Some Empirical Evidence. In *Nations and Households in Economic Growth: Essays in Honor of Moses Abramowitz*; P. A. David, M. W. Reder, Eds.; Academic Press: New York, NY, 1974; pp 89–125.

Easterlin, R. A. Income and Happiness: Towards a Unified Theory. *Econ. J.* **2001,** *111*, 465–484.

Field, T. M.; Hernandez-Reif, M.; Quintino, O.; Schanberg, S.; Kuhn, C. Elder Retired Volunteers Benefit from Giving Massage Therapy to Infants. *J. Appl. Gerontol.* **1998,** *17*, 229–239.

Frederick, S.; Loewenstein, G. Hedonic Adaptation. In *Well-Being: The Foundations of Hedonic Psychology;* Kahnemann, D., Diener, E., Schwarz, N. Eds.; Russell Sage Foundation: New York, NY, 1999; pp 302–329.

Harbaugh, W. T.; Myer, U.; Burghart, D. R. Neural Responses to Taxation and Voluntary Giving Reveal Motives for Charitable Donations. *Science* **2007,** *316*, 1622–1625.

Hagerty, M. R.; Veenhoven, R. Wealth and Happiness Revisited: Growing National Income Does Go with Greater Happiness. *Soc. Indic. Res.* **2003,** *64*, 1–27.

Helliwell, J. F.; Haifaing F. H. How's the Job? Well-Being and Social Capital in the Workplace. *Ind. Labor Relat. Rev.* **2008,** *63* (2), 205–227.

Kasser, T. *The High Price of Materialism*. The MIT Press: Cambridge, MA, 2002.

Lane, R. E. *The Loss of Happiness in Market Democracies.* Yale University Press: New Haven, CT, 2000.

Lucas, R. E.; Schimmack, U. Income and Well-Being: How Big is the Gap Between the Rich and the Poor? *J. Res. Personal.* **2009,** *43* (1), 75–78.

Lyubomirsky, S.; Diener, E.; King, L. The Benefits of Frequent Positive Affect: Does Happiness Lead to Success? *Psychol. Bull.* **2005,** *131*, 803–855.

Lykken, D.; Tellegen, A. Happiness is a Stochastic Phenomenon. *Psychol. Sci.* **1996,** *7*, 186–189.

Meier, S.; Stutzer, A. Is Volunteering Rewarding in Itself? *Economica* **2008,** *75*, 39–59.

Rasciute, S.; Downward, P. Health or Happiness? What Is the Impact of Physical Activity on the Individual? *Kyklos*. **2010,** *63* (2), 256–270.

Samanni, M.; Holmberg , S. *Quality of Government Makes People Happy*. Gothenburg: The Quality of Government Institute, University of Gothenburg. 2010 (Working Paper 2010:1). www.qog.pol.gu.se, access date 18.03.2017

Scitovsky, T. *The Joyless Economy: An Inquiry into Human Satisfaction and Consumer Dissatisfaction*; Oxford University Press: New York, NY, 1976.

Stewart, J. C.; Janicki, D. L.; Muldoon, M. F.; Sutton-Tyrrell, K.; Kamarck, T. V. Negative Emotions and 3-Year Progression of Subclinical Atherosclerosis. *Arch. Gen. Psychiatry* **2007,** *64* (2), 225–233.

Shi, G.-X.; Yang, H.-R. Subjective Well-Being of Middle School Students. *Chin. Mental Health J.* **2006,** *20*, 238–241.

Tatarkiewicz, W. *Analysis of Happiness*. Martinus Nijhoff: The Hague, The Netherlands, 1976.

UNDP. *Human Development Report 2002—Bosnia and Herzegovina*. United Nation Development Program: New York, NY, 2002.

Uslaner, E. M. *The Moral Foundation of Trust*. Cambridge University Press: New York, NY, 2002.

Veenhoven, R. (2000). Freedom and happiness: A comparative study in forty-four nations in the early 1990s. In E. Diener & E. M. Suh (Eds.), *Culture and subjective well-being* (pp. 257-288). Cambridge, MA, US: The MIT Press.

Veenhoven, R. Healthy Happiness: Effects of Happiness on Physical Health and the Consequences for Preventive Health Care. *J. Happiness Stud.* **2008,** *9*, 449–469.

Wang, L.; Graddy, E. Social Capital, Volunteering, and Charitable Giving. *Voluntas* **2008,** *19*, 23–42.

Wilkinson, R.; Pickett, K. *The Spirit Level: Why More Equal Societies Almost Always Do Better*. Allen Lane: London, UK, 2009.

CHAPTER 26

Critical Periods in the Historical Process in Turkish Higher Education and Effectiveness on Turkey

BELGİN TURA*

Department of Educational Sciences, Hacettepe University, Ankara, Turkey

*E-mail: belgintura@gmail.com

ABSTRACT

Turkey is engaged in acquis adaptation studies in the field of education as well as in other fields in the negotiation process as a candidate country to the European Union. In this context, the higher education system in Turkey, which is included in the Bologna process, has developed for many years in the shadow of politics and at the center of social events. Today, the Turkish higher education system has been striving to fulfill its responsibility to produce science and to play a role in the development of the country at the expense of current problems and reform debates. Higher education debates in Turkey usually emerge on the basis of political and ideological axis of power quarrels. This situation interrupts the changes that need to be made in higher education. To understand the discussions about the Turkish higher education system, it is necessary to understand the process the system has been through since its establishment and to make today's assessment in the light of the past. The understanding of the critical periods in the Turkish higher education system and its short- and long-term effects on the system is also important in terms of the identification and implementation of strategies for the solution of existing problems of higher education and the success of restructuring efforts. In this study, which was conducted for this purpose, critical periods and the day-to-day effects of the Turkish higher education system have been examined since the foundation of the Republic of Turkey.

In the study, firstly, the higher education reforms carried out were handled on the historical line, and then other events that had an impact on the higher education system were evaluated. Finally, in the study, the regulations that Turkey has taken in relation to the Bologna process which has participated in the European Union candidacy have been discussed.

26.1 INTRODUCTION

Universities have been the institutions that have influenced societies and shaped the future of countries with their role in the development of science and technology throughout history. It has been perceived as the responsibility of the universities to educate new generations from the past to the present day, to develop scientific thinking, to disseminate critical thinking skills, and to earn high-quality workforce with industrialization. Today, the rapid change in globalization and information technologies has brought universities to face the responsibility of educating global citizens of knowledge societies beyond the training of the business force needed by the industry. For this reason, universities continue their aim of being a driving force in the development of societies, by transforming themselves in order to fulfill the mission of becoming the pioneer of change beyond adaptation to change.

Higher education debates in Turkey usually emerge on the basis of the political and ideological axis of power quarrels. This situation interrupts the changes that need to be made in higher education. Despite there has been a social consensus on the need for restructuring the higher education system, comprehensive and radical steps have not been taken in the name of restructuring of higher education, because a common framework cannot be established by different sectors as to what strategy to follow and what to do (Küçükcan and Gür, 2009, pp 15–16).

The educational system in Turkey has a strong centralized tradition. This is true not only for the curriculum and financing of schools but also to staff recruitment and student-selection exams. The most criticism of the higher education system in the context of centralization and autonomy in Turkey is the regulation of the higher education system and the aim of directing the higher education institutions is to establish pressure on the universities with the extensive authority of the Higher Education Council.

The most criticism of the higher education system in the context of centralization and autonomy in Turkey is made due to the fact that the Higher Education Council which was established with the aim of organizing the higher education system and managing higher education institutions have been making pressure on the universities with its large authorities. Although the regulation of the authority of the Council of Higher Education was included in all development plans and government programs published after 1990, all attempts on restructuring Council of Higher Education have failed in the midst of political and ideological debates (Çelik and Gür, 2013, p 160).

According to the understanding of discussions about the Turkish higher education system, it is necessary to understand the process of the system that has been followed since its establishment and to make today's assessment in the light of the past. The understanding of the critical periods in the Turkish higher education system and its short- and long-term effects on the system is also important in terms of the identification and implementation of strategies for the solution of existing problems of higher education and the success of restructuring efforts.

In this study which was conducted for this purpose, critical periods and the day-to-day effects of the Turkish higher education system have been examined since the foundation of the Republic of Turkey. In the study, firstly, the higher education reforms carried out were handled on the historical line, and then other events that had an impact on the higher education system were evaluated. Finally, in the study, the regulations that Turkey has taken in relation to the Bologna process which has participated in the European Union candidacy have been discussed.

26.2 FROM THE EMPIRE TO THE REPUBLIC, FROM DARÜLFÜNUN TO UNIVERSITY: 1933 REFORM

Darülfünun, which was the first university idea of the Ottoman Empire was established in 1845. Darülfünun obtained the name of Darülfünun-ı Osmani in 1909 and the name of Darülfünun in 1912 after attaining a regular structure and continuity at the end of 65 years of dealing. The institution gained scientific autonomy with the regulation in 1912 and the fiscal and administrative autonomy in 1919.

The Republic of Turkey was established in 1923 with the national struggle given by the Turkish nation under the leadership of Mustafa Kemal Atatürk with the aim of establishing the Turkish state and successive revolutions were made for the foundation of a civilized and democratic structure.

The Republic of Turkey has the legacy of an autonomous Darülfünun that has the power to elect the rector. Scientific freedom was granted to

Darülfünun with the law issued in the Republican period in 1922, and academic liberty of Darülfünun was made untouchable and legal personality was acquired by the law issued in 1924 (Widmann, 2000, p 74). In 1932, there were 240 academic staff and 2469 students in schools of medicine, literature, theology, science and law, pharmacists and dentists in İstanbul Darülfünün (Hirsch, 1998a, p 236).

The revolutionary administrators of the young Republic had many expectations for Darülfünun on the success of the revolutions and in reaching the level of civilization the country intended. However, Darülfünun is far from satisfying the expectations of the administrators and his academic staff. While scientific research was only based on the personal efforts of the faculty members and they have not reached sufficient numbers, on the other hand, the curricula were not sufficiently comprehensive enough but the curricula were also behind the new scientific developments. The problem becomes unsolvable when the scientific inadequacies of many academic staff are combined with the inadequacies of the students of Darülfünun (Bahadır, 2007, p 55).

Darülfünun failed to fulfill expected responsibilities as a higher education institution in the development of democracy, modernity, and scientific thought that were targeted by the republican regime; this situation caused dissatisfaction between Darülfünun and country administrators and other intellectuals. Minister of National Education Reşit Galip expressed this dissatisfaction with the speech that he made at the inauguration of Istanbul University in 1933 (Hirsch, 1998a, s. 312):

> *There were political, social, great revolutions in the country. Darülfünun was left with an exhausted counsel against these. There were great movements based on the economic scene. Darülfünun seemed unaware of these. There have been radical changes in the law. Darülfünun only took the new laws into the education program. The letter was revived; the movement of Turkish began, Darülfünun never smitten. A new history keeps all the country in a national movement. It took me about three years to deal with this for Darülfünun. Istanbul Darülfünun had stopped, closed to itself, pulled own hand from the external realm in a vestigial trial.*

The criticisms on Darülfünun revealed the need for institutional reform and Albert Malche, who was a professor at the University of Geneva, was invited to Turkey in 1931 by the government to conduct a review and prepare a report on the conditions of the reform. Prof. Malche came to İstanbul on January 16, 1932, after his studies in Darülfünun, Malche came to Ankara and presented his report to Esat Sagay who was the Minister of National

Education (Bahadır, 2007, p 60). Among the noteworthy items in the Malche report is the fact that Darülfünun's autonomy tends to distance it from the society and the government, the inadequacy of Turkish scientific publications, the use of encyclopedic books as books and obsolete teaching methods, the inadequacy of foreign language knowledge of Turkish students and the inability of Darülfünun on educating professors of future under Darülfünun conditions so the need for educators to be educated outside, the payment of low wages to professors (Widmann, 2000, pp 75–76).

The insensibility of the Istanbul Darülfünun to the revolutions of the newly established Republic, as well as the inadequacies that revealed in the Malche report and passive resistance, constituted the basic justification for the reform of 1933. Istanbul Darülfünun was closed with the Law no. 2252, which was accepted on May 31, 1933, and Istanbul University was established on August 1, 1933.

University autonomy was also abolished with the closure of Darülfünun and the newly opened university was reduced to a "department" affiliated to the Ministry of National Education. About 157 of 240 faculty members in Darülfünun were terminated and only 83 faculty members and deputies were able to move to the newly established university (Hatiboğlu, 2000, p 120). The aim of 1933 University Reform is to bring research and education activities in universities to the level of modern Western universities and to bring the management system of these universities to Turkish universities (Doğramacı, 2007, p 15).

Universities were given scientific autonomy, but the administration of the university was left to the rector with the 1933 Reform. The rector was appointed by the recommendation of the Ministry of National Education with a triple decree and with the approval of the President and the dean was appointed by the recommendation of the rector and the approval of the Minister of National Education. The funding of the newly established university has been provided on the basis of general budget rules. The university which was aimed for education and training on the basis of nationalism and revolutionary principles has eight institutes and faculties of medicine, law, science, and literature. These institutes were The Institute of Turkish Revolution, Institute of National Economics and International Relations, Turkic Institute, Institute of Geography, Institute of Morphology, Institute of Chemistry, Electro-Mechanic Institute, Institute of Islamic Studies (Ataünal, 1998, pp 32–35).

The university has found its academic staff from three sources. These are selected faculty members from the Darülfün, are students who have returned from abroad who have been sent abroad for postgraduate study, and refugees who had seeking refuge in Turkey from Nazi Germany. Among these German scientists were valuable professors such as Philipp Schwartz, Ernst E. Hirsch, Fritz Neumark, and Ernst von Aster. Foreign scientists provided contributions on the improvement of teaching programs and methods in higher education, increasing the number and quality of textbooks, the development of university faculties and libraries, and the establishment of institute libraries. They had done scientific researches such as field, laboratory, clinic and had served Turkish science life and educated young Turkish scientists. These scientists also played an important role in the establishment of the Institute of Higher Agriculture in Ankara in 1933 (Taşdemirci, 2000, pp 199–201).

Law School in 1925, Higher Agriculture Institute in 1933 at Ankara, Faculty of Language and History-Geography in 1935, School of Political Science in 1936, and Faculty of Medicine and Science in 1940 were established. Ankara University, which is one of the outstanding higher education institutions of the Republic of Turkey, was formally established.

When the Republic of Turkey was founded in 1923, the population was greatly reduced due to the major losses that had been experienced in the Balkan wars, the First World War, and the War of Independence. Population growth in the first year of the Republic emerged as a necessity in order to restore social and economic life in the country after long years of war (Koç et al., 2010, p 49). In this period, the country's economy was based on agriculture. Agriculture in the countryside continued to expand. The share of agricultural production in total production was around 50%. Industrialization which was led by the state, created new employment areas in an appropriate rate to the growth of the urban population without causing migration from rural to urban (DİE, 1995, p 43). According to the first population census made in 1927, the population of the country was determined as 13,648,270. By the year 1935, the population of the country reached 16,158,018 with a rate of 21.1%. In 1935–1936 academic year, 7277 students were enrolled in higher education institutions out of which, 6162 were male and 1115 were female. This situation which appeared to be against the students of the girls, continued in 1945–1946 academic year, and 19,273 higher education students consisted of 15,688 male and 3585 female students (TÜİK, 1923–2013).

The higher education institution has gained a regular structure in administrative and academic terms with the 1933 University Reform, modern Turkish scientists are educated and the number of scientific publications has increased. Darülfünun, which was the last educational institution of the

Ottoman Empire, has been closed down and replaced by a modern university structure in line with the requirements of the times, and the transformation of the Turkish nation towards modernization with the younger republic has also become a demonstration.

26.3 1946 UNIVERSITY REFORM AND AUTONOMY DISCUSSIONS

The Turkish higher education system has undergone radical changes with the University Reform and with the multiparty life transition in Turkey in 1946. Reasons for 1946 University Reform are briefly explained as follows (Hirsch, 1998b, p 816).

Universities do not have a general system of management and there are similarities and common features among different universities, with different structures due to their governance, historical, national, economic, and even religious, sectarian reasons. The common features of these universities are their scientific and administrative autonomy, their ability to select teaching members and organs. Faculties and universities must be in a structure that is not similar to other State institutions and their management needs a separate expertise; these faculties and universities are apart from the other administrative organization, have a separate administration and a separate budget which are autonomous and legal entities and are directly connected to the Minister of National Education.

According to this new structure, duties and responsibilities will be scattered among all academic staff and the heavy responsibility of this work based on knowledge and expertise will be raised through the rectorate and the ministry, and a wider and stronger foundation will be imposed and quickness and expertise will be provided in the works.

For the reasons stated above, the "Law on Universities" no. 4936 was adopted and enforced on June 13, 1946. The purpose of the law is to ensure that the universities are autonomously managed and to determine the conditions of university academic staff's upbringing, selection, and promotion on specific principles.

According to the Law no. 4936, universities defined as science, research, and education associations that have become autonomous and legal entities and consist of faculties, institutes, schools, and scientific institutions. The faculties constituting the university are also determined to have autonomy of science and management and legal personality according to the law (Article 1).

The faculties consist of The General Assembly, the Professors Board, the Board of Directors, and the Dean. Faculty General Assemblies consist of ordinaryus professors, professors, associate professors and instructors of courses that do not have associate professors (Article 5). The Professors Board consists of the ordinaryus professors and professors of the Faculty, associate professors who are independently responsible for the management of a course, and two associate professors to be elected separately (Article 6). The Board of Directors consists of the dean, former dean, and three professors to be elected by the Professors Board from ordinaryus professors or professors for 2 years (Article 7). According to the Law, the Dean is elected by the Professors' Board for two years from among the ordinaryus professors or professors of the faculty who are monthly salaried, and the dean who finishes his term cannot be elected dean again before 4 years (Article 8).

Until the 1960s, universities accepted candidates who applied to them without examination. Over time, the universities that faced an exceeding demand over their current quota realized student selection in three ways (www.osym.gov.tr).

- Registration according to the application order and stopping the registration after the quota is filled,
- Acceptance of graduates of science or literature department of high school according to the quality of education given in faculties,
- Selection of students from the top row by sorting according to the grade of high school graduation.

The administration of autonomy in the universities introduced by the 1946 Reform caused the professors who were supposed to produce science to stay between election campaigns and political polarization and as a result the colleges were dragged into a chaotic environment.

While the chaos arising from administrative mechanisms in universities has been prevailing, Hacettepe University was founded in Ankara on July 8, 1967. Instead of the implementation that the rector provided to the universities is replaced by a simple majority provided by all professors in every two years with the Law no. 4936, Hacettepe University has had an opportunity to appoint rector through elections for 5 years by the senate of the university where a limited number of professors participated and this period can be removed in 8 years with the Hacettepe University Establishment Law no. 892. The establishment of Hacettepe University has been evaluated as an effort to bring a new regulation to the system beyond the participation of a new university in the higher education system (Doğramacı, 2007, p 19).

After the foundation of Hacettepe University with the Law no. 892, Istanbul and Ankara universities sued for the cancellation of Hacettepe University Law on the grounds that Hacettepe University is not autonomous and therefore, this is contrary to the Constitution. As a result of the lawsuit, the cancellation order was rejected by the Constitutional Court (Güvenç, 2009, p 31). This situation further exacerbated the chaotic environment in the universities.

The military coup was executed in Turkey on May 27, 1960, then the State Planning Organization affiliated to the Prime Ministry was established on September 30, 1960 and a planned period was made in 1962 with the preparation of the first 5-year development plan (1963–1967). At the beginning of the planned period, there were seven universities in the country. In 1973, five more universities were established and the number of universities in the country increased to 12. Rapid population growth has created an intense demand for higher education, and since 1962 private colleges have started to meet this demand. In the 1966–1967 academic year, 14,000 students enrolled in private colleges (Second Five Year Development Plan).

In total, 21,531 students were enrolled in higher education institutions in 1946–1947 academic year and 17,417 of these students were male and 4114 were female students. At the beginning of the planned period, in the 1962–1963 academic year, 56,193 of the 70,649 higher education students were male and 14,356 of them were female students. Since the establishment of the Republic, the appearance of schooling that has taken place against girls in higher education has continued in the 1960s (TÜİK, 1923–2013).

In the 1950s, Turkey's socioeconomic characteristics began to change. As a result of the industrialization policies of the governments, the industrial sector began to need more workforce. In rural areas, reaching the limit of cultivable area has caused the rural population to start migration to the cities. Especially in the field of education such as health, the establishment of universities in cities has made the cities an attraction center (Koç et al., 2010, p 52). In 1950, the population of the country was 20,947,188 and the rate of population growth was 21.7 per thousand. In the beginning of the planned period in 1960 the population of the country was 27,754,820 and the population growth rate reached the highest level with 28.5 per thousand (TÜİK, 1923–2013). Problems such as uneven urbanization, unemployment, and economic stagnation which resulted as an output of the population growth rate led to the questioning of the population policies that valid up to that date and with the increase in the number of educated individuals in the country, the rate of population growth started to show irreversible trends (Koç et al., 2010, p 52).

Regulations made with the University Reform in 1946 provided an artificial immunity to academic staff members of the universities, unlike the intended ones (Doğramacı, 2007, p 16). The election of the rectors and deans who already have limited authority by all the professors led to the election campaigns in universities and led to politicized of them. The struggles within the university have been interrupted by the Republican People's Party between 1946 and 1950 and the Democratic Party's administration between 1956 and 1960. So, the expected development and change in the way of modernization in the higher education system has not been achieved at the expected level.

26.4 1973 REFORM AND HACETTEPE UNIVERSITY WHICH WAS LOST SELF-FREEDOM WHILE EARNING AUTONOMY

The student events that first started in France in May 1968, with the butterfly effect in a chained way, leapt to Turkey and the reform initiated by the students of Ankara and İstanbul University had already spread to other universities in the chaos and turned into a social turmoil. The military memorandum issued to the President Cevdet Sunay by the Turkish Armed Forces on March 12, 1971 and the subsequent amendments to the 1971 Constitution also required a new higher education law in order to increase the violence gradually and to control the situation.

The Law on Universities no. 1750 was adopted on June 20, 1973 and was enforced on July 7, 1973 with the publication of the Official Gazette. For the first, time together with the law, the Higher Education Council was established as a senior committee to provide coordination between higher education institutions and follow up implementations and make proposals to the competent authorities (Article 4).

The Higher Education Council is composed of the members to be elected by the authorized body of each university among the professors for 2 years and the members appointed by the Council of Ministers on the proposal of the Minister of Education with the same duration as the number of university representatives on the board. In this way, the requirement to have at least one member has been found at least one member among those assigned by this method from official higher education institutions, the Ministry of Finance, National Education, Youth and Sports, and the State Planning Organization and the Scientific and Technological Research Council of Turkey (TUBITAK) (Article 5). The tasks and authorities of the Higher Education Board are as follows (Article 6):

- The establishment of new higher education institutions, the development of existing ones, the fulfillment of human power needs, the effective use of investments and resources, the short and long term planning of educating academic and research staff in the country and abroad and a balanced arrangement of the positions of the relevant institutions determined by senates,
- Submission of opinions to the Ministry of National Education on the design of laws and regulations related to universities and other higher education institutions,
- Increasing the number of semesters in the branches of education considered necessary for the characteristics, capacities, human power, and material needs of the universities, making proposals for parallel education or night education,
- Providing assistance to universities in the academic direction of higher education institutions within the framework of the integrity of higher education and the monitoring of the results of these services together with the relevant institutions,
- Publishing the results of the studies related to the tasks in a report every year,
- Making in cooperation with universities on the research topics required by the development plan and the allocation of private research funds to the universities,
- Determining the principles and conditions for the fees and fees received from the students,
- Establishment of National Committees to participate in International Scientific Institutions, supervision of these committees, and provision of financial support to them,
- To fulfill other duties that was given by the law.

Universities' autonomy continued with the Law no. 1750, but the University Supervisory Board affiliated to the Prime Ministry was established to provide supervision and control over the universities. The council under the chairmanship of the Prime Minister consists of Minister of Education, Minister of Justice, three members elected for 3 years by lot from the rectorate faculty members of universities subject to this law, a member from

Undersecretary of the State Planning Organization, a member to be elected for 3 years among the deans by the National Security Council (Article 7).

The power of coordination and control over the universities of the State on universities has been tried to be increased without touching the autonomy of the universities with the Law no. 1750. However, as a result of the application made by the Ankara University to the Constitutional Court, the provisions of the Law no. 1750 on the Higher Education Council (YÖK) were cancelled due to the fact that the Higher Education Council was contrary to the university autonomy guaranteed by the Constitution (Constitutional Court Decision, 1973/37, Decision Number: 1975/22, Decision Day: February 11, 12, 13, 14, and 25, 1975, official Gazette date/issue: December 3, 1975/15431). The initiative to increase coordination and control over universities by State has failed with the cancelled decision.

Another effect of the Law no. 1750 is the abolition of the original statute of Hacettepe University which was established by the Establishment Law no. 892, and makes the University subject to the same law as other universities. Hacettepe University has achieved a more effective working system than other universities with its management model. This management model, also known as the Hacettepe model, has given an effective management and supervisory authority to the region. According to the model, the rector, who is the chief director of the University, may be in charge of deans, school, and institute directors or general secretary, if she/he considers this authority necessary. Faculty and assistant representative members, faculty board, and university senate were attending, while the instructors were in full-time working order. Financial provisions related to university expenditures have also been determined by a regulation that was prepared by the University Senate (Doğramacı, 2007, pp 20–21). While university was winning autonomy, the institution has lost its management freedom which has achieved international success.

Institutions that are most affected by student actions that have sprouted and spread around the universities and created a chaotic environment within the society have become universities again. Among the challenges and livelihoods of governments trying to control universities and the universities that are trying to protect their autonomy, higher education institutions have failed to adequately demonstrate their expected development in the process of modernization and have struggled to survive and lose their status as universities in the midst of social events.

In this period, social and economic changes in the country have reduced the demand for families to have children, and the traditional large family structure has begun to be resolved and its core families taken up. In this process, family types (single person and single parent) scattered in social life began to be encountered and these family types started to grow rapidly in the country (Koç et al., 2010, p 54). While the population of Turkey was 35,605,176 and the rate of increase was 25.2 per thousand in 1970, when the population of Turkey reached to 44,736,957 in 1980, the rate of increase decreased to 20.7 per thousand. It is thought that the demographic transformation of the living is the effect of the inclusion of the educated female population in the labor force. In 1970, the total labor force in the country was 20,726,989 while the total female l labor force was determined as 10,354,309. By the year 1980, 27,303,045 female labor force was realized as 13,590,999 (TÜİK, 1923–2013).

There were 19 universities in Turkey as one in 1974, five in 1975 and one in 1978. In the 1968–1969 academic year when the student events started, there were 140,000 students in the higher education institutions and a total of 346,476 students reached to higher education in 1977–1978 academic year. Of these students, 247,145 were male and 79,538 were female students (TÜİK, 1923–2013). Gender inequalities among higher education students in the country still exist as a problem in the country at the end of the 1970s. In 1968–1969, a total of 9906 faculty members and staff members were employed in higher education institutions while in the year 1976–1977; 17,607 faculty members and members were employed (Fourth Five Year Development Plan).

As the number of high school graduates increased as a result of the increase in the population in the country and therefore the demand for education, the selection methods of the universities applied to high school education graduates were not able to respond adequately. For this reason, on November 19, 1974, the Universities' Student Selection and Placement Centre (ÖSYM) was established by the Council of Universities in accordance with Article 52 of the Law no. 1750. For the first time in 1974, students were taken to the universities by ÖSYM and 37,271 of the 229,994 students who applied for the exam were placed in a higher education institution (www.osym.gov.tr).

The irregularity that emerged in 1968 with the actions that started in Turkey increased the violence by being bifurcated with another expression through a transformation toward the whole society. The armed conflicts that have taken place in some places created a chaotic environment in the country and a military coup was carried out on September 12, 1980, which resulted in the seizure of the army in the direction of the instability.

26.5 1981 HIGHER EDUCATION REFORM AND FOUNDATION OF YÖK

One year after the September 12 military coup, the Higher Education Law no. 2547 was adopted on November 4, 1981 and entered into force upon publication in the Official Gazette dated November 6, 1981. The Law on Universities no. 1750 does not contain any qualification change in higher education. The Law on Higher Education no. 2547 initiated a reform process that radically changed the higher education system in Turkey.

All universities, academies, and other colleges were organized together in one unity for the first time with the 1981 Higher Education Reform. The university-academy dilemma in the higher education system has come to an end and colleges and conservatories affiliated to different ministries have been gathered under the roof of universities and a minimum partnership has been established in the planning, coordination and training programs among higher education institutions (Doğramacı, 2007, p 25). Administrative autonomy was abolished by law and universities were defined as higher education institutions with scientific autonomy and public legal personality (Article 3d).

With the Law no. 2547, the Higher Education Council which was previously cancelled by the Constitutional Court because of contrariness to the Constitution was reestablished. Higher Education Council consist of seven members which were elected by the President by giving priority to professors who have successfully served as rectors and faculty members, seven members which were elected by the Council of Ministers and elected among senior State officials or pensions, and seven members which were elected and seven members which were selected from professors who are not board members by Interuniversity Board (Article 6b). The Higher Education Council has undertaken tasks related to planning, coordination, and control in universities.

The rectors are appointed by the President of the Higher Education Council (Article 13a) and the deans by the Higher Education Council by the recommendation of the rector with Law no. 2547 (Article 16a). The institute directors are appointed by the rector with the proposal of the dean (Article 19b), heads of department are appointed by the dean in the faculties (Article 21) and the professors are appointed by the rector (Article 26b). With the 1981 reform, the structure of organization from top to bottom was established by appointment at universities rather than the principle of university administrators being elected by the boards.

The procedure for appointment of the rector of the University has been changed with "Law on the Amendment of Article 13 of the Higher Education Law no. 2547, dated November 4, 1981" no. 3826, which was adopted in July 1, 1992 and published in the Official Gazette dated July 7, 1992 with the Law no. 3826. With the Law no. 3826, the right of the university to speak was increased but the last election was left to the President of the Republic.

With the Law no. 2547, the titles of doctorate, associate professorship, and professorship that were previously acquired according to the laws of the universities and academies were accepted as equivalent (Tekeli, 2009, p 72). For the first time in the law, assistant professorship staff has been defined and the assignment authority has been left to rector (Article 23a).

With the 1981 Reform, the selection of students for higher education was based on a central examination principle. The Universities Student Selection and Placement Center (ÖSYM), which was established in 1974, is connected to the Council of Higher Education by Law no. 2547 and is named as Student Selection and Placement Centre (Article 45). In 1981, 420,855 candidates applied for university entrance examination organized by ÖSYM and 55,037 of these candidates were placed in a higher education program. In the year 1990, 112,865 students from 898,975; in 2000, 262,857 of 1,414,827 were placed in a structured higher education program. By the year 2016, 792,249 of the 2,255,386 candidates who applied for the university exam were placed in a structured higher education program (www.osym.gov.tr).

With the 1981 Higher Education Reform, the number of universities opened in the country has increased. There were 8 universities in 1982, 22 universities in 1992, 15 universities in 2006, and 17 universities in 2007. By 2017, there are a total of 183 universities in the country, 118 of which are state, and 65 are foundation universities (www.yok.gov.tr).

After the 1981 reform with the increase of higher education institutions, the number of students also increased rapidly. In the academic year of 1981–1982, there were 240,403 students that including 174,345 male and 66,058 female students in higher education institutions. In the 1990–1991 academic year, there were a total of 705,409 student that 468,406 of them were male and 237,003 were female students. In the 2000–2001 school year, the total number of students reached 888,945 male and 621,260 female students that totaled to 1,508,205 (TÜİK, 1923–2013). The number of girls and boys in higher education indicates an increase in favor of female students over the years, as can be understood from the number of students studied over 10 years. In the year 2015–2016, a total of 6,689,185 students in higher education came out as 3,621,517 male and 3,067,668 female students (www.

istatistik.yok.gov.tr) indicate that gender equality was almost guaranteed among students in higher education.

During the academic year of 1981–1982, 2080 professors and 2647 associate professors served in higher education institutions. In 1990–1991 education year, 4775 professors and 2433 associate professors are served in higher education institutions; during the 2000–2001 academic year, there were 8804 professors and 5249 associate professors. In the 2015–2016 education year, the number of professors who have been working in higher education institutions reached 22,416 and the number of associate professors reached 15,023.

According to the 1981 reform in Turkey and the development of scientific publications produced by higher education institutions, 386 articles were published in 1981, 1152 in 1990, and 6423 articles in 2000, according to the number of articles published in scientific journals scanned in international citation indexes (Doğramacı, 2007, p 84). Between the years of 2010 and 2015, Turkey settled in the 18th place in the world rankings with 204,216 scientific publications. 36,007 (17.63%) of 204,216 publications were written with international cooperation. Turkey cooperate the most in the preparation of scientific publications with the United States (13,819), Germany (5892), and the United Kingdom (5586). The most international cooperation fields of Turkey in scientific publications are material sciences, common disciplines (1727), engineering, electrical and electronics (1646) and mathematics, applied (1382) science branches (www.cabim.ulakbim.gov.tr).

Following the 1981 reform, the number of higher education graduates employed in the labor market increased in parallel with the increase in the number of higher education graduates in the country. 870,000 of the 17,754,000 individuals employed in the country in 1988; in 2000, 1,894,000 of 21,581,000 individuals were graduated from higher education (TÜİK, 1923–2013). By the year 2016, a total of 30,535,000 workforces in the country were identified as 6,892,000 graduates of higher education (TÜİK, 2016).

For the first time in the 1980s, the concept of "family planning" took place in the Constitution. Population growth in the country is now completely out of necessity and Turkey has shifted to the last stage of its demographic transformation. While the population of Turkey in this period was 44,736,957 in the year 1980 and the rate of population growth was 20.7 per thousand (TÜİK, 1923–2013), as a result of the population policies applied by years when the population reached to 79,814,871 in 2016, the growth rate decreased to 13.5.

26.6 DEVELOPMENTS THAT AFFECT HIGHER EDUCATION SYSTEM IN TURKEY

Other developments that have emerged during the reform processes in Turkey and which have had significant implications for the development of the higher education system are discussed below.

26.6.1 FOUNDATION OF THE TURKISH SCIENTIFIC AND TECHNOLOGICAL RESEARCH COUNCIL (TÜBİTAK)

The Turkish Scientific and Technological Research Council (TÜBİTAK), was established on July 24, 1963 with no. 278 "the Law on the Establishment of a Research Institution," with the aim of providing guidance to governments in the development of science policies and encouraging the development of scientific studies and technology in the country. The name of the Scientific and Technical Research Council of Turkey has been changed to the Scientific and Technological Research Council of Turkey by Law no. 5376, which was enacted on July 7, 2005.

The provisions of TÜBİTAK's Establishment Law, which limits the field of activity to natural sciences, were expanded by Law no. 5376 and social sciences were included in the duty field of the Institution.

Four research groups in the fields of basic sciences, engineering, medicine, agriculture, and animal (Presidency of Research Support Programs which now includes 10 research groups) and Scientist Training Group (today, the Department of Supporting Scientists) have been established in order to support basic and applied scientific studies, especially in natural sciences and to encourage young researchers was established (www.tubitak.gov.tr).

The Scientist Training Group established in TÜBİTAK has started with high school and university scholarship programs in order to educate and encourage young scientists; today, activities have been continuing with national undergraduate, postgraduate, doctoral scholarship programs; international doctoral programs, postdoctoral fellowships and research scholarship programs, and support programs for participating in national and international scientific events. In the book of "25 the annual TÜBİTAK" that was published in 1989, it was reported that TÜBİTAK gave high school scholarship to 2626 students, license scholarship to 2381 students, graduate scholarship to 607 students, domestic doctoral scholarship to 506 students, abroad scholarship to 506 doctorate students between the years of 1965 and 1988 (Erim, 2009, p 358).

By the year 2013, TÜBİTAK provided 567 national graduate students and 982 students a scholarship at the level of doctoral degree. About 324 doctoral students and 669 postdoctoral researchers were benefited from overseas research scholarships. The total expenditure realized by the Scientific Humanity Support Program in 2013 was 105.1 million TL. About 94.3 Million TL of 105.1 Million TL is allocated to adult scientists, 7.9 million TL is transferred to primary and secondary students, and 2.9 million TL is expenditures made under administrative/financial services. TÜBİTAK supported 5122 AR-GE projects in public and private sector in 2015 and transferred 7009 million TL to these projects (www.tubitak.gov.tr).

TÜBİTAK has played an important role in the development of research culture in Turkey since the day it was founded and in the realization of researches on project basis. Turkey provides support and guidance services to the governments through the coordination duties that Turkey undertakes in the development of science policies, international cooperation, and especially in the process of harmonization with the European Union. It has undertaken a locomotive role in the development of industry and technology in the country with the support of AR-GE that provided by the research institutes, the private sector, and the public. TÜBİTAK continues to make contributions in the name of combining the supportive powers of the industry with the driving force of education through the scholarship opportunities provided to young scientists and university–industry cooperation support.

26.6.2 DEVELOPMENT OF DISTANCE LEARNING IN TURKISH HIGHER EDUCATION

Despite the developments in Turkey with radical reforms carried out in all areas of the country since the establishment of the Republic in 1923, the gross enrolment rates in the country in 1961 are 70% in primary school, 23% in junior high school, 9% in high school and 4% in higher education (First Five Year Development Plan). The educational problems in the country were dealt with in a holistic approach with all dimensions, and in addition to the existing applications, the search for alternative methods was made to increase access to education with the transition to the planned economy period in 1962. For this purpose, distance learning applications have started to take place in the Turkish Education System in order to provide access to education and at the same time to meet the needs to the teachers in the formal education.

The most defined and organized distance education application within the country's education system is the Letters Teaching Centre that was established in 1961 under the Ministry of National Education's Vocational Technical Education Under secretariat. The aim of the center is to prepare the students who want to finish primary and secondary schools out of the finishing examinations, to train certificate in the arts types and to arrange vocational courses by letter in the fields required for the workers employed in the industry. The Letters Teaching Centre was transformed into the Centre for Letters to Higher Education in order to provide higher education opportunities to students who did not have the opportunity to enter a higher education program and graduated from high school and balanced schools in 1974 and to meet the need to teachers in the country (Özkul, 2009, p 682–685). "Higher Education with Letters" that was active in 1974–1975 academic year was limited to teacher education program only. Due to the negative attitude toward teaching the letter in public, the application did not see the expected disposition; and was left the place to the Trial Higher Teacher School in 1975, which was established within the Ministry of National Education. The trial Higher Teacher School project was finalized by the order of the Ministry of National Education in 1975 without going into practice.

After the failure of the Letters Teaching Centre and the Trial Higher Teacher School, the Common Higher Education Institution (YAYKUR) was established by the Ministry of National Education in 1975. The aim of YAYKUR is to provide equality of opportunity to educated young people who are left out of higher education due to the inadequacy of organized higher education and to raise the intermediate human power that the industry needs. In the YAYKUR curriculum published by the Ministry of National Education, education is planned as formal and open education and the courses are planned to be applied 2 years in the theoretical second year (Öztürk, 1979, p 37–39). In 1975–1976 academic year, 3962 students were trained in 24 Vocational Schools within the scope of YAYKUR (DİE, 1980). YAYKUR programs started to be closed from 1978 to 1979 academic year and completely abolished in 1981. Distance education programs failed to institutionalize as a result of unstable attitudes of political governments until the 1981 Higher Education Reform.

In 1981, with the Law no. 2547 on Higher Education, the task of arranging distance education activities at the level of higher education was transferred to universities. A year later, the distance education practices were left to Anadolu University, which was transformed and restructured in 1982 by the Eskişehir Economic Commercial Sciences Academy, which was established in 1958. Anadolu University has formed the first example of bilateral education in Turkey with the faculties of formal education and the faculty of open education.

Anadolu university open education faculty in 1982–1983 education started with the undergraduate programs of Economics and Business Administration. In total, 29,500 students enrolled in these programs. There were 1,835,563 undergraduate students enrolled in Anadolu University open education programs during the 2015–2016 academic year and 24,811 undergraduate students enrolled in distance education programs of other universities (www.istatistik.yok.gov.tr).

26.6.3 PRIVATIZATION IN HIGHER EDUCATION: ESTABLISHMENT OF FOUNDATION UNIVERSITIES

In Turkey, since the 1960s, in order to meet the increase in demand for higher education, private higher schools are being established in accordance with the goals of schooling in higher education determined in the Development Plans. The number of private schools that were first established in 1962 reached 13 in the first 3 years and 41 in 1968. According to the Ministry of National Education Planning-Research and Coordination Department, 1799 students were educated in 1963; 36,006 students were educated in 1968 and 50,246 students were educated in 1972 (Ataünal, 1998, p 51). Private schools that were operated in the 1960s were transferred to universities and academies in accordance with Law no. 1472 enacted on June 2, 1979.

The Higher Education (Law no. 2547) that was enacted in the 1981, Higher Education Reform introduced a provision allowing the establishment of nonprofit foundation universities in Turkey (Additional Article 2) and provisions on how to establish these universities were regulated in Additional Article 3 of this Law. According to the Constitution of the Republic of Turkey, Foundation universities are subject to the same provisions as state universities for academic studies outside the financial and administrative subjects, provision of teaching staff and security aspects (Article 130).

The Foundation has three separate sources of funding: the universities, contributions of founding foundations, student fees, and the state aid. The state aid provided to foundation universities is made in accordance with the provisions of Additional Article 18 of Law no. 2547. In 2005, 10 foundation universities received state aid, and at these universities, the ratio of state aid within the total income of the university varied between 0.9% and 3.6% (Vakıf Üniversiteleri Raporu, 2007).

Bilkent University, which was founded in 1984, is the first foundation university in Turkey. This university was followed by Koç University in 1992 and Başkent University in 1994. The number of foundation universities reached 8 in 1996 and 23 in 2001. The number of students in the foundation universities increased from 426 in 1986–1987 year to 5846 in 1990–1991, to 46,022 in 2000–2001 academic year. In the foundation universities, 67 academic staff worked in 1986–1987 academic year, whereas 442 and 3721 academic staff attended universities in 1990–1991 and 2000–2001 academic year , respectively (Vakıf Üniversiteleri Raporu, 2007). In 2017, there were 65 foundation universities in the country. In 2015–2016 academic year, 534,177 students have been studying at foundation universities while 23,174 academic staff have been working (www.istatistik.yok.gov.tr). Benefits provided by the foundation universities to Turkish higher education are summarized as follows by Berkman (2009);

- Reducing quota pressure on state universities,
- Transfer of foundation resources to higher education besides public resources,
- The fact that state universities feel compelled to conduct high-level research on competition and to attract students by improving existing conditions for academic staff and students,
- New options and training opportunities for students,
- Offer alternate opportunities for students who want to go abroad for education,
- Increased employment opportunities for academic staff,
- Increasing the qualified human power in the country,
- Cultural and artistic activities that carried out by higher education institutions and enrichment of social life.

26.7 BOLOGNA PROCESS, QUALITY, AND ACCREDITATION STUDIES

Following the signing of the partnership agreement with the European Economic Community in 1963, Turkey became a member of the European Union (EU) in 1987 and was recognized as a candidate country by EU members in 1999 (www.tr.wikipedia.org). The Bologna process originally contained six action areas. These are:

1. Create easily understandable and comparable higher education diploma or diploma grades,
2. To move to a two-stage degree system, including undergraduate and graduate degrees in higher education,
3. Apply European Credit Transfer System (ECTS),
4. Ensuring and promoting the mobility of students and academic staff,
5. Establish and expand the network of quality assurance systems,
6. Develop the European dimension in higher education.
   On May 19, 2001, 32 European countries, including Turkey, gathered in Prague to evaluate the Bologna process and further three action areas were added to the Bologna process.
7. Promote life-long learning,
8. Ensuring active participation of students and higher education institutions to the process,
9. Making the European Higher Education area attractive.

In 2003, in Berlin, the 10th action area included a synergy between the European Research Area and the European Higher Education Area and added doctoral studies (Bologna Süreci Durum Değerlendirme Raporu 2005–2007).

Turkey officially joined the Bologna process in 2001. Since its inclusion in the Bologna process, Turkey has implemented various arrangements within the higher education system. These arrangements are discussed below.

The Decision of the General Assembly of YÖK dated March 11, 2005 and the application of the Diploma Supplement and the European Credit Transfer System have become compulsory for all higher education institutions since the end of 2005–2006 academic year (Türkiye'nin Yükseköğretim Stratejisi, 2007).

Regulations of "Higher Education Institutions Student Councils and National Education Councils of Higher Education Institutions" prepared by the Council of Higher Education for the purpose of establishing national student unions which will ensure the active participation of students from the main action headings of the Bologna Process have been put into action with the publication of Official Gazette dated September 20, 2005 and numbered 25,942.

Regulation on "Academic Evaluation and Quality Improvement in Higher Education Institutions," prepared by YÖK, was published in the Official Gazette dated September 20, 2005 and no. 25942 within the scope of establishment of quality evaluation and development systems fully compatible with the principles determined at European level in higher education institutions.

The Turkish Quality Assurance System is also open to international evaluations. Up to now, a total of 42 engineering programs at four universities have been considered equivalent in "substantial equivalence" rating by the Accreditation Board for Engineering and Technology (USA) (ABET) over various time periods. Higher Education Quality Assurance Regulation dated July 23, 2015 was published in the Official Gazette no. 29423 and entered into force. With this regulation, YÖDEK continues to serve as the Higher Education Quality Board (http://www.yok.gov.tr/web/uluslararasi-iliskiler/kalite-guvencesi).

The work of the Bologna Process in relation to the framework of higher education qualifications is carried out by the Council of Higher Education. Turkey Higher Education Qualifications Framework (TYYÇ) Basic Field Competencies was approved by the General Assembly of The Council of Higher Education dated January 31, 2011.

Within the context of Bologna process alignment, Turkey has taken the steps to internationalize its universities by making the higher education system a contemporary structure. In this context, the Bologna process is seen as an opportunity for the development of the Turkish higher education system, for the enhancement of qualifications at universities and for the gain of internationally recognized advantages.

26.8 CONCLUSION

When the history of Turkish higher education is examined, it can be seen that it is actually an intervention history. The higher education system, which develops in the shadow of politics and at the center of social events, continues its existence with the present debate with a history of nearly 100 years.

The adventure of Turkish higher education, where political actors played an active role, first started with the 1933 reform. After the foundation of the republic in 1923, the concept of establishing a modern university for the settlement of western lifestyles and success of the revolutions developed and the Ottoman heritage Darülfünun which was an autonomous institution for this purpose was closed. One of the reasons for the failure of Darülfünun was that the electoral appointment of the administrators was not included in the republic universities and was determined by the appointment procedure. Istanbul University which was the first university in the Republic and Ankara University, which was established in the aftermath, have been in important services in the development of scientific thinking and modernization in the country.

Turkey, for the first time in 1946, had gone into a multiparty system and the democratization movement that took place has also been influential in educational institutions. With the 1946 reform, self-management was introduced to higher education institutions and the universities began to be managed by their own elected management units. Because of the inability to fully establish the planning, coordination, and control systems in the higher education system, university autonomy has turned away from the supervisory control of university administrators, and universities are slowly becoming politicized. Some bureaucrats who were disturbed by the emerging chaotic situation entered into the search for alternative universities and founded Hacettepe University in 1967. The establishment of Hacettepe University has enabled the development of a successful management model in higher education. Despite the removal of the university's status in the following years, the University has also developed its presence internationally.

The 1968 student events that shook the whole world deeply affected Turkey. The state that had difficulty in controlling the social complexity that was living was hit by the military coup and this brought the 1973 University Reform with the amendments of the Constitution. Though the 1973 reform and the state tried to increase the control over the universities, the political debate and the legal resistance of the universities led to the failure of the 1973 Reform.

The military coup was carried out on September 12, 1980 on the ground that the social turbulence that started with the student events in the country could not be controlled despite all the preventions taken and the violence of the time progressed further into a chaos. For the first time after the military coup, the higher education system was considered as a whole. With the 1981 Higher Education Reform, radical changes were made in the higher education system. Under the Reform, the Council of Higher Education was established and the appointment procedure was restored in the administration abolished by the 1946 Reform. Although the solid administration procedure in the university administration has been softened by the laws passed in the following years, the autonomy of the university and the authority of the Higher Education Board have continued to be a debate issue in the following years. The reforms realized for the solution of the problems in the Turkish higher education brought new problems together with the developments provided and opened new chaotic areas.

The Turkish higher education system has been struggling to close the 900-year gap between successive reforms in the shadow of politics and social turbulence in the adventure of nearly 100 years of university experience and much earlier than Europe's nearly 1000 years of university experience, and the EU Bologna process continues to be shaped by the impact of global developments and market dynamics.

KEYWORDS

- **Bologna process**
- **higher education**
- **university reform**
- **Turkey**
- **universities**

REFERENCES

1472 Sayılı Kanunla Açılmış Bulunan Resmi Yüksekokulların Üniversite ve Akademilere Devredilmesi Hakkında Kanun. Kanun No: 2237.

2547 Sayılı Yükseköğretim Kanununun 13 üncü Maddesinin Değiştirilmesi Hakkında Kanun. Kanun No: 3826.

Açık Yükseköğretim Yönetmeliği. Resmi Gazete Sayısı: 17860.

Anadolu Üniversitesi. https://www.anadolu.edu.tr (accessed Mar 25, 2017).

Ataünal, A. *Türkiye'de Yükseköğretim (1923–1998)*. Ankara: Milli Eğitim Bakanlığı Yükseköğretim Genel Müdürlüğü, 1988.

Bahadır, O. 1933 Üniversite Reformu Niçin Yapıldı? In *Türkiye'de Üniversite Anlayışının Gelişimi (1861–1961),* Aras, N. K., Dölen, E., Bahadır, O., Eds; Ankara: Türkiye Bilimler Akademisi, 2007; pp 52–85.

Berkman, Ü. Vakıf Üniversiteleri Modeli ve Bilkent Üniversitesi. In *Türkiye'de Üniversite Anlayışının Gelişimi II (1961–2007)*, Çelik, T., Tekeli, I., Eds., Türkiye Bilimler Akademisi, 2009: Ankara, pp 743–763.

Birinci Beş Yıllık Kalkınma Planı (1962–1967), http://www.sbb.gov.tr/wp-content/uploads/2018/11/Birinci-Be%C5%9F-Y%C4%B1ll%C4%B1k-Kalk%C4%B1nma-Plan%C4%B1-1963-1967%E2%80%8B.pdf

Bologna Süreci Durum Değerlendirme Raporu 2005–2007. T. C. Yükseköğretim Kurulu. http://www.yok.gov.tr/documents (accessed Mar 20, 2017).

Cahit Arf Bilgi Merkezi. http://cabim.ulakbim.gov.tr/bibliyometrik-analiz/turkiye-bilimsel-yayin-performans-raporlari/ (accessed Mar 24, 2017).

Çelik, Z.; ve Gür, B. Turkey's Education Policy During the AK Party Era (2002–2013). *Insight Turk.* **2013,** *15* (4), 151–176.

Devlet İstatistik Enstitüsü (DİE). Milli Eğitim İstatistikleri Yükseköğretim 1974–1978. *MEB,* Ankara, 1980.

Devlet İstatistik Enstitüsü (DİE). Türkiye Nüfusu, 1923–1994. Demografi Yapısı ve Gelişimi. 21. Yüzyıl Ortasına Kadar Projeksiyonlar. Başbakanlık Devlet İstatistik Enstitüsü Matbaası: Ankara, 1995.
Doğramacı, İ. *Türkiye'de ve Dünyada Yükseköğretim Yönetimi.* Ankara: Meteksan Matbaacılık ve Teknik Sanayi Ticaret A. Ş, 2007.
Dördüncü Beş Yıllık Kalkınma Planı (1979–1983), http://www.sbb.gov.tr/wp-content/uploads/2018/11/D%C3%B6rd%C3%BCnc%C3%BC-Be%C5%9F-Y%C4%B1ll%C4%B1k-Kalk%C4%B1nma-Plan%C4%B1-1979-1983%E2%80%8B.pdf.
Erim, R. Kuruluşundan Bugüne Türkiye Bilimsel ve Teknik Araştırma Kurumu. In *Türkiye'de Üniversite Anlayışının Gelişimi II (1961–2007)* Çelik, T., Tekeli, I., Eds.; 2009; pp 351–388).
Güvenç, B. Yükseköğretim Sisteminin Geçirdiği Değişimler YÖK Öncesi: 1961–1981. In *Türkiye'de Üniversite Anlayışının Gelişimi II (1961–2007),* Dönemi. T., Çelik, İ., Tekeli, Eds.; 2009; pp 19–54)
Hatiboğlu, M. T. *Türkiye Üniversite Tarihi.* Selvi Yayınevi: Ankara, 2009.
Hirsch, E. *Dünya Üniversiteleri ve Türkiye'de Üniversitelerin Gelişmesi. (I. cilt).* Ankara Üniversitesi Yayınları No: 211: Ankara, 1998a.
Hirsch, E. *Dünya Üniversiteleri ve Türkiye'de Üniversitelerin Gelişmesi. (II. cilt).* Ankara Üniversitesi Yayınları No: 211: Ankara, 1998b.
Kasım 1981-Kasım 1988 Döneminde Yükseköğretimdeki Gelişmeler. http://www.yok.gov.tr/documents (accessed Mar 24, 2017).
İkinci Beş Yıllık Kalkınma Planı (1968–1972), http://www.sbb.gov.tr/wp-content/uploads/2018/11/%C4%B0kinci-Be%C5%9F-Y%C4%B1ll%C4%B1k-Kalk%C4%B1nma-Plan%C4%B1-1968-1972%E2%80%8B.pdf
Türkiye İstatistik Kurumu (TÜİK) http://tuik.gov.tr/Kitap.do?KT_ID=0&KITAP_ADI=Genel&metod=AnaKategori
Koç, İ.; Eryurt, M. A.; Adalı, T.; ve Seçkiner, P. *Türkiye'nin Demografik Dönüşümü. Doğurganlık, Aile Planlaması, Anne-Çocuk Sağlığı ve Beş Yaş Altı Ölümlerdeki Değişimler: 1968–2008.* Hacettepe Üniversitesi Nüfus Etütleri Enstitüsü: Ankara, 2010.
Küçükcan, T.; ve Gür, B. *Türkiye'de Yükseköğretim. Karşılaştırmalı Bir Analiz.* Seta Yayınları: Ankara, 2009.
Ölçme, Seçme ve Yerleştirme Merkezi Başkanlığı (ÖSYM). http://www.osym.gov.tr (accessed Mar 20, 2017)
Özdil, İ. Uzaktan Öğretim Uygulamasının Türkiye'deki Durumu. *Eğitim ve Bilim* **1982,** *6* (36), 43–48.
Özer, B. Türkiye'de Uzaktan Eğitim: Anadolu Üniversitesi Açıköğretim Fakültesi'nin Uygulamaları. 1989; pp 1–24. http://www.academia.edu (accessed Mar 25, 2017).
Özkul, A. E. Türkiye'de Açık ve Uzaktan Yükseköğretim. In *Türkiye'de Üniversite Anlayışının Gelişimi II (1961–2007),* Çelik, T., Tekeli, I., Eds.; 2009; pp 677–741).
Öztürk, D. YAYKUR'un Türk Eğitim Sistemine Getirdiği Yenilikler. *Eğitim ve Bilim* **1979,** *4* (22), 37–48.
Taşdemirci , E. Türkiye'de Yükseköğretim Kurumlarının Gelişmesinde Avusturya ve Alman Bilim Adamlarının Katkısı. *Erciyes Üniversitesi Sosyal Bilimler Enstitüsü Dergisi* **2000,** *1*(9), 189–213.
Tekeli, İ. Türkiye'de Üniversitelerin YÖK Sonrasındaki Gelişme Öyküsü (1981–2007). In *Türkiye'de Üniversite Anlayışının Gelişimi II (1961–2007),* T. Çelik, T.; Tekeli, I., Eds.; 2009; pp 55–226.

Türkiye'nin Avrupa Birliği Üyelik Süreci. https://tr.wikipedia.org/ (accessed Mar 25, 2017).
Türkiye Bilimsel ve Teknolojik Araştırma Kurumu (TÜBİTAK). http://www.tubitak.gov.tr (accessed Mar 25, 2017).
Türkiye Cumhuriyeti Anayasası. Kanun No: 2709.
Türkiye İstatistik Kurumu (TÜİK). İstatistik Göstergeler. 1923–2013.
Türkiye İstatistik Kurumu (TÜİK). İşgücü İstatistikleri, 2016.
Türkiye Yükseköğretim Yeterlilikler Çerçevesi. http://www.tyyc.sakarya.edu.tr (accessed Mar 25, 2017).
Türkiye'nin Yükseköğretim Stratejisi. T. C. Yükseköğretim Kurulu. 2007. http://www.yok.gov.tr/documents (accessed Mar 20, 2017).
Üniversiteler Kanunu. Kanun No: 4936, http://www.resmigazete.gov.tr/arsiv/6336.pdf
Üniversiteler Kanunu. Kanun No: 1750 http://www.resmigazete.gov.tr/arsiv/14587.pdf. Accessed date (accessed Mar 25, 2017).
Vakıf Üniversiteleri Raporu, 2007. http://www.yok.gov.tr (accessed Mar 24, 2017).
Widmann, H. *Atatürk ve Üniversite Reformu,* Kazancıgil, A., Bozkurt, S., Çev., Eds.; Kabalcı Yayınevi; İstanbul, 2000.
Yükseköğretim Bilgi Yönetim Sistemi. https://istatistik.yok.gov.tr/. (accessed Feb 22, 2017)
Yükseköğretim Kurulu. http://www.yok.gov.tr (accessed Mar 20, 2017).
Yükseköğretim Kanunu. http://www.mevzuat.gov.tr/MevzuatMetin/1.5.2547.pdf
Yükseköğretim Kurulu Denklik Birimi. http://www.yok.gov.tr/web/guest/icerik (accessed Mar 25, 2017).
Yükseköğretim Kurulu Uluslararası İlişkiler Birimi. http://www.yok.gov.tr/web/uluslararasi-iliskiler/kalite-guvencesi (accessed Mar 25, 2017).

Glossary

Darülfünun: Darülfünun which was the first university idea at the beginning of Ottoman Empire. Darülfünun obtained a name of Darülfünun-ı Osmani in 1909 and a name of Darülfünun in 1912 after attaining a regular structure and continuity at the end of 65 years of dealing.

Drop of honey effect: Generally a chaotic behavior is characterized by its extreme sensitivity to the initial conditions, meaning this sensitivity that a very small perturbation of the system in an initial condition may lead it to an exponentially type divergent final state. The trajectories may behave in a very different way in neighboring points, approaching and moving away one from the other in a really unpredictable way: consider for illustration either the Lorenz' metaphor of the "butterfly effect" or the "drop of honey effect."

Economic Vulnerability Index: Economic vulnerability index (EDI) measures the vulnerability of the countries against economic and environmental shocks. By looking at this index, it is possible to compare economic vulnerability conditions of different countries.

Fuzzy ANP: Saaty (1996) developed fuzzy analytical network process (ANP) method to make decision in complex situation.

TOPSIS: Chen and Hwang (1992) developed TOPSIS (Technique for Order Preference by Similarity to an Ideal Solution) method as a multicriteria decision-making approach. While using this method, it can be possible to compare the values of different alternatives.

Index

U

W

www.ingramcontent.com/pod-product-compliance
Lightning Source LLC
LaVergne TN
LVHW021003300926
849160LV00007B/236

* 9 7 8 1 7 7 4 6 3 4 3 0 1 *